D0208879

Hazardous and Industrial Waste Treatment

Hazardous and Industrial Waste Treatment

CHARLES N. HAAS
L. D. Betz Professor of Environmental Engineering
Drexel University

RICHARD J. VAMOS
DAI Environmental, Inc.

PRENTICE HALL, Englewood Cliffs, New Jersey 07632

Library of Congress Cataloging-in-Publication Data

Haas, Charles N.
Hazardous and industrial waste treatment / Charles N. Haas,
Richard J. Vamos.
p. cm.
Includes bibliographical references and index.
ISBN 0-13-123472-2
1. Hazardous wastes—Purification. 2. Factors and trade waste-
-Purification. 2. Waste minimization. I. Vamos, Richard J.
II. Title.
TD1060.H33 1995
628.4—dc20 94-26977
CIP

Acquisitions editor: *William Stenquist*
Editorial/production supervision: *Raeia Maes*
Cover design: *Cathy Mazzucca*
Manufacturing buyer: *William Scazzero*

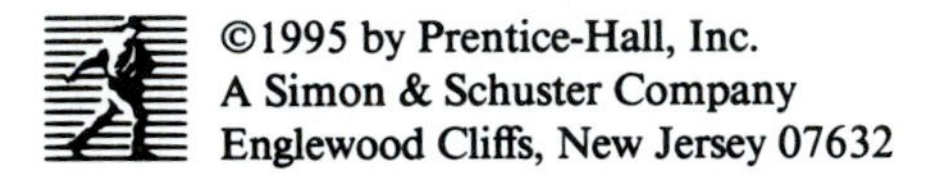

A Simon & Schuster Company
Englewood Cliffs, New Jersey 07632

Printed in the United States of America

10 9 8 7 6 5 4 3 2 1

ISBN 0-13-123472-2

Prentice-Hall International (UK) Limited, *London*
Prentice-Hall of Australia Pty. Limited, *Sydney*
Prentice-Hall Canada Inc., *Toronto*
Prentice-Hall Hispanoamericana, SA, *Mexico*
Prentice-Hall of India Private Limited, *New Delhi*
Prentice-Hall of Japan, Inc., *Tokyo*
Simon & Schuster Asia Pte. Ltd., *Singapore*
Editora Prentice-Hall do Brasil, Ltda., *Rio de Janeiro*

Contents

PREFACE ***xi***

1 SOURCES AND CLASSIFICATION OF HAZARDOUS WASTES 1

Sources of Hazardous Wastes 1
Quantitative Waste Generation Estimates 4
Waste Generation Factors 5
Comparisons of Geographic Regions and Various Industries 6
Heterogeneity in Scale of Waste Generation 8
Application of Generation Factors 9
Characteristics of Hazardous Waste 11
Domestic Hazardous Waste Generation 13
Problems and Discussion Questions 13
References 14
Bibliography 14

2 ASSESSMENT OF EXPOSURE POTENTIAL: TRANSPORT PROCESSES 20

Introduction 20
Routes of Exposure 21
Direct Contact 21
The General Contaminant Mass Balance 21
Contaminant Transport Equations 23
Unreactive Contaminants, 23
Reactive Contaminants, 28
Dispersion Estimates, 29
Estimates of Sorption Equilibria, 30
Estimates for Reaction Rate Due to Volatilization, 31
Estimates for Chemical Reaction Rates, 33
Estimates for Photolysis Rates, 33
Estimates for Biodegradation Rates, 35

Examples 36
Atmospheric Dispersion 39
Groundwater Transport 45

Problems 51
References 53

3 *OVERVIEW OF THE WASTE MANAGEMENT PROBLEM* **56**

Ongoing Versus "Old" Sites 56
Source–Receptor Paradigm 57
Matching Applicable Wastes with Applicable Processes 57
Levels of Design 59

4 *PHYSICAL WASTE TREATMENT PROCESSES* **60**

Aim of Presentation 60
Process Classification 60
Separation by Differential Action of a Force 61
Gravitational Separation Processes, 62
Centrifugation Processes, 67

Separation by Differential Passage Through Fixed Media 70
Granular Filtration, 70
Vacuum Filters, 75
Pressure Filters, 77
Common Features of Pressure and Vacuum Filters, 78
Macromolecular and Molecular Filtration, 78

Separation by Differential Phase Partitioning 86
Liquid–Liquid Separation, 86
Vapor–Liquid Separation, 90

Solid–Liquid and Solid–Vapor Separation 105
Problems 112
References 114

5 *CHEMICAL WASTE TREATMENT PROCESSES* **118**

Process Classification 118
Neutralization 118
Introduction, 118
Solution pH, 119
Acids and Bases, 120
Acidity, Basicity, and Buffering, 122
Neutralization Processes, 125
Neutralization Process Design and Control, 130

Precipitation 132
Introduction 132
Nucleation, 137
Activity Coefficient Calculations, 138

Precipitant Dosage, 140
Coagulation/Flocculation, 144
Coprecipitation, 145

Design of Precipitation Processes 145
Common Precipitation Applications, 147

Chemical Oxidation and Reduction 152
Introduction, 152
Redox Reactions in Hazardous Waste Treatment, 158
Applications of Redox Processes in Hazardous Waste Treatment, 159
Cyanide Destruction, 160
Oxidation of Organic Compounds, 163
Wet Oxidation, 163
Supercritical Water Oxidation, 164
Ozone and Advanced Oxidation Processes, 167
Molecular Ozone, 169
Ozone Mass Transfer, 170
Molecular Ozone Process Design, 171
Ozone AOPs, 172
Ozone AOPs Process Design, 177
AOP Treatment Costs, 179
Applications of Ozone and Other AOPs, 180

Ion Exchange 181
Ion Exchange Types, 185
Physical Properties of Ion Exchangers, 187
Classification of Ion Exchangers, 190
Ion Exchange Equilibria, 195
Ion Exchange Kinetics, 205
Ion Exchange Processes, 207
Columnar Ion Exchange Processes, 209
Local Equilibrium Model, 214
Limitations of Binary Local Equilibrium Model, 217
Column Design and Operation, 218
Applications of Ion Exchange in Hazardous Waste Treatment, 226
Ion Exchange Process Costs, 230

Solidification/Stabilization 230
Problems 231
References 233

6 BIOLOGICAL TREATMENT PROCESSES 240

Why and How Microorganisms Grow 240
Stoichiometry, 241
Energetics of Microbial Growth, 244
Kinetics, 246

Physical Classification of Microbial Processes 250
Suspended Growth Processes, 250

Design Equations for Processes 255
Batch Suspended, 255
Batch Film, 257
Continuous Suspended, 260
Continuous Film, 271

Oxygen Transfer Considerations 280
Operational Stability of Anaerobic Processes 281
Biological Degradation of Xenobiotics 283
Genetic Alterations, Acclimation, and Selection 284
Volatilization and Physical Removal Mechanisms 285
Coupled Physical and Biological Processes 286
Problems 287
References 288

7 *THERMAL PROCESSES* *292*

Energetics and Stoichiometry of Combustion Processes 293
Development of a Combustion Equation, 293
Incineration Enthalpy Balances, 296

Equilibria and Kinetics of Combustion Reactions 303
Equilibrium Constant Estimation, 303
Kinetics, 306
Turbulent Conditions during Incineration, 308
Formation of NOx, CO, and Acid Gases, 308
Products of Incomplete Combustion, 310

Physical Configurations 311
Liquid Injection, 311
Rotary Kilns, 311
Hearth Incinerators, 314
Fluidized Bed Systems, 316
Miscellaneous Systems, 317

Air Pollution Equipment 317
Conduct of a Trial Burn 318
Operations and Control 321
Problems 322
References 323

8 *WASTE ELIMINATION OPTIONS* *326*

Introduction 325
Strategies for Minimization 326
Improving Plant Operations, 326
Altering Process Technology, 328

Recycle/Recovery/Reuse, 329
Changing Raw Materials, 330
Product Reformulation, 331

Waste Audits 331
Pre-Audit Phase, 332
Audit Phase, 333
Feasibility Analysis, 334
Implementation Phase, 339

Institutional Disincentives 339
Problem 339
References 339

9 SYSTEMS ANALYSIS FOR REGIONAL PLANNING OF HAZARDOUS WASTE MANAGEMENT OPTIONS 342

Introduction 342
Problem Formulation 342
System Constraints, 343
Formulation of the Objective Function, 344

Solution Methodology 345
Problem Generalizations 348
Nonlinearity of Cost Functions, 348
Multiple Waste Types and Treatment/Disposal Facilities, 349
Multiperiod Problems, 350
Impact Assessment of Alternatives, 350
Sensitivity Analyses, 352

References 353

INDEX 354

Preface

Environmental engineers have, from the inception of the discipline, been concerned with the mitigation of adverse impacts of people's activities. This originated with control of domestic pollution of the air and water. With industrialization and urbanization, the field of concern has expanded to control of discharge of hazardous wastes. The recognition of the hazardous waste problem came to public attention with catastrophic events at "Valley of the Drums" and Love Canal. In the United States, this resulted in the passage of the Resource Conservation and Recovery Act (RCRA) and the Comprehensive Environmental Restoration Cleanup and Liability Act (CERCLA, or "Superfund"). These legislative initiatives resulted in substantial need to train individuals to assess, design, and implement methods for hazardous waste control.

Hazardous waste problems may be divided, albeit imperfectly, into the management of ongoing sources of hazardous and industrial waste and the remediation of prior activities. The focus of this book is on the former problem, although many technologies described here may be applied to remediation activities. Furthermore, the focus of this book is specifically on treatment and minimization of these wastes, rather than their ultimate disposal. The latter material is, in its own right, the locus of a distinct course, and often those involved in ultimate disposal issues, for example, landfills, must possess a quite distinct variety of skills from those most directly involved in the former issues.

This book was written to serve as a textbook and reference for those engaged in finding solutions to the management of ongoing sources of hazardous waste. These individuals may have prior training in environmental science and engineering or in one of the other science or engineering disciplines. This text can serve as the foundation for one or more courses in the field of hazardous waste management.

In developing this material, the authors have assumed a limited basic knowledge of environmental chemistry and in pollution control. This prior background may have been obtained in the typical undergraduate or graduate introductory courses. The instructor who uses this book can supplement this material with current journal and regulatory references and with case studies of particular local interest.

This book was developed by the authors from teaching at Rensselaer Polytechnic Institute, Illinois Institute of Technology, the University of Illinois at Urbana-Champaign, and Drexel University. The authors thank the many students and professional colleagues who contributed to the evolutionary development of this material. Among those whose ideas assisted in this development are James W. Van Nortwick, James W. Patterson, Kenneth E. Noll, William W. Shuster, and Kirankumar Topudurti. The first author is also particularly grateful to Professor Richard S. Engelbrecht, of the University of Illinois at Urbana-Champaign for hosting a sabbatical during which much of the present material was refined. The second author is particularly grateful to his colleagues at DePaul and Associates for their thoughts, ideas, and insights.

The staff at Prentice Hall, particularly Doug Humphrey and Bill Zobrist, are acknowledged for their forbearance in allowing this effort, which developed through a prolonged gestation, to come to a fruitful conclusion. Finally, both authors thank their professional colleagues and their families for their understanding in allowing time and resources to be brought to bear in this effort.

Like most other environmental activities, there is never a final answer, but only a process that ever more closely approaches the ideal solution. Clearly the future will bring new problems and uncover better or alternative methods for solution of these problems. The authors hope that, in contributing to the education of those who will advance the field, some modest effort toward environmental protection has been made.

Charles N. Haas
Richard J. Vamos

Hazardous and Industrial Waste Treatment

1

Sources and Classification of Hazardous Wastes

SOURCES OF HAZARDOUS WASTES

As a rough estimate, 1 metric ton of hazardous wastes is generated per capita per year in the United States (Conservation Foundation, 1984). Since 1980 and the legislative enactment of the Resource Conservation and Recovery Act (RCRA) there have been diverse estimates of the volume of hazardous waste generated. Table 1-1 presents a summary of various national surveys on waste generation.

There was a sixfold increase between the 1980 and the 1983 estimate, due to a more complete coverage of manufacturing operations. One unknown is an estimate of the degree to which manufacturing operations that generate hazardous waste actually comply with applicable regulations and therefore are included under national surveys and compilations of data.

Hazardous wastes originate from diverse sources. Even if the whole superstructure of RCRA did not exist, classification of the sources and types of hazardous wastes would be necessary to allow us to efficiently handle these materials.

The standard system of categorization used by the Census Bureau is called the Standard Industrial Classification (SIC Code). SIC codes are analogous to a Dewey Decimal system for industries in the United States. This system of classifying sources provides a relatively straightforward method of categorizing industrial and commercial operations in the United States by means of their *principal economic activity.* It also provides the basis by which the Census Bureau ascertains the various economic activities of industries and businesses.

Table 1-2 enumerates the two digit codes and provides some examples of three and four digit codes. The Census Bureau defines SIC codes up to six digits, providing a high degree of precision for describing industrial waste sources. For most purposes, two and three digit subdivisions are the most useful for waste categorization.

TABLE 1-1 SUMMARY OF SURVEYS ON NATIONAL HAZARDOUS WASTE GENERATION. (Adapted from Conservation Foundation, 1984.)

Year	Estimate in million metric tons	Comment
1978	56	Environmental Protection Agency (EPA) estimate in 1978 of generation to be expected in 1980. 61 percent expected to come from industrial sources.
1980	41	EPA contractor estimate in 1980 of likely waste generation in that year, with expected range 28–54 million metric tons.
1983	250	Office of Technology Assessment estimate based on surveys of all states, with range expected between 255–275 million metric tons.
1984	264	EPA contractor estimate of waste that was generated in 1981 based on manifest data.

As an example of the use of these codes, SIC 28 refers to the entire realm of industries that deals with chemical manufacture—pesticides, inorganic chemicals, organic chemicals, petrochemicals, and so on. The U.S. Census of Manufacture (U.S. Department of Commerce, 1981) includes information on the total employees, dollar value of sales, and dollars of value added by manufacture for that particular industry in geographic subdivisions of the United States. Table 1-3 is a summary of information on this particular industry taken from U.S. census data.

There is a disadvantage to the use of SIC codes for categorizing waste management. Since industries are classified by the census on the basis of their primary end-product, different industries that generate hazardous waste in common intermediate operations are not necessarily classified together. For example, an automobile plant would be classified as SIC code 3711. However, that same auto-

TABLE 1-2 STANDARD INDUSTRIAL CLASSIFICATION (SIC) WITH EXAMPLE SUBDIVISIONS. (U.S. Department of Commerce, 1981.)

Code	Industry
01–09	Agriculture
10–14	Mining
15–17	Construction
20–39	Manufacturing
	28 Chemical manufacture
	286 Industrial organic chemicals
	2862 Explosives
40–49	Transport, communication, and utilities
50–51	Wholesale trade
52–59	Retail trade
60–67	Financial, insurance, real estate
70–89	Services
91–97	Governmental
99	Nonclassifiable

TABLE 1-3 CENSUS INFORMATION FOR SIC 28. (Chemical Manufacture; U.S. Department of Commerce, 1981.)

Statistic	Value for 1977 (Nationwide)
Purchased fuels and electrical energy	2.978.5 trillion BTU
Employees	880,200
Payroll	$13,839 million
Value added by manufacture	$56,721 million
Value of shipments	$118,154 million
Cost of materials	$62,294 million

mobile plant contains electroplating and metal finishing operations, degreasing processes, and so forth. Therefore, in classifying an industry on the basis of the end-product, the fact that industry contains a number of particular processes that have more in common with other industrial categories may be overlooked.

Another example of the potential misclassification errors obtained by strict reliance on SIC codes occurs in the computer industry. The census code for computer manufacturers is 3573. However, the computer industry manufactures semiconductor chips and circuit boards. In such processes, operations of electroplating, washing and degreasing the circuit boards, and metal forming occur. These have much more in common with the automotive and electroplating industries in terms of waste products produced than is implied by the end-product or the associated SIC code. Unfortunately, the SIC is the only realistic alternative for classifying waste, since it is the only readily available scheme by which economic data are classified.

Of the two digit SIC categories, the primary sources of hazardous waste are contained in the broad series between 20 and 39, and, to a somewhat lesser degree, in the range between 91 to 97 which, although described as governmental, include industries such as waste treatment plants, universities, and hospitals.

Certain industries generate wastes which are, under the legislative or regulatory definitions of hazardous waste, specifically excluded. These include mining, and perhaps more importantly, agricultural wastes (40 CFR 261.4[1]). Thus, the farmer who has used a drum of pesticide, despite the fact that pesticide is a hazardous material, may not be generating a hazardous waste in a regulatory sense; agricultural waste is specifically excluded under RCRA.

Within the various industries, hazardous wastes originate from a relatively common variety of specific operations. These can be divided into four very common sources, with a number of special cases other than the four. The four major categories may be described as:

1. Off specification products and excess raw materials;
2. Spent catalysts and purification residues;
3. Sludges and other residuals from waste treatment operations within the manufacturing facility;
4. Contaminated solvents and solvent residues.

[1] CFR is the abbreviation for "Code of Federal Regulations" in which all regulatory matters, including environmental protection, are codified. The first number refers to the title and the second number refers to the section within the regulations.

Probably the most common source of hazardous waste by all of the industries is off specification products and excess raw materials. For example, a pharmaceutical plant may be carrying out a fermentation to produce a given drug. The plant gets a contaminated organism within the fermentation liquor, and therefore must discard the entire batch of that product. That, in combination with what went into the liquor to begin with, may now become a hazardous waste. As a second example, a fine chemical manufacturing operation may buy a slight surplus of a chemical that is relatively cheap in order to be sure that the other, more expensive, chemicals which are used by its manufacturing operation are used to a higher degree. If now discarded, the excess raw material may become a hazardous waste.

The second broad category consists of spent catalysts and purification residues. Activated carbon is used very commonly in a variety of chemical industries such as solvent manufacturing operations, and manufacture of other chemicals (Broughton, 1978). In such cases, the spent adsorbent or—in the case of a catalyst, where a catalyst might be used for oxidation—the spent catalyst, now may become hazardous because of its flammability or ignitability, for example. In the oil industry, many of the spent catalysts contain substantial concentrations of trace metals, which are at least partially responsible for their catalytic activity (Mills and Cusumano, 1978). When the catalysts become spent or become poisoned, they are discarded and thus may become a hazardous waste.

The third category consists of waste treatment sludges. For example, an electroplating firm (a firm that plates metal on various surfaces) will very often precipitate its wastewater with lime or sodium hydroxide to remove metals from the bulk liquid (Patterson, 1985). The result is a sludge which is a hazardous waste (based on the "derived from" rule) and must be disposed. This rule stipulates that any waste arising from treating or processing a specifically listed hazardous waste is, itself, a hazardous waste.

The fourth and final category consists of contaminated solvents, and solvent residues. Solvents are used in a variety of operations, such as degreasing. In a number of cases the spent solvents are then purified such as by extraction. The spent solvents themselves, or the residue from the repurification operations, become waste to the industry that generated them.

These four sources are probably the most common sources of waste from all the industries of all the SIC codes, and each of the industries in each of the groups might have quite specific types of processes from which each of four broad categories of waste originate.

QUANTITATIVE WASTE GENERATION ESTIMATES

Much of the early data on hazardous waste generation in the United States comes from before 1976—the date of RCRA passage. Indeed, estimates of the amount of hazardous waste were formulated as early as the mid-1960s. In the early 1970s, there were a series of studies done under federal sponsorship in which hazardous waste disposal practices were assessed by industry. These are referred to as "industrial assessments." The bibliography at the end of this chapter enumerates the industrial assessments that were part of this series.

In these industrial assessments, flowcharts were developed for a single industry and, from this, the total waste flow to all media was assessed on a per unit product production basis. Since many of these studies were done even before the contemplation of RCRA, and, therefore, before the term "hazardous waste" was carefully defined, the data of these industrial assessments must be interpreted carefully.

After the passage of RCRA in 1976, but prior to its implementation in 1980, a number of individual states prepared an estimate of the amount of hazardous waste being generated within their boundaries. These "state-wide assessments" were formulated on the basis of the RCRA legislative definition of hazardous waste. These statewide assessments come a bit closer to the current definition of hazardous waste, but they still may differ to a degree. The bibliography to this chapter enumerates the statewide assessments prepared during this timeframe.

The best data currently available arises from the requirements, under RCRA, for each state to prepare an annual or biannual report of all the hazardous waste that has been generated, treated, stored, or disposed within that state. All manifests originating within the state are tallied and then categorized by industry (according to SIC), by type of waste (US EPA waste category), by fate, and so forth. For example, the Illinois report for 1983 indicates that about 6,000 metric tons of hazardous waste were accepted into Illinois that originated in California (Illlinois EPA, undated).

WASTE GENERATION FACTORS

Using the waste generation data, it is desirable to reduce it in some manner that would permit the development of rules of thumb for estimating the magnitude of new sources. Using all industries of a given category, one could determine the average waste production by region and assess geographic variability. If such a rule of thumb existed, then for a new manufacturing operation one could estimate how much hazardous waste it might be expected to generate. If one wanted to develop a management plan for an entire region, in the absence of actual data, one could use such rules of thumb to estimate, on the basis of census data, the waste generation profile.

The following rule of thumb has been used with classical domestic wastewater treatment: 100 gallons of wastewater is generated per capita per day (Water Pollution Control Federation, 1977). Within at least a factor of two, the population of a locality can be multiplied by 100 gallons per capita per day to get something fairly close to the wastewater generation rate. For example, in the city of Chicago, using 2 million as a round number for the population times 100 gallons per capita per day, one gets 200 million gallons per day.

Using such generation factors, one could assess systematic differences among states or geographical regions in hazardous waste generation, treatment, or disposal practices. For example, if a leather tanning operation in one state produces an above average amount of waste, certain deductions might be made about that state's system for dealing with hazardous waste, or about relative manufacturing conditions. There might be geographic variability in generation rates. Perhaps the regulatory officials in one state exert a different type of pressure than the regulatory officials in other states. Maybe there are fewer land disposal facilities available, at greater cost, in one state. Perhaps the cost of water or sewage disposal is different, and this encourages more volume and more waste. Can these differences be assessed systematically?

To produce such generation factors, the various sources of hazardous waste generation data may be collated with census information, or some other source of economic activity. One must then decide the most appropriate measure of industrial scale by which to divide waste generation. In the case of domestic wastewater generation, cited earlier, the scale factor is traditionally population, so that the generation factors are reported as gallons per capita per day.

What is the proper unit to divide the overall amount of hazardous waste by in order to get a number that might properly reflect the scale of that industry? Based on the readily available census information, there are three possibilities. The first follows the practice of ordinary industrial wastewater treatment and scales waste generation by unit of product produced. For example, in the case of leather tanning one would use pounds of waste generated per thousand square feet of leather processed. The second possible divisor is waste generated per size of facility, measured by number of manufacturing employees. The third divisor that might be used is value added by manufacture. Value added by manufacture is defined as the value of the finished product shipped minus the value of the energy and materials used to manufacture that product in the enterprise (U.S. Department of Commerce, 1981).

Now, the argument in favor of this third divisor is as follows: If the traditional wastewater generation rate (waste produced/unit product) is used, a different ratio results for each industry, and it becomes difficult to compare industries since they are computed on incommensurable scales.

The second and third divisors both produce common scales of measurement. Both will give the same units (waste produced/employee or waste produced/dollar, respectively) for all industries. The argument in favor of dollar value added by manufacturing, rather than per manufacturing employee, is obtained by considering for a moment what would happen when a manufacturing facility automates its production process. Using a waste generated per employee ratio, if the output did not change, the generation factor would increase. However, if the value added by manufacturing divisor was used, the generation factor would remain essentially constant. Hence, the latter approach appears to be less sensitive to variations in degree of automation than the former approach. Therefore, all the generation factors reported here will employ the latter approach.

COMPARISONS OF GEOGRAPHIC REGIONS AND VARIOUS INDUSTRIES

Table 1-4 summarizes waste generation factors estimated from a variety of data sources for diverse industries and for various geographical regions. Table 1-5 summarizes, for the various databases, the relative proportion of hazardous waste from

TABLE 1-4 HAZARDOUS WASTE GENERATION FACTORS BY SIC FROM SEVERAL DATA BASES.

		Gallons generated/$ value added by manufacture			
		Chicago	New York State	Nationwide	
SIC	Industry	(1)	(2)	(3)	(4)
26	Paper and allied products	0.0225	0.00183	0.0208	—
27	Printing and publishing	0.00001	0.00006	0.000379	—
28	Chemicals and allied products	0.00856	0.0529	0.102	0.50
29	Petroleum and coal products	0.0176	0.0756	0.0237	0.079
30	Rubber and misc. products	0.0002	0.003	0.00369	—
31	Leather and leather products	0.0284	0.0193	0.013	—
32	Stone, clay and glass	0.00071	0.00568	0.012	—
33	Primary metals	0.00848	0.0355	0.0713	0.027
34	Metal products	0.00283	0.00165	0.0168	0.018
35	Nonelectrical machinery	0.00301	0.00221	0.00938	0.034
36	Electrical and electronics	0.0023	0.00107	0.00145	0.014
37	Transportation equipment	0.0142	0.00654	0.00641	0.036
38	Instruments and related products	0.00017	0.002	0.00129	—
39	Miscellaneous manufacturing	0.00053	0.00147	0.00236	—

(1) Patterson and Haas (1982).

(2) Original data from Iannotti et al. (1979).

(3) Computed from national extrapolations of Jennings (1982).

(4) Computed from USEPA (1983).

each region attributable to that industry. Based on this table, several observations can be made.

First of all, the major sources of hazardous waste are the chemical industry (SIC code 28) which accounts for roughly 70 percent of the hazardous waste nationwide, followed by primary metals (SIC 33), which account for approximately 20 percent. All other industries make up the remaining 10 percent. However, this balance will dramatically alter in smaller geographical areas. For example, in the Metropolitan Water Reclamation District of Greater Chicago (MWRDGC), there is an approximate reversal of those two industrial categories. This indicates the necessity of characterizing the precise mix of contributors to the hazardous waste stream in a given geographic area prior to developing a management plant.

The second broad observation is that, if the generation factor itself (waste produced/value added) is accepted as an index of the relative propensity of waste to be produced, industries differ in the degree to which they produce waste. In the table, there is an enormous difference among generation rates, with, for example, pulp and paper the low example and the chemical industry much higher.

A third observation arises from comparing SIC generation rates for the various geographic regions. For example, the generation factors for the primary metals industry (SIC 33) among the different data sets differ by a factor over 8. This may be real, or it may reflect various dates of some of the surveys, or the changing definitions

TABLE 1-5 PROPORTION OF HAZARDOUS WASTES ATTRIBUTABLE TO VARIOUS SIC's IN DIVERSE SURVEYS COMPARED TO PROPORTION VALUE ADDED BY MANUFACTURE.

		Percentage of Total					
		Chicago[1]		New York State[2]		Nationwide[3]	
SIC	Industry	Waste	VAM	Waste	VAM	Waste	VAM
26	Paper and allied products	13.7	2.8	0.7	2.7	3.8	3.8
27	Printing and publishing	0.2	10.6	0.1	14.3	0.1	5.5
28	Chemicals and allied products	13.4	8.0	55.4	7.0	47.8	9.7
29	Petroleum and coal products	2.4	0.6	3.3	0.3	3.2	2.8
30	Rubber and misc. products	0.1	2.6	0.8	1.8	0.6	3.4
31	Leather and leather products	17.7	0.3	2.8	1.0	0.4	0.6
32	Stone, clay and glass	0.2	1.6	1.9	2.2	1.9	3.3
33	Primary metals	13.5	8.0	19.6	3.7	22.1	6.4
34	Metal products	6.9	12.5	0.9	5.1	6.3	7.9
35	Nonelectrical machinery	6.8	10.4	3.5	10.8	5.2	11.5
36	Electrical and electronics	5.0	13.4	1.6	10.0	0.6	8.6
37	Transportation equipment	15.1	5.2	4.8	5.5	3.4	11.0
38	Instruments and related products	0.1	4.1	3.6	12.2	0.2	3.2
39	Miscellaneous manufacturing	0.3	2.8	0.7	3.6	0.2	1.8

[1] Patterson and Haas (1982).

[2] Original data from Iannotti et al. (1979).

[3] Computed from national extrapolations of Jennings (1983).

VAM—Value Added by Manufacture (1977)—as % of total in region.

of what may have been hazardous, or differences among states in implementing regulations.

HETEROGENEITY IN SCALE OF WASTE GENERATION

The additional major factor which must be considered is the wide range in size of the various classes of hazardous waste generators. Figure 1-1 plots the frequency distribution of hazardous waste management (generators, and treatment, storage, and disposal facilities) by the magnitude of their annual waste generation. The largest number of waste generators is responsible, collectively, for the smallest amount of waste generated, while the largest fraction of total waste generated is from a relatively small number of large sources. Essentially, 90 percent of the waste is produced by 10 percent of the generators. Thus, if the concern were simply about managing overall volume, the effort might be best concentrated on managing the larger generators. Therefore, in the evolution of EPA regulations, first consideration was given to larger sources, and subsequent concern was extended to the so-called "small quantity generators."

This volume frequency distribution depends upon the individual industry being considered. For example, in the Chicago area, the average annual waste generated by paper and allied products (SIC 26) is 34,000 gallons, while in the rubber

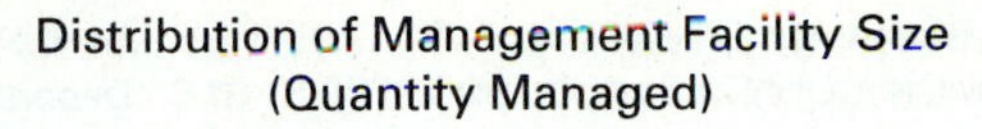

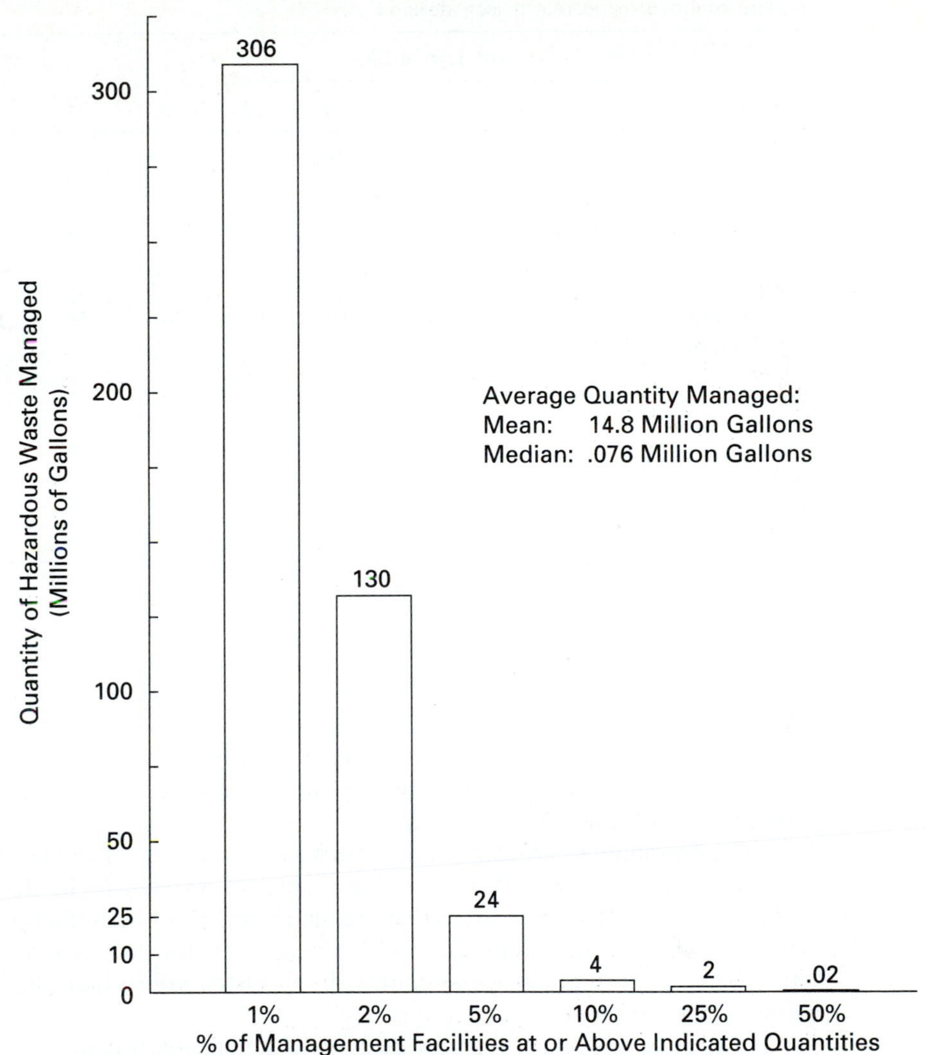

Figure 1-1 Histogram of frequency of generators of size (U.S.). (U.S. EPA, 1984.)

industry (SIC 30), the average is 1,100 gallons. This implies a much different set of handling facilities and modes of transportation by the diverse industries.

APPLICATION OF GENERATION FACTORS

The estimates of generation factors that are produced can be used to estimate the geographical distribution of waste generation. Of course, actual data will yield more precise estimates. Generation factors, however, can be used to make an initial ap-

TABLE 1-6 DOLLAR VALUE ADDED BY MANUFACTURE IN NORTH CAROLINA FOR MAJOR COUNTIES AND INDUSTRIES (U.S. Department of Commerce, 1981). (*Note:* Figures rounded to nearest million dollars.)

	10^6 \$/yr in SIC								
County	26	27	28	30	32	33	34	35	39
Catawba				16					
Durham		16							
Forsyth							107		
Gaston		17	38				12	113	11
Guilford		22	34				9		
Iredell							32	26	
Mecklenberg	17	81	137	108		16	59	195	
New Hanover			246						
Pitt								22	
Rowan				11					
Rutherford				12					
Union							17		
Wake							72		
Wayne							5		
TOTAL	17	136	455	147	11	16	313	355	11

Note: SIC's with insignificant activity excluded.

proximation of the nature of the waste generation pattern, and to pinpoint possible areas where further attention must be directed. This will be illustrated by example.

Suppose we are interested in estimating hazardous waste generated in the State of North Carolina. This example is selected since there is a publication of the actual North Carolina data that can be used to check the precision of the method of estimation (Pierce and Davidson, 1982).

Data on manufacturing by county are needed to do the estimate. The census figures for the state of North Carolina are listed in Table 1-6. In this table, for simplicity, only the major counties that are major centers of manufacturing activity and the major SIC codes are listed. Within the table are the dollars added by manufacture per year, in that particular county, by industry within that SIC code. This represents the data base.

Now, using any one of the previously tabulated generation factors, along with the census data, one can determine the amount of hazardous waste anticipated from that industry in that county.

For example, using the U.S. EPA national average data base, one can construct the county-by-county waste estimates in Table 1-7. (Table 1-7 also compares the estimates to actual generation rates noted by Pierce and Davidson (1982).) By adding all industries in a given county for any set of the generation factors in Table 1-4, one can obtain estimates of the geographic dispersion of sources. In the case of the national or the overall EPA generation factors, one arrives at an estimate of 62 million gallons per year from all sources in all counties. The actual data indicates 41 million gallons per year. Using the other generation factors, one obtains 6.4 million gallons per year and 26.5 million gallons per year from the MSDGC and New York

TABLE 1-7 PROJECTED HAZARDOUS WASTE GENERATION RATES IN COUNTIES OF NORTH CAROLINA USING GENERATION FACTORS FROM USEPA (1983) COMPARED TO ACTUAL.

County	Million gallons/Year Estimated	Actual
Mecklenberg	18.673	10.724
Guilford	3.578	7.631
Gaston	5.183	3.797
Wake	6.367	2.157
Rowan	0.137	2.271
New Hanover	25.133	3.551
Cumberland	—	2.112
Forsyth	1.803	0.852
Other	1.537	8.113
Total	62.411	41.195

State surveys repectively. The average of the estimates from all of the generation factors is 34 million gallons per year, which is within 25 percent of actual.

At the level of individual counties, this approach is much less precise. From the table, the estimates may be no better than an order of magnitude, looking in particular at New Hanover county, where the national generation factors project 25 million gallons per year versus actual generation of 3.5 million gallons. This is not surprising since, for smaller geographical regions, the average is taken over fewer industries, which magnifies the potential for error.

CHARACTERISTICS OF HAZARDOUS WASTE

From a management point of view, the EPA waste classification numbers are not useful. Instead, one needs a categorization scheme that is less finely differentiated than that, so as to enable the development of preliminary plans for management and treatment systems. Table 1-8 presents one type of reasonable categorization framework which will be used in this book.

Initially, Table 1-8 divides waste into two categories, according to whether it is primarily organic or primarily inorganic. Within the inorganic category, materials are subdivided according to a chemical property: acidic, alkaline, neutral. Within the organic category, the concentration of the organic material is judged. At this second level, subcategories are not made on the basis of acidity, because acidity is not as relevant to organic waste as it is for inorganic waste.

The implications of acidity and alkalinity are straightforward. Defining organic concentration is more difficult. For our purposes, concentrated organic is that level of organic material that will support self-sustaining combustion—generally regarded as 10–15 percent organic carbon. This level of organic matter also generally corresponds to the point where, physically, the material starts to resemble a cake.

TABLE 1-8 WASTE CATEGORIZATION SCHEME MODIFIED FROM SHUSTER ET AL. (1979).

INORGANICS

Acid Wastes

100 relatively uncontaminated
101 containing metals
102 containing organics
103 containing minor contaminants other than the above

Alkaline Wastes

110 relatively uncontaminated
111 containing metals
112 containing organics
113 containing nonmetallic cyanides
114 containing metals and cyanides
115 miscellaneous cleaning solutions
116 alkaline sludges

Other Inorganic Wastes

120 salt solutions
121 solids
122 metals
124 nonmetals

ORGANICS

Concentrated Liquids

200 lightly contaminated halogenated
201 lightly contaminated nonhalogenated
202 lightly contaminated mixed solvents
203 highly contaminated halogenated
204 highly contaminated nonhalogenated
205 highly contaminated mixed solvents

Dilute Aqueous Solutions

210 readily oxidized, halogenated
211 readily oxidized, nonhalogenated
212 difficult to oxidize, halogenated
213 difficult to oxidize, nonhalogenated

Organic Solids

221 salts
222 tars and residues
223 sludges

Organic Gases/Vapors

231 combustible
232 noncombustible

SPECIAL WASTES

301 strong oxidizers
302 strong reducers
303 explosives
304 biological wastes

At a third level of classification, one can determine more specific chemical characteristics. If it is inorganic, does the material contain metals or dilute organics, or does it contain cyanide?

In subsequent chapters, as various treatment and management systems are

TABLE 1-9 PER HOUSEHOLD GENERATION OF HAZARDOUS WASTES IN ALBUQUERQUE, NEW MEXICO (Adapted from Brown, 1987).

	Pounds/household-year Disposal methods				
Waste Type	Landfill	Sewer	On/off site	Recycle	TOTAL
Antifreeze	0.940	0.605	1.642		3.186
Cleaners		0.049	0.011		0.059
Drain openers	0.005	0.005	0.005		0.016
Fertilizers	0.081	0.005			0.086
Motor oils	5.217	0.346	3.737	1.674	10.974
Paints	1.685	0.043	0.022		1.750
Pesticides	0.221	0.007	0.011		0.239
Poisons	0.011	0.005			0.016
Polishes		0.005			0.005
Solvents/thinners	0.124	0.005	0.130		0.259
TOTAL	8.339	1.033	5.547	1.674	16.592

considered, the individual types of wastes in terms of this classification which are amenable to treatment by various processes will be indicated.

DOMESTIC HAZARDOUS WASTE GENERATION

In the foregoing discussion, the focus has been on hazardous wastes generated by industrial and commercial operations. It is important to recognize that households can also generate hazardous wastes. There is less quantitative data on this source, but the evidence suggests that the magnitude of domestic hazardous waste generation is far less than from commercial and industrial activities. Nevertheless, for the operation and management of domestic waste treatment facilities (either wastewater or solid waste), the presence of hazardous materials in the waste stream must always be considered. Table 1-9 summarizes the waste generation reported in a recent survey in Albuquerque, NM (Brown, 1987). Comparing the overall hazardous waste generation rate of about 1 metric ton/capita-year to the rate of 16.6 lb/household-year (7 kg/household-year), it becomes clear that the household generation of hazardous wastes is under 1 percent of the total.

PROBLEMS AND DISCUSSION QUESTIONS

1.1. In 1982, Illinois reportedly generated 629 million gallons of hazardous waste. How does this compare to estimates that can be made using census information?

1.2. Consider the estimates for waste generation from North Carolina. If the objective is to refine the estimates for geographic distribution of waste generators, from what industries would it be most cost-effective to attempt to collect more precise information?

1.3. Use the most recent census information available for your state. In what counties is the most waste generated? Which are the largest waste generation industries?

REFERENCES

BROUGHTON, D. B., "Adsorptive Separation," *Kirk Othmer Encyclopedia of Chemical Technology,* 1:531–81, John Wiley, New York (1978).

BROWN, T. S., "Household Hazardous Waste: The Unresolved Water Quality Dilemma," *Journal of the Water Pollution Control Federation,* 59:120–24 (1987).

CONSERVATION FOUNDATION, *State of the Environment: An Assessment at Mid-Decade,* Washington, D.C. (1984).

IANNOTTI, J., et al., *An Inventory of Industrial Hazardous Waste Generation in New York State,* New York State Department of Environmental Conservation (1979).

ILLINOIS EPA, "Computer Printouts for the 1983 Annual Hazardous Waste Report" (1985).

JENNINGS, A. A., "Profiling Hazardous Waste Generation for Management Planning," *Journal of Hazardous Materials,* 8:69–83 (1983).

MILLS, G. A., AND CUSUMANO, J. A., "Catalysis," *Kirk Othmer Encyclopedia of Chemical Technology,* 5:16–61, John Wiley, New York (1978).

PATTERSON, J. W., *Industrial Wastewater Treatment Technology,* Butterworth, Boston (1985).

PATTERSON, J. W., AND HAAS, C. N., *Management of Hazardous Wastes—An Illinois Perspective,* Report to the Illinois Institute of Natural Resources (1982).

PIERCE, J. J., AND DAVIDSON, G. M., "Linear Programming in Hazardous Waste Management," *ASCE Journal of the Environmental Engineering Division,* 108: 10–14, (1982).

SHUSTER, W. W., et al., *Technology for Managing Hazardous Wastes,* Report to the New York State Environmental Facilities Corporation (1979).

U.S. DEPARTMENT OF COMMERCE, *1977 Census of Manufacturers,* Washington, D.C. (1981).

USEPA, "National Survey of Hazardous Waste Generators, Treatment, Storage and Disposal Facilities Regulated Under RCRA in 1981: Preliminary Highlights of Findings" (1983).

USEPA, "National Survey of Hazardous Waste Generators and Treatment, Storage and Disposal Facilities Regulated Under RCRA in 1981," EPA/530-SW-84/005 (1984).

WATER POLLUTION CONTROL FEDERATION, *Wastewater Treatment Plant Design,* Manual of Practice 8, Washington D.C. (1977).

BIBLIOGRAPHY

Industrial Assessments

Assessment in Industrial Hazardous Waste Management Petroleum Re-Refining Industry, John W. Swain, EPA OSWMP Report No. (SW-144c), June, 1977.

Assessment of Industrial Hazardous Waste Practices, Inorganic Chemicals Industry, Versar, Inc., EPA, OSWMP Report No. (SW-104c), March, 1975.

Assessment of Industrial Hazardous Waste Practices—Electronic Components Manufacturing Industry, Wapora, Inc., EPA OSWMP Report No. (SW-140c), 1977.

Assessment of Industrial Hazardous Waste Practices, Electroplating and Metal Finishing Industries Job Shops, Battelle Columbus Labs, EPA OSWMP Report No. (SW-136c), Sept., 1976.

Assessment of Industrial Hazardous Waste Practices, Leather Tanning and Finishing Industry, SCS Engineers, EPA OSWMP Report No. (SW-131c), Nov. 1976.

Assessment of Industrial Hazardous Waste Practices, Organic Chemicals, Pesticides, and Explosives Industry, TRW Systems Group, EPA OSWMP Report No. (SW-118c), April 1975.

Assessment of Industrial Hazardous Waste Practices, Paints and Allied Products Industry, Contract Solvent Reclaiming Operations, and Factory Application of Coatings, Francis Scofield et al., Wapora, Inc., EPA OSWMP Report No. (SW-119c), Sept. 1975.

Assessment of Industrial Hazardous Waste Practices, Rubber and Plastics Industry, Appendices. Foster D. Snell, Inc., EPA OSWMP Report No. (SW-163c.4), March, 1978.

Assessment of Industrial Hazardous Waste Practices, Rubber and Plastics Industry, Executive Summary, Foster D. Snell, Inc., EPA OSWMP Report No. (SW-163c.1), 1978.

Assessment of Industrial Hazardous Waste Practices, Rubber and Plastics Industry, Rubber Products Industry, Foster D. Snell, Inc., EPA OSWMP Report No. (SW-163c.3), 1978.

Assessment of Industrial Hazardous Waste Practices, Special Machinery Manufacturing Industries, Wapora, Inc., EPA OSWMP Report No. (SW-141c), March, 1977.

Assessment of Industrial Hazardous Waste Practices, Storage and Primary Batteries Industries, Versar, Inc., EPA OSWMP Report No. (SW-102c), January, 1975.

Assessment of Industrial Hazardous Waste Practices, Textiles Industry, US EPA OSWMP Report No. (SW-125c), (1976).

Assessment of Industrial Hazardous Waste Practices in the Metal Smelting and Refining Industry. Volume I. Executive Summary, Calspan Corp., EPA OSWMP Report No. (SW-145c.1), April, 1977. Volume II. Primary and Secondary Nonferrous Smelting and Refining. Calspan Corp., EPA OSWMP Report No. (SW-145c.2), April, 1977. Volume III (SW-145c.3), April, 1977. Volume IV (SW-145c.4), April, 1977.

Assessment of Hazardous Waste Practices in the Petroleum Refining Industry, Jacobs Engineering Co., EPA OSWMP Report No. (SW-129c), June 1976.

Industrial Hazardous Waste Practices, Rubber and Plastics Industry, Plastic Materials and Synthetics Industry, Foster D. Snell, Inc., EPA OSWMP Report No. (SW-163c.2), 1978.

Pharmaceutical Industry, Hazardous Waste Generation, Treatment, and Disposal, EPA OSWMP Report No. (SW-508), 1976.

Statewide Assessments

Alabama Waste Management Plan, prepared by Roy F. Weston, Inc., for the Alabama Water Improvement Commission, Nov. 1, 1978.

Alaska Summary of Hazardous Waste Sources in Alaska: FIT Task Report, prepared by Ecology and Environment, Inc. for the U.S. EPA, EFSR 8011-0120 (10), Nov. 1980.

Arizona A Report on Industrial and Hazardous Wastes, prepared by Behavioral Health Consultants, Inc. for the Arizona Department of Health Services, June 20, 1975.

Report to the Arizona State Legislature Regarding Siting of a Statewide Hazardous Waste Disposal Facility "Executive Summary", Arizona Department of Health Services, Division of Environmental Health Services, Bureau of Waste Control, Jan., 1981.

The Hassayampa Landfill Hazardous Waste Disposal Site: Disposal Analysis (Apr. 20, 1979–Oct. 28, 1980), Arizona Dept. of Health Services.

Arkansas Solid Waste Management Plan, State of Arkansas, Dept. of Pollution Control and Ecology, Solid Waste Management Division, Jan. 1981.

California Handbook of Industrial Waste Compositions in California 1978, David L. Storm, California Dept. of Health Services, Hazardous Materials Management Section, Nov. 1978.

Summary of Off-Site Hazardous Waste Generation for 1979, Unpublished Data Summary, Dept. of Health Services, Hazardous Materials Management Section.

Colorado Hazardous Wastes in Colorado: A Preliminary Evaluation of Generation and Geologic Criteria for Disposal. J. L. Hynes and C. J. Sutton, Colorado Geological Survey Dept. of Natural Resources and Colorado Dept. of Health, 1980.

Connecticut Connecticut 1979 Industrial Waste Generation and Disposal Alternatives Inventory, prepared by TRC Environmental Consultants, Inc., for the Connecticut Areawide Waste Treatment Management Board, Aug. 1979.

Florida Hazardous Waste Survey for the State of Florida, prepared by the Florida Resources and Environmental Analysis Center for the State of Florida Department of Environmental Regulation, Oct. 1977.

Hawaii Hazardous Waste Management Problem Assessment and Strategy Formulation for Hawaii, Guam, American Somoa, Northern by Gerretson, Elmendorf, Zinov, Reibin Architects and Engineers for the U.S. EPA. April, 1978.

Idaho Idaho Solid Waste Management Industrial Survey Report, Idaho Dept. of Environmental and Community Services, June, 1973.

Illinois Hazardous Waste Management in Illinois, prepared by Patterson Associates, Inc. for the Illinois Institute of Natural Resources, Document No. 79/32, October 1979.

Illinois Industrial Waste Survey, Data Summary, Illinois Environmental Protection Agency, October 1980.

Iowa Hazardous Waste Management in Iowa: Interim Report, The Environmental Quality Commission and the Iowa Dept. of Environmental Quality, 1981.

Kansas A Survey of Hazardous Waste Generation and Disposal Practices in Kansas, Vols. I and II, State of Kansas, Dept. of Health and Environment, March 1977.

Summary of Bulletins 4-8.1 and 4-8.2, Survey of Hazardous Waste Generation and Disposal Practices in Kansas, Dept. of Health and Environment, March 1977.

Kentucky Hazardous Waste Generated in Kentucky, Preliminary Report, Dept. of Natural Resources and Environmental Protection, Bureau of Environmental Protection, Division of Waste Management, Jan., 1981.

Hazardous Waste Survey of Kentucky, Dept. of Natural Resources and Environmental Protection, Bureau of Environmental Protection, Division of Hazardous Materials and Waste Management. Jan., 1978.

Louisiana Preliminary Quarterly Report of Hazardous Waste for the First Quarter of 1980, Dept. of Natural Resources, Office of Environmental Affairs, Hazardous Waste Management Division, 1980.

Maine Initial Hazardous Waste Survey Report Findings and Analysis, Maine Department of Environmental Protection, Sept. 1980.

State of Maine Hazardous Waste Survey, Industrial Sludge Quantities, Industrial Waste Survey, 1977.

Maryland Maryland's Hazardous Waste Manifest Systems Two Years After. Department of Natural Resources, May 1980.

Report on Hazardous Waste Practices, State of Maryland, Department of Health and Mental Hygiene and Department of Natural Resources, May 1977.

Massachusetts Hazardous Waste Management in Massachusetts, Draft Environmental Impact Report, Massachusetts Department of Environmental Management, Bureau of Solid Waste Disposal, Jan. 1981.

Minnesota Hazardous Waste Management, Minnesota's Issues/Options, Barr Engineering Co., Tenech Environmental Consultants, Inc., Noble and Associates, Peat Marwick, Mitchell, & Co., and Dayton Herman Graham & Getts, Aug., 1979.

Mississippi Technical Report on Hazardous Waste Management in Mississippi, Mississippi Hazardous Waste Management Council, January 1981.

Montana Hazardous Wastes in Montana—A Survey of Waste Generation and Management Practices, Dept. of Health and Environmental Sciences Division, Solid Waste Management Bureau, Dec. 1977.

New Jersey Hazardous Solid Waste Manifest System Waste Pickups, in Pounds for the Period 07/01/80 to 09/30/80, Statewide Totals, New Jersey Dept. of Environmental Protection, Nov. 21, 1980.

Hazardous Solid Waste Manifest System Waste Pickups, in Pounds for the Period 1/01/80 to 6/30/80, Statewide Totals, New Jersey Department of Environmental Protection, Aug. 2, 1980.

Hazardous Solid Waste Manifest System Waste Pickups, in Pounds for the Period 10/01/80 to 12/31/80, Statewide Totals, New Jersey Department of Environmental Protection, Feb., 3, 1981.

New York An Inventory of Industrial Hazardous Waste Generation in New York State, New York State Department of Environmental Conservation, Bureau of Hazardous Waste, June, 1979.

	Hazardous Waste Disposal Sites in New York State, First Annual Report, New York State Dept. of Environmental Conservation, Division of Solid Waste, June, 1980.
	Toxic Substances in New York's Environment, An Interim Report, New York State Dept. of Environmental Conservation and New York State Dept. of Health, May, 1979.
North Carolina	Hazardous or Difficult to Handle Waste Survey Report, North Carolina Dept. of Human Resources, Oct., 1978.
Ohio	State of Ohio Hazardous Waste Management Plant, Part I and Part II, prepared by Battelle for the State of Ohio Environmental Protection Agency, Dec., 1978.
Oklahoma	Oklahoma Controlled Industrial Waste Projected Inventory—Annual Basis, Oklahoma State Department of Health, Industrial and Solid Waste Division, March 1978.
Oregon	Hazardous Waste Management Planning (1972–73), State of Oregon, Department of Environmental Quality, March 1974.
Pennsylvania	Activity 10.03—Residual Waste Projections, Bureau of Land Protection, Division of Solid Waste Management (undated).
Rhode Island	Generation of Hazardous Waste in Rhode Island—1980, Rhode Island Dept. of Environmental Management, Division of Air and Hazardous Materials, Jan. 1981.
South Carolina	Hazardous Waste Survey by Activity and County, (Computer Printout L50260EA), South Carolina Dept. of Health and Environmental Control, Bureau of Solid and Hazardous Waste Management, 1981.
	In State Waste Disposed in South Carolina SCA Services as of Jan. 31, 1979, South Carolina Dept. of Health and Environmental Control, Feb., 1980.
	In State Waste Disposed in South Carolina SCA Services as of Jan. 31, 1980, South Carolina Dept. of Health and Environmental Control, Feb., 1980.
	In State Waste Incinerated in South Carolina for 1979 and 1980 (to March, 1980), South Carolina Dept. of Health and Environmental Control.
	Waste Disposal in South Carolina—First Quarter, Jan. 1—March 31, 1980. South Carolina Dept. of Health and Environmental Control, June, 1980.
Texas	Draft Solid Waste Management Plan for Texas, Volume 2—Industrial Solid Waste, Texas Department of Water Resources, November, 1980.
Vermont	Hazardous Waste in Vermont, Agency of Environmental Conservation, State of Vermont, Spring 1979.
Washington	An Industrial and Hazardous Waste Inventory of Washington's Manufacturing Industries (1980), Department of Ecology, Hazardous Waste Section, Feb. 1981.

West Virginia Industrial Waste Survey, State of West Virginia Department of Natural Resources, 1980.

Hazardous Waste Survey Results, State of West Virginia, Department of Health, Solid Waste Division, 1978.

Wisconsin The Current Status of Hazardous Waste Management in Wisconsin, Department of Natural Resources, Bureau of Solid Waste Management, Hazardous Waste Section, May 1979.

Wyoming Survey and Analysis of Hazardous Waste Generation and Management in Wyoming, Department of Environmental Quality, Solid Waste Division, Aug., 1980.

2

Assessment of Exposure Potential: Transport Processes

INTRODUCTION

In order for the disposal of a hazardous waste to pose a problem, it must have the potential to come into contact with some population (human or otherwise) at concentrations sufficient to exert an undesirable biological effect. In the previous chapter, the relationships between chemical concentrations and biological effects were discussed. In this chapter, the mechanisms for transport will be considered. Given either an ongoing source of waste (an active incinerator or landfill, for example) or a preexisting closed facility (a site on the National Priorities List, for example), to what extent might such biologically-significant contamination occur? Alternatively, to what degree should barriers or treatment be installed in order to prevent or minimize the extent of such exposures? In this discussion, the exposed population (human or nonhuman) will be collectively denoted "receptors."

First, the relationship between the amount of material at a source (management facility or disposal site), and the extent of its transport to an impacted receptor must be quantified. Then the decision as to what types of intermediate barriers are needed can be made. Such barriers might be simply a physical distance between the source and the receptor, some type of engineering structure or treatment system designed to attenuate the source of pollution and its transport from the source for the receptor, or simply removal—shutting down that generator, removing the dump site, cleaning up the source of that disturbance.

ROUTES OF EXPOSURE

The following are the routes by which exposure to hazardous waste occurs:

- direct contact and dermal exposure or ingestion
- contamination of surface water and subsequent ingestion
- contamination of groundwater and subsequent ingestion
- contamination of the overlying atmosphere and transport, followed by inhalation, or deposition to the soil or water (with possible exposure via one of the foregoing routes)

It is also possible for exposure to occur after some concentration or transformation involving other organisms that have been exposed via one of the above routes. Perhaps one of the more dramatic examples of this food chain contamination involved polybrominated biphenyls (PBB's). A number of years ago, in Michigan, the accidental incorporation of PBB's into cattle feed resulted in direct exposure of under 1,000 persons at the affected farms. However, due to the contamination of the milk supply before action could be taken, more than 10,000 persons were exposed to the chemical (Archer et al., 1979).

However, hazardous waste can only contaminate the food chain when transported through the physical environment. Hence, by considering physical transport mechanisms, one can assess the propensity for such contamination to occur.

DIRECT CONTACT

The mechanism of direct contact, by definition, does not involve a transport step. In this mechanism, the receptor or the potential receptor comes in close proximity to the source of pollution. As a result, it is unnecessary to estimate risks associated with pollutant transport. The remaining portions of risk assessment associated with direct contact focus only on the dose-response relationships for the particular toxic materials.

Direct contact exposure can cause significant effects. There has been substantial controversy regarding the significance of dioxin exposures by direct contact at certain disposal sites. Uncontrolled application of waste oils containing dioxin resulted in widespread contamination in Missouri, including at Times Beach (Piontek, 1983). In the case of dioxin, the data on dioxin exposure, with a couple of exceptions, is mainly from major industrial accidents where the primary effect was a direct contact, in fact, rather than inhalation, ingestion, or food chain contamination.

THE GENERAL CONTAMINANT MASS BALANCE

To evaluate the other transport processes, it is necessary to examine the basic transport equations for contaminants in the environment. Consider an infinitesimal volume in space, bounded by the planes x, y, z, $x + dx$, $y + dy$, $z + dz$ (Figure

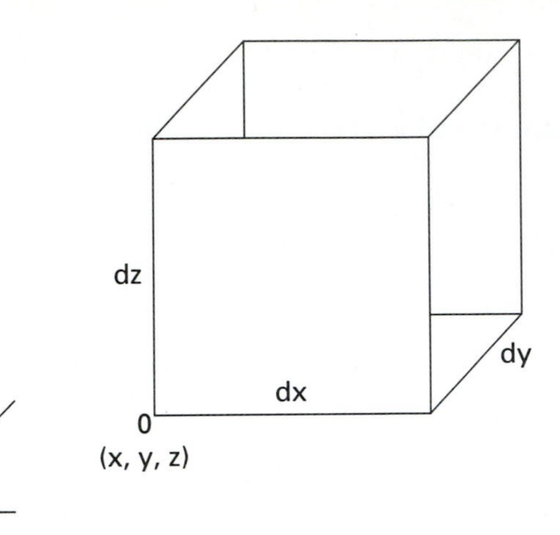

Figure 2-1 Infinitesimal volume in space.

2-1). The volume of the element is $(dx\ dy\ dz)$. There may be a velocity of bulk fluid characterized by a vector function with components v_x, v_y, v_z. Additionally, due to spatial gradients of concentration (C), additional transport may occur due to either molecular or eddy diffusion. As an approximation (generally good in all but extremely anisotropic materials), this process can be characterized by a vector of fluxes in each of the three coordinate directions, that is,

$$N_i = -D_i(\partial C/\partial i) \tag{2.1}$$

with i taking the indices x, y, and z, D_i being the diffusion coefficient (either molecular or eddy) in that particular direction, and N_i being the component of diffusive flux (mass/area-time) in that direction. In this volume, there may also be processes occurring which produce or consume some materials, and these will be described by reaction rates, r, denoting the mass produced or consumed by reaction per unit volume per unit time.

Providing that the fluid is incompressible, a mass balance can be written in words as:

$$\begin{aligned}\text{Rate of accumulation} = {} & \text{rate in via flow} + \text{rate in via diffusion} \\ & - \text{rate out via flow} - \text{rate out via diffusion} \\ & + \text{rate of reaction}\end{aligned}$$

Using a Taylor series expansion for evaluation of quantities at differential increments, this can be written mathematically in the following form:

$$\begin{aligned}dx\ dy\ dz(\partial C/\partial t) = {} & C(v_x dy\ dz + v_y dx\ dz + v_z dx\ dy) \\ & - [D_x dy\ dz(\partial C/\partial x) + D_y dx\ dz(\partial C/\partial y) + D_z(\partial C/\partial z)] \\ & - \{dy\ dz[Cv_x + (\partial Cv_x/\partial x)\ dx] + dx\ dz[Cv_y + (\partial Cv_y/\partial y)dy] \\ & + dx\ \ dy[Cv_z + (\partial Cv_z/\partial z)\ dz]\} \\ & + \{dy\ dz[D_x(\partial C/\partial x) + (dx)\partial(D_x \partial C/\partial x)/\partial x] \\ & + dx\ dz[D_y(\partial C/\partial y) + (dy)\partial(D_y \partial C/\partial y)/\partial y] \\ & + dx\ dy[D_z(\partial C/\partial z) + (dz)\partial(D_z \partial C/\partial z)/\partial z]\} + r\ dx\ dy\ dz\end{aligned} \tag{2.2}$$

where r is the net rate of formation (mass/volume-time) of all reactions (chemical, physical, or biological) that occur in the infinitesimal volume. Cancellation of like terms results in the following:

$$(\partial C/\partial t) = -[(\partial Cv_x/\partial x) + (\partial Cv_y/\partial y) + (\partial Cv_z/\partial z)] + \{[\partial(D_x\partial C/\partial x)/\partial x] + [\partial(D_y\partial C/\partial y] + [\partial(D_z\partial C/\partial z)/\partial z]\} + r \quad (2.3)$$

Finally, if it is assumed that both the velocity vector and the diffusivities are spatially uniform, then one has:

$$(\partial C/\partial t) = -[v_x(\partial C/\partial x) + v_y(\partial C/\partial y) + v_z(\partial C/\partial z)] + [D_x(\partial^2 C/\partial x^2) + D_y(\partial C^2/\partial y^2) + D_z(\partial C^2/\partial z^2)] + r \quad (2.4)$$

Solution of the transport equation requires the specialization and simplification of the problem so that analytical or numerical solutions can be employed. In particular, one needs to know the dimensionality of the problem (i.e., 1, 2, or 3 spatial dimensions), the homogeneity and isotropy of the velocity and diffusive transport fields (whether the velocities and dispersions are constant in space and time, and whether the dispersions are independent of direction), the existence of reactions or partitioning into other phases, and whether or not the system may be regarded as being at steady state.

Particular cases of interest in predicting the dispersal of hazardous wastes in the environment will be considered below. For more general discussions on the solution of the transport equation, one should consult Bird et al. (1960), Crank (1975), or Thibodeaux (1979).

CONTAMINANT TRANSPORT EQUATIONS

Environmental transport processes are functions of the source (i.e., surface water, groundwater, air). In the case of surface waters, the velocities at various positions are generally known. However, for groundwaters, the velocities are not generally subject to direct measurement, but must be inferred. Consideration of transport in groundwaters is deferred to a later section.

The simplest example of a surface water is a river or stream, in which flow is one dimensional. In such a situation, the concentration of material in the river varies solely as a function of distance downstream, and perhaps time. Only for very large rivers with considerable heterogeneity in the bottom is there significant variation along the width or the depth.

Unreactive Contaminants

If a pollutant is nonreactive, by definition there are no mechanisms available for chemical or biological transformation, or for loss due to volatilization or adsorption to sediments. In this system, where there are no gradients in other than the axial direction, the general mass balance is written in the following form:

$$\partial C/\partial t = D_z \partial^2 C/\partial z^2 - v_z \, \partial C/\partial z \tag{2.5}$$

where

z = the linear distance downstream
t = the chronological time
C = the pollutant concentration at distance z and time t
D_z = the axial dispersion coefficient for the stream
v_z = the axial velocity of the stream

Given initial and boundary conditions which describe the nature of the contamination, this equation can be solved, either numerically or analytically. There are two important cases in which the solution of this equation is analytic and has application to the problems under discussion.

If, for example, a single pulse of material is released at zero distance and time (z, $t = 0$), as, for example, in the case of a sudden and rapid loss of a chemical from a storage facility located on a river, the appropriate conditions for solution of Equation 2.1 become:

At $t = 0$ for $z \geq 0$, $C = 0$ (before the spill, there is no contamination). This condition need not hold precisely. If, for example, there is a constant average concentration in the river, the contribution due to the dispersion of the spill is additive above the baseline average.

At $z = \infty$, for $t \geq 0$, $C = 0$ (if one goes far enough downstream, the stream becomes clean). This condition is termed the "semi-infinite" boundary condition, and does not adequately describe the situation; the solution to the problem may be an infinite series (Crank, 1975).

The mass per unit area which passes the stream at any z is identical to the mass per area discharged in the spill. Mathematically, if M is the mass per area released, then the following is a third condition for this problem:

$$\frac{M}{v_z} = \int_0^\infty C(z, \tau) \, d\tau \tag{2.6}$$

for all z.

Given these conditions, the downstream concentration as a function of distance and time can be written as

$$C = \frac{M}{(4\pi D_z t)^{1/2}} \exp\{-(z - v_z t)^2/4D_z t) \tag{2.7}$$

Equation 2.7 describes what happens to the single slug as it moves downstream. The form of the equation is identical to a Gaussian, or "normal," distribution found in statistics. Figures 2-2 and 2-3 show the concentration downstream at a fixed point as a function of time, and at a fixed time as a function of distance. Note that only the latter curve is symmetric about and has a maximum at the point where $z = v_z t$. Only

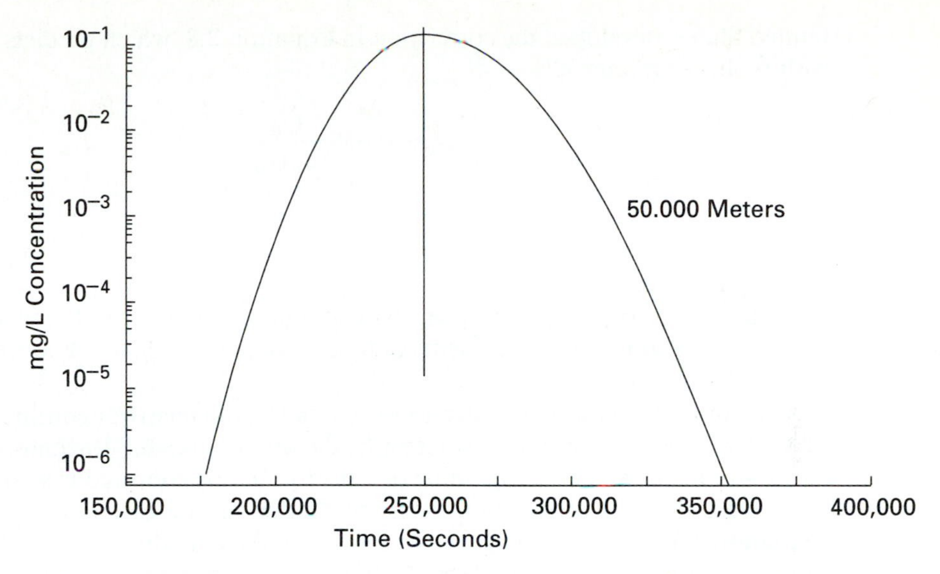

Figure 2-2 Plume dispersion for a slug release without reaction, versus time ($M = 1000$ g/m^2, $D_z = 50$ m/s^2, $v_z = 0.2$ m/s). Vertical lines indicate the theoretical hydraulic residence time.

when samples are taken along a transect at a fixed time is this relationship exactly true. In the more common case of "watching" a contaminant plume pass a fixed point, neither of these observations is exactly correct.

To utilize Equation 2.7, the value of the axial dispersion in a river must be known. McQuivey and Keefer (1974), based on a study of a number of rivers in the

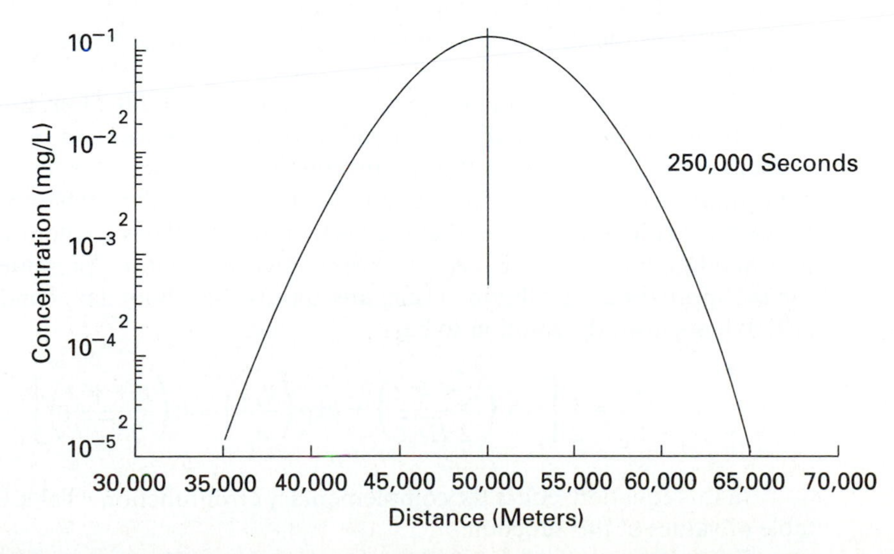

Figure 2-3 Plume dispersion for a slug release without reaction, versus distance ($M = 1000$ g/m^2, $D_z = 50$ m/s^2, $v_z = 0.2$ m/s). Vertical lines indicate the theoretical hydraulic residence time.

United States, developed the correlation in Equation 2.8, which predicts dispersion within plus or minus 30%.

$$D_z = 0.058 \frac{Q_0}{S_0 W_0} \tag{2.8}$$

where

Q_0 = volumetric flow rate
W_0 = width of the river
S_0 = slope of the specific energy gradient (as a practical matter, in all but very rapid rivers, this is identical to the slope of the channel bottom)

Instead of the single sudden release, one can have a relatively continuous source of contamination. Consider, for example, the onset of landfill leachate discharging into a surface water at the beginning of a storm. In this continuous-source case the same basic differential equation as in the sudden-release case remains applicable (Equation 2.5). All that changes is the third boundary condition. The initial condition ($t = 0$) and the first boundary condition ($z = \infty$) remain as previously, and the new boundary condition stipulates that the flux of material passing the point $z = 0$ is identical to the flux introduced. This is a so-called "open" boundary condition, and is written as:

$$\text{at } z = 0, \text{ and } t \geq 0, -D_z(\partial C/\partial z) + v_z C = J \tag{2.9}$$

where J is the mass/area-time released in the discharge.

The quantity J/v_z has units of concentration, and it is physically equal to the concentration that would be attained merely by dilution into a continuously flowing stream. It will be convenient in subsequent discussion to define the dimensionless ratio, $v_z z/D_z$ as the Peclet number, representing the ratio of advective to dispersive transport.

The boundary condition represented by Equation 2.9 often produces great difficulty in solving the applicable equations without resorting to infinite series or numerical approximation. Although this condition is "exact," the condition that the concentration at $z = 0$ (for time > 0) is fixed at C_0 provides a suitable approximation to the exact solution provided that the Peclet number (Pe) is greater than 16 (Nauman and Buffham, 1983). Unless otherwise stated, all solutions presented below will use the approximate condition. Using this approximate boundary condition, Ogata (1970) has shown the solution to be:

$$\frac{C}{C_0} = \frac{1}{2}\left[\operatorname{erfc}\left(\frac{z - V_z t}{2\sqrt{D_z t}}\right) + \exp\left(\frac{v_z z}{d_z}\right)\operatorname{erfc}\left(\frac{z + V_z t}{2\sqrt{d_z t}}\right)\right] \tag{2.10}$$

In this equation, erfc is the complementary error function. Table 2-1 is a short table of values of this function.

For computer calculations, it may be approximated by a rational function. For example Equation 2.11 provides precision better than 0.000025 (Gautsch, 1964).

TABLE 2-1 SELECTED VALUES OF erf(x). (Reproduced from Tuma, 1987.)

x	0	1	2	3	4	5	6	7	8	9
0.	0	0.1125	0.2227	0.3286	0.4284	0.5205	0.6039	0.6778	0.7421	0.7969
1.	0.8427	0.8802	0.9103	0.9340	0.9523	0.9661	0.9764	0.9838	0.9891	0.9928
2.	0.9953	0.9970	0.9989	0.9994	0.9996	0.9998	—	—	—	—

(Note: erf(x) = 1 − erf(x))

$$\text{erfc}(|x|) = \exp(-x^2)(0.3480242W - 0.0958798W^2 + 0.7478556W^3) \quad (2.11)$$

where $W = 1/(1 + 0.47047x)$. Note that $\text{erfc}(-x) = 2 - \text{erfc}(|x|)$.

Mathematically, the error function is related to the area under the Gaussian curve, and is thereby related to the solution from the point slug source case. The continuous case can be regarded as resulting from an integrating of an infinite number of minute slug sources over time. Particular care is necessary in evaluation of Equation 2.6 for large values of the Peclet number since the exponential in the second term will be large and the value of *erfc* will be small. In such circumstances, the following approximation, valid for large values of x, may be useful (Lapidus and Amundson, 1952):

$$\text{erfc}(|x|) = \exp(-x^2)\pi^{-1/2}\left[\frac{1}{x} - \frac{1}{2x^3} + \frac{3}{4x^5} - \frac{15}{8x^7} \cdots\right] \quad (2.12)$$

So far, only situations involving spills into streams, that is, one dimensional systems, have been considered. A more complicated possibility is an open body of water in which it is important to consider what happens in three dimensions. In this case, at each point the velocity is a vector with components in the three cartesian coordinates. This situation may arise in large lakes, oceans, or estuaries. And, as will be noted later, the groundwater dispersion problem involves, in part, a case of this type.

In the case of a large body of open water such as a major lake, the velocity of water may be regarded as essentially constant with respect to both time and position. This is referred to as "steady" (i.e., time invariant) and "uniform" flow. This is not the most general case that could be considered and it does not describe all open bodies of water. For example, it does not describe estuaries which generally have a cyclical (tidal) variation in velocities.

For the steady, uniform situation, if a single slug of pollutant is released having a mass of contaminant "W," providing that the release is sufficiently away from the shoreline to minimize the effects of the surface, then the concentration profile is described as a three-dimensional Gaussian curve as in Equation 2.13. The velocity of the water body is described using the cartesian coordinate system in the axial (x), transverse horizontal (y), and vertical (z) directions.

$$C = \frac{W}{(2\pi)^{2/3}(8D_x D_y D_z t^3)^{1/2}} \exp\left\{-\frac{(x - v_x)^2}{4D_x t} - \frac{(y - v_y)^2}{4D_y t} - \frac{(z - v_z)^2}{4D_z t}\right\} \quad (2.13)$$

Reactive Contaminants

If a contaminant is nonconservative ($r < > 0$, hence total mass in the fluid phase changes with respect to time, generally decreasing) and thus capable of reacting, the previous transport models must be modified before the fate of the material is adequately described. There are, however, many mechanisms available for reactions of pollutants in natural systems. These include the following:

- physical association with particulates and removal from the water column by sedimentation
- volatilization of materials into the atmosphere
- chemical, photochemical, or biological decomposition of material in the water column

To a first approximation, each of these reactions can be modeled assuming either a linear equilibrium relationship or a first-order irreversible reaction. With either of these approaches, the dispersion models lead to analytical solutions. For more complex models, numerical solutions to the relevant transport equations are required.

In the case of the one-dimensional or stream problem, simplification of Equation 2.4 to account for the transformation process leads to:

$$\partial C/\partial t = D_z \partial^2 C/\partial z^2 - v_z\, \partial C/\partial z - r_d \tag{2.14}$$

where r_d is the rate (mass/volume-time) at which the contaminant of interest is removed from the aqueous phase.

In the case of a first-order irreversible reaction, $r_d = k_d C$, where k_d is a rate constant (units time^{-1}). For the continuous discharge case with reaction, under steady-state conditions, the solution to Equation 2.14 was shown by Wehner and Wilhelm (using "exact" boundary conditions), (1956), to be:

$$\frac{C}{C_0} = \frac{4\, a\, e^{\mathrm{Pe}/2}}{(1 + a)^2 \exp(a\mathrm{Pe}/2) - (1 - a)^2 \exp(-aPe/2)} \quad \text{with } a = \sqrt{1 + \frac{4k_d z}{v_z Pe}} \tag{2.15}$$

For the one-dimensional system with a first-order decay process and a continuous source of pollution the analogous equation under unsteady state is (Ogata, 1970):

$$\frac{C}{C_o} = \frac{1}{2}\left\{\exp\left[\frac{(1 - a)\mathrm{Pe}}{2}\right]\mathrm{erfc}\left(\frac{z - av_z t}{2\sqrt{D_z t}}\right) + \exp\left[\frac{(1 + a)\mathrm{Pe}}{2}\right]\mathrm{erfc}\left(\frac{z + av_z t}{2\sqrt{D_z t}}\right)\right\} \tag{2.16}$$

In the case of equilibrium sorption to solids (such as in transport of material through the ground or in a sediment-laden river) any one of these equations can be used following a simple modification of the time and the distance axes. If, as above,

C is the concentration of pollutant in the water (mass/volume) and S is the concentration of the material on the solid phase (in units of mass of contaminant per mass of adsorbing solid phase) then the equilibrium relationship can be written as:

$$S = K_{sed} C \tag{2.17}$$

K_{sed}, the Henry's law constant for adsorption has units of volume/mass (of adsorbing solid phase) and is a property of both the contaminant and the solid phase itself. The larger the value of K_{sed}, the greater the affinity for the solid phase. For this situation, when a general mass balance is derived over a contaminant, the total mass rate of change of pollutant (mass/volume-time) is written as

$$\epsilon \partial C/\partial t + \rho \partial S/\partial t \tag{2.18}$$

where ρ is the mass of solid phase present in a given total system volume and ϵ is the volume of fluid-filled pore space per total system volume (i.e., porosity, which for sediment-laden streams is virtually unity). However, from differentiation of Equation 2.17 this total rate of change may also be found to equal:

$$(\epsilon + \rho K_{sed}) \partial C/\partial t \tag{2.19}$$

This term appears on the left-hand side of any mass balance (e.g., Equation 2.14). By dividing any constants on the right-hand side by the term $(1 + \rho K_{sed}/\epsilon)$ and substituting the "corrected" constants in an appropriate integrated solution (such as Equation 2.10), the time history of the pollutant transport wave is obtained. The term $\rho K_{sed}/\epsilon$ is often written as R and defined as the retardation. The higher the value of the retardation, the more strongly associated the pollutant is with the sediment or with the solid phase. If v_{H_2O} is the velocity of water through a system, and v_p is the velocity of the center of mass of a pollutant, then the retardation may also be defined as:

$$R = (v_{H_2O}/v_p) - 1 \tag{2.20}$$

So, for example, in Equation 2.10 quantity v_z is replaced by $v_z/(1 + R)$ and the diffusion coefficient, D_x is replaced by $D_x/(1 + R)$ Similarly, if there was a reaction-rate constant in a mass balance (characterizing a reaction that occurred only in the aqueous phase) then it (k_d) would be replaced by $k_d/(1 + R)$.

In order to use these equations we need to estimate dispersion coefficients (D_x, D_y, D_z), retardation values, and values for any reaction rates. One example of the estimation of longitudinal dispersion (D_z) in streams was previously given. In multiple-dimensional cases in particular, such as groundwater migration, it is also necessary to know the velocity vector (possibly as a function of position). Once these parameters are estimated, then the appropriate equation can be used to calculate the pollutant attenuation versus time and/or location.

Dispersion Estimates

The three types of dispersion that appear are distinguished by the coordinate direction. In a river, the major type of dispersion process that is of concern to us is axial dispersion—dispersion along the axis of the flow of the river (D_z) for which a correlation has been presented. For multi-dimensional systems, in addition to D_z, two other dispersion terms may be of significance. First, there is horizontal transverse dispersion (D_x) perpendicular to the direction of bulk flow. In the case of a river, this would be dispersion along the width, from channel to channel. One correlation for D_x in rivers is presented by Thibodeaux (1979):

$$D_x = 6.25 \times 10^{-6} \frac{t^{2.6}}{D_z} \tag{2.21}$$

This equation is not dimensionally consistent; therefore t, the time for which the plume concentration is sought, and both D_z and D_x are in cm^2/sec. Although this relationship fits the data on dispersion in waters, it violates one of the major assumptions used in developing the general transport equation—namely, the constancy of the dispersion parameters. Hence, the adoption of Equation 2.21 carries with it an implicit recognition of the failure of a Gaussian dispersion model with constant dispersion parameters to describe the data on contaminant transport.

The vertical-dispersion coefficient in waters can be estimated since it is primarily due to the presence of vertical-density gradients from differential solute or particle concentrations or differential temperatures. If ρ is the density of the water as a function of depth (y, in meters), then, to within an order of magnitude, D_y (in cm_2/s) can be given by (Thibodeaux, 1979):

$$D_y = \frac{0.0001}{\left| \frac{1}{\rho} \frac{\partial \rho}{\partial y} \right|} \tag{2.22}$$

Estimates of Sorption Equilibria

For sorption of materials to sediments, it is necessary to estimate the K_{sed} value of solid-water partitioning for a pollutant. For organic pollutants, the major factor governing association with the solid phase is the organic carbon content of the solid, sorbing, phase. In general, therefore, one can write:

$$K_{sed} = f_{oc} K_{oc} \tag{2.23}$$

where f_{oc} is the decimal fraction organic carbon content of the solid phase and K_{oc} is the partition coefficient of the pollutant into the organic carbon of the solid phase with units of volume/mass. Given an estimate for K_{oc}, one can determine K_{sed} and hence the retardation.

The solid phase organic carbon consists of some unknown mix of organic material. However, there have been a number of laboratory studies done on the partitioning of organic contaminants between water and immiscible organic phases. Most of the available data base, which is quite broad, uses octanol as the immiscible organic phase and consists of measurements of K_{ow}—the octanol:water partition coefficient (dimensionless). It has been determined that the partitioning between octanol and water can be used in a predictive model for K_{oc} using the following log-log correlation:

$$\log(K_{oc}) = A \log(K_{ow}) + B \tag{2.24}$$

Given the two constants A and B, which are available from prior work, and K_{ow}, which is available in various compilations (Leo and Hansch, 1971), or may be estimated using a number of group contribution methods (Lyman et al., 1982), K_{oc} can be determined. It should be noted that this equation is not dimensionally consistent, and specific units of K_{oc} must be used—generally mL/g. Other correlations of a similar nature exist between the aqueous solubility (S) of an organic pollutant and its K_{oc} value. Lyman et al. (1982) have critically reviewed these correlations.

Estimates for Reaction Rate Due to Volatilization

The final information needed to analyze pollutant transport processes are the quantitative values of the reaction rate constants. For physical processes, particularly volatilization, it is possible to come up with estimates based on general correlations on mass transport. For chemical and biological processes, only broad ranges for reaction rates may be given.

For volatilization, providing that the concentration of contaminant in the air overlying a body of water is low relative to the equilibrium concentration which would be sustained over such an aqueous concentration, and that the water column at any location is well mixed vertically (with respect to the contaminant concentration) the rate of water–air transport can be characterized as a first-order reaction in soluble contaminant. This may be seen by the following analysis. Transport between a bulk liquid and a bulk vapor phase is generally described by the following formulation:

$$J = K_L(C_L^* - C_L) \tag{2.25}$$

where J is the flux (mass/area-time) from gas to liquid, C_L is the soluble concentration of volatile species, C_L^* is the hypothetical concentration of volatile pollutant which would be in the liquid if the liquid were in equilibrium with the overlying gas phase, and K_L is the overall mass transfer coefficient (length/time) with liquid phase driving forces. If the actual gas phase partial pressure of contaminant was given by P_G and the gas-liquid equilibrium was described by Henry's law (i.e., $K_H C_L^* = P_G$ with K_H being the Henry's law constant with units of volume-pressure/mass), then (2.25) could be rewritten as:

$$J = K_L(P_G/K_H - C_L) \tag{2.26}$$

Clearly, however, providing that $P_g/K_H < C_L$,

$$J = -K_L C_L \tag{2.27}$$

The flux on an area basis (J) can be transformed into a pseudo-reaction rate (r_V, rate of volatilization in mass/volume-time) by multiplying by the area/volume ratio. For uniform rectangular cross sections, this ratio is the reciprocal of the water depth (d) and hence,

$$r_V = -(K_L/d)C_L \tag{2.28}$$

It can thus be seen that the ratio K_L/d has units of time t^{-1}—identical to a first-order decay constant.

The value of the overall mass transfer coefficient can be estimated by applying the two film theory of mass transfer which hypothesizes the existence of a liquid resistance and a gas phase resistance in series, characterized by the respective film transfer coefficients k_L and k_G, each with units of length/time. The three coefficients are related by the following equation:

$$\frac{1}{K_L} = \frac{1}{k_L} + \frac{RT}{K_H k_G} \tag{2.29}$$

In this equation, R is the ideal gas law constant, and T is the absolute temperature (Kelvin).

The most commonly used expression to estimate k_L is the O'Connor and Dobbins (1956) correlation relating the mass transfer coefficient to the stream depth (d), the linear velocity (v_z), and the molecular diffusivity of the volatile species in water (D_L):

$$k_L = \sqrt{\frac{D_L v_z}{d}} \tag{2.30}$$

Note that D_L is the *molecular* diffusivity of the pollutant in water, and not the turbulent dispersion (D_z). For k_G, for the gas film transfer coefficient, the correlations are somewhat more complicated because in general, the physical interaction of the wind with the overlying water surface strongly influences transport. Thibodeaux (1979) provides the following correlation:

$$k_G = \frac{0.036\, D_G\, (Sc_G)^{1/3}\, (\mathrm{Re})^{0.8}}{L} \tag{2.31}$$

where D_G is the molecular diffusivity of the volatile species in air (*not* the same as D_L), L is the "fetch" across the water body—defined as the length of the water body measured along an axis parallel to the wind and perpendicular to the water surface, and Sc_G and Re are, respectively, the Schmidt number and Reynolds' number defined by the relationships below.

$$Sc_G = \frac{\mu_G}{\rho_G D_G} \tag{2.32}$$

$$\mathrm{Re} = \frac{v_R L \rho_G}{\mu_g} \tag{2.33}$$

In these equations, ρ_G is the density of the atmosphere, μ_G is the viscosity of the atmosphere and v_R is the vector difference between the wind velocity and the component of the stream velocity in the direction of the wind. From trigonometry, L and v_R are related to the wind velocity (v_W), the stream velocity (v_z), the width of the stream (w) and the angle between the wind direction and the stream direction (Θ) by the relationships:

$$L = w/\sin\Theta \tag{2.34}$$

$$v_R = |v_W - v_z \cos\Theta| \tag{2.35}$$

From the information above, k_G can be computed for a water body and used to determine K_L. In the case of a lake, surface impoundment or ground accumulation, the previous relationships for k_G remain applicable. In such cases, there would be no velocity associated with the flow of water, ($v_z = 0$). However, k_G could still be computed.

Estimates for Chemical Reaction Rates

In solutions or in suspension, many organic compounds are subject to degradation due to chemical hydrolysis reactions. In these processes, the molecule reacts with either water or other dissolved constituents to produce transformed materials. The reactions may be acid or base catalyzed.

In general, the rate of hydrolysis may be considered to be first order in the organic compound, with an overall rate constant, k_T, given as a summation of various individual processes:

$$k_{hyd,T} = k_H[H^+] + k_{OH^-}[OH^-] + \Sigma\, k_{HA_i}[HA]_i + \Sigma\, k_{B_j}[B]_j \qquad (2.36)$$

In this equation, k_0 is the uncatalyzed hydrolysis rate constants, k_H and k_{OH} are, respectively, the second-order rate constants (units L/mol-s) for catalysis by hydronium and hydroxyl ions, respectively, and k_{HA_i} and k_{B_j} are, respectively, the second-order rate constants for catalysis by other acids and bases. Square brackets indicate molar concentrations of the individual chemical species.

The significance of Equation 2.36 is that the rate of hydrolysis may be significantly influenced by the pH of a system, and by the presence of acids or bases. At present, however, it is difficult to estimate quantitatively (Lyman et al., 1982) the individual values of the hydrolysis rate constants. However, certain generalizations are possible with regard to the compound half-life due to hydrolysis. These are presented in Figure 2-4. The half-life ($t_{1/2}$) due to hydrolysis is related to the first-order rate constant by the relationship $k_{hyd,T} = 0.69/t_{1/2}$.

In some cases, soils may catalyze the chemical degradation of organic compounds. This appears to occur by a free radical mechanism in which the organic is relatively reduced and the soil has a high oxidation potential. The soil moisture and the presence of trace metals in the soil mineral phases may be influential. Little information is available for the quantitative or qualitative prediction of these processes (Dragun and Helling, 1982).

Estimates for Photolysis Rates

Many chemicals are capable of absorbing light in the UV or near UV wavelength bands. In such cases, a portion of the absorbed light may result in the degradation of molecular structure due to photochemical reactions. In addition, the irradiation of natural waters with UV or near UV (including sunlight) can result in the production of free radicals and oxidizing agents (such as hydrogen peroxide) that can cause chemical transformations. The rate of this process (particularly where direct photo-

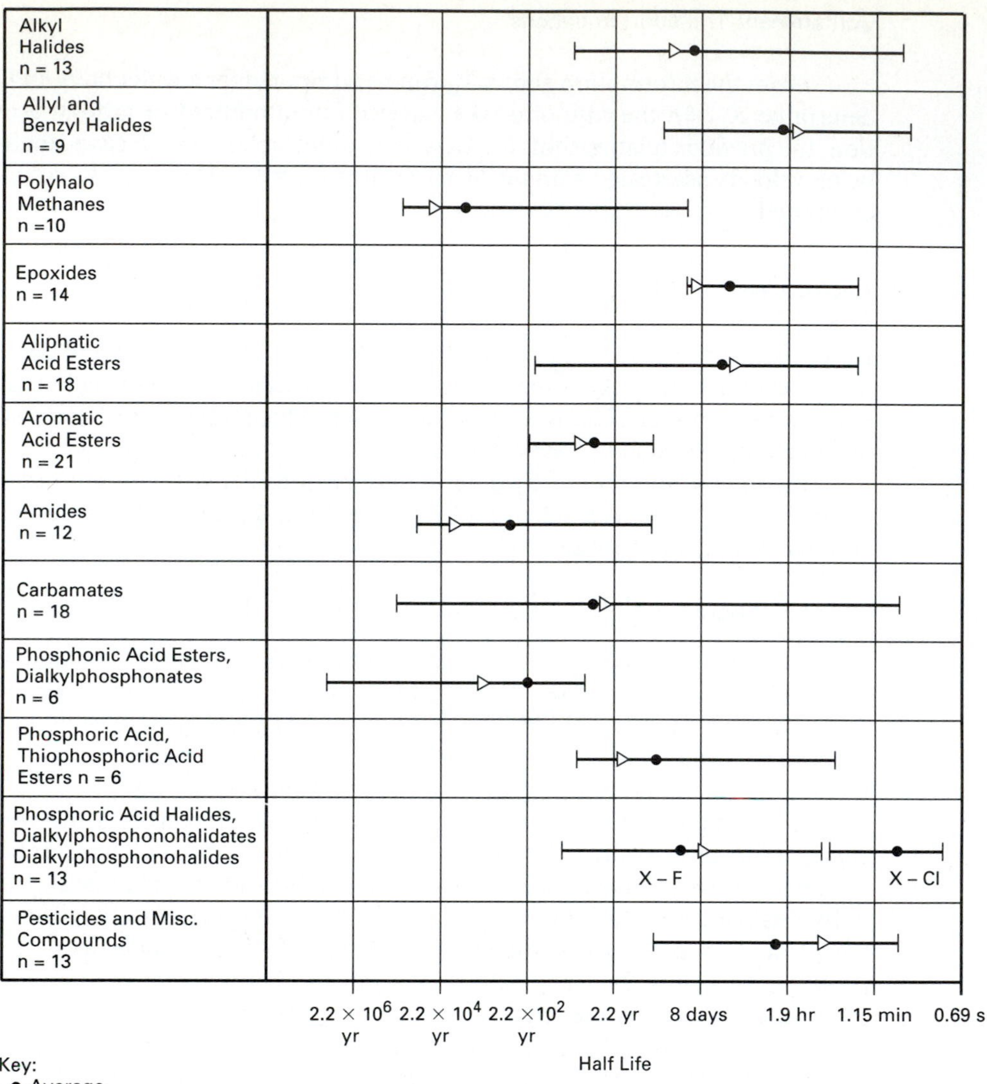

Figure 2-4 Half lives for hydrolysis at pH 7 and 25°C. (Reprinted from Lyman et al., 1982 by permission of Arthur D. Little, Inc.)

chemical reaction is considered, rather than the intervention of oxidizing intermediates) is described by first-order kinetics.

The expression for the first-order photolysis rate may be given by:

$$k_{\text{phot}} = 2.303 \frac{\Phi_{\text{solar}}}{j} \int_{\lambda} I_{\lambda} \epsilon_{\lambda} d\lambda \tag{2.37}$$

In Equation 2.37, I_λ (photons/cm^2-s-nm) and ϵ_λ (L/mol-cm) are, respectively, the irradiation received at the point at which the rate constant is sought and the molar absorptivity of the compound, both as functions of the wavelength λ (nm), j is Avogadro's number multiplied by 1,000 (to account for the conversion between cm^3 and L, i.e., 6.023×10^{20}) and Φ_{solar} is the mean quantum yield—that is, the proportion of photons which are absorbed that result in a photochemical reaction for irradiation received in the solar spectrum.

I_λ may be estimated from meterological conditions and the physics of light transmission. ϵ_λ is known once the compound of interest is specified. The quantity Φ_{solar} is quite compound specific and has only been measured for a limited set of materials. The range in the quantum yield exhibited is 0.00015–0.3 (Lyman et al., 1982).

Estimates for Biodegradation Rates

Few rules of thumb exist on which to formulate quantitative estimates of biodegradation rates of organic materials found in hazardous wastes *a priori* (Grady, 1986), although a large number of synthetic organics have been found to be amenable to biodegradation (Kobayashi and Rittman, 1982). However, a number of qualitative generalizations can be made (Lyman et al., 1982):

1. Highly branched compounds are more resistant to biodegradation.
2. Short chain substituents are less rapidly degraded than long chain substituents.
3. Highly oxidized materials, such as halogenated or oxygen-rich materials may resist further oxidation under aerobic conditions (in the presence of oxygen) but may be more rapidly degraded under anaerobic conditions.
4. More highly polar compounds tend to be more biodegradable than less polar compounds.
5. Unsaturated aliphatic compounds are more readily biodegraded than saturated aliphatic compounds.
6. Alcohols, aldehydes, acids, esters, amides and amino acids are more susceptible to biological decomposition than their analogous alkanes, olefins, ketones, dicarboxylic acids, nitriles and chloroalkanes.
7. Increased substitution on aliphatic compounds impedes biodegradation.
8. Increased halogenation impedes biodegradation.
9. Methyl, chloro, nitro, and amino substituents on aromatic rings inhibit biodegradation of the ring.

The biodegradation process, particularly in relatively dilute systems, can often be defined by a second-order rate constant (k'_{biol}) with units volume/mass-time. From this, the first-order rate constant (k_{biol}) with units 1/time can be obtained by

$$k_{biol} = (k'_{biol})(X) \tag{2.38}$$

where X is the concentration of microorganisms (mass/volume) in the system which are capable of degrading the specific compounds in question. This formulation assumes that the system is relatively dilute, and that all other necessary nutrients for microbial growth are present in adequate amounts (including the presence or absence, as necessary nutrients for microbial growth are present in adequate amounts (including the presence or absence, as necessary, of oxygen). In general, the magnitude of k_{biol} may be as high as 1 g/g-d, with values decreasing (by orders of magnitude) for compounds of lesser biodegradability.

EXAMPLES

For the first set of problems, a river will be considered. The river has a velocity of 0.3 m/s, and a cross section of 3 m (wide) and 2 m (depth). The river slope is 0.0004.

Example 2-1

First, consider a slug discharge of 100 kg of nickel cyanide, $Ni(CN)_2$—a nonvolatile, but highly toxic material. It can be assumed that there is no sorption or any other reaction mechanism, and that the contaminant behaves conservatively. The river is to be cleaned up by pumping and treating the entire flow at a point 50 km downstream. When should pumping be initiated and how long should it continue in order to ensure that no concentration in excess of 1 mg/L (0.001 mg/cm³) passes? Equation 2.7 will be used to solve this problem (slug discharge, unreactive, unsteady state, one dimensional).

The axial dispersion is estimated from Equation 2.8. Since the flow (Q_0) equals the velocity times the cross section ($Q_0 = 0.3 \text{ m/s} \times 3 \text{ m} \times 2 \text{ m} = 1.8 \text{ m}^3\text{/s}$),

$$D_z = 0.058 \times 1.8 \text{ m}^3\text{/s} /(0.0004 \times 3 \text{ m}) = 87 \text{ m}^2\text{/s or } 8.7 \times 10^5 \text{ cm}^2\text{/s}$$

The mass of material discharge per unit area is given by:

$$M = (100 \text{ kg} \times 10^6 \text{ mg/g})/(3 \text{ m} \times 2 \text{ m} \times 10^4 \text{ cm}^2\text{/m}^2) = 1667 \text{ mg/cm}^2$$

By substitution into Equation 2.3, the following is obtained, where t are the unknown times (in seconds) at which the concentrations are 1 mg/L:

$$0.001 \text{ mg/cm}^3 = \frac{1667 \exp\{-[(50 \text{ km})(10^5 \text{ cm/km}) - (0.3 \text{ m/s})(100 \text{ cm/m})t]^2/[4(8.7 \times 10^5 \text{ cm}^2\text{/s})t]\}}{[4(3.14159)(8.7 \times 10^5 \text{ cm}^2\text{/s})t]^{1/2}}$$

or

$$1.98 \times 10^{-3}\, t^{1/2} = \exp[-(7.184 \times 10^6/t) + 86.21 - 2.586 \times 10^{-4}\, t]$$

This equation can be solved either by trial and error or by Newton–Raphson iteration. There are two roots—that is, the time at which the concentration rises above 1 mg/L and the time at which it falls below 1 mg/L. Their numerical values are 154,000 and 178,000 seconds, respectively (or 42.78 and 49.44 hours). Therefore, a total volume of water to be pumped and treated would be given by:

$$1.8 \text{ m}^3\text{/s} \times (178{,}000 - 154{,}000 \text{ s}) = 43{,}200 \text{ m}^3$$

Example 2-2

Consider the same river as in Example 2-1. However, in this case, there is a continuous (illegal) discharge of benzene which is both volatile and adsorbable. First, in this example, the significance of the sorption will be considered ignoring volatilization. Subsequently, the volatilization process will be examined ignoring sorption. The sediment concentration is assumed to be 5 mg/L of organic carbon (plus other inorganic sediment components).

For adsorption to sediment, Lyman et al. provide the following correlation for K_{oc} (in mL/g) for a set of organic compounds:

$$\log K_{oc} = -0.55 \log (\text{solubility in mg/l}) + 3.64$$

The solubility of benzene in water at 20–25° C is 1.8 g/l or 1800 mg/l. Therefore,

$$\log K_{oc} = -0.55 \log (1800) + 3.64 = 70 \text{ mL/g}$$

The retardation (R) is obtained by multiplying K_{oc} and the concentration of sediment in the water column, hence $R = 70\ (\text{mL/g}) \times (5\ \text{mg/L}) \times (10^{-3}\ \text{L/mL}) = 3.55 \times 10^{-4}$. Since this number is substantially less than 1, the transport of benzene down the river will be substantially unimpeded by the presence of sediment.

To analyze the effect of volatilization (in the absence of adsorption) on the transport of benzene downstream, a number of transport properties are required. These are summarized below.

Temperature	25°C		
D_L	8.0×10^{-6} cm²/sec	D_G	0.088 cm²/sec
ρ_L	1 g/cm³	ρ_G	1.18×10^{-3} g/cm³
μ_L	1 g/cm-s	μ_G	1.8×10^{-4} g/cm-sec
K_H	5.49 L-atm/mol		

The wind speed across the stream is 10 miles/hour, and it is at a perpendicular direction to the flow of the water (i.e., $\Theta = 90°$). It is necessary to estimate the concentration of benzene downstream of a steady discharge from a source that has been in existence sufficiently long for the system to come to steady state.

From Equation 2.34, the fetch is equal to the stream width (i.e., $L = w = 3$ m) since $\Theta = 90°$.

From Equation 2.35, the relative velocity between the stream and the air can be computed:

$$v_R = |(10 \text{ miles/hr})(0.447 \text{ m/s/mile/hr}) - (0.3 \text{ m/s}) \cos(90°)| = 4.47 \text{ m/s}$$

Preparatory to computing the individual mass transfer coefficients, the (gas phase) Reynolds and Schmidt numbers can be computed:

$$\begin{aligned} Re &= 4.47 \text{ m/s}(100 \text{ cm/m})(3 \text{ m})(100 \text{ cm/m})(0.00118 \text{ g/cm}^3)/(1.8 \times 10^{-4} \text{ g/cm-s}) \\ &= 8.79 \times 10^5 \end{aligned}$$

$$\begin{aligned} Sc_G &= \frac{1.8 \times 10^{-4} \text{ g/cm-s}}{(1.18 \times 10^{-3} \text{ g/cm}^3)(0.088 \text{ cm}^2/\text{s})} \\ &= 1.73 \end{aligned}$$

From the above, the gas phase mass transfer coefficient can be computed using Equation 2.31:

$$\begin{aligned} k_G &= \frac{(0.088 \text{ cm}^2/\text{s})(0.036)(8.79 \times 10^5)^{0.8}(1.73)^{0.33}}{3 \text{ m } (100 \text{ cm/m})} \\ &= 0.72 \text{ cm/s} \end{aligned}$$

k_L can be computed directly from Equation 2.30:

$$k_L = \left[\frac{(8 \times 10^{-6}\ \text{cm}^2/\text{s})(30\ \text{cm/s})}{200\ \text{cm}}\right]^{1/2} = 1.095 \times 10^{-3}\ \text{cm/sec}$$

The overall mass transfer coefficient can be obtained by the combination rule in Equation 2.29:

$$\frac{1}{K_L} = \frac{1}{1.095 \times 10^{-3}\ \text{cm/s}} + \frac{(0.082\ \text{L-atm/mol-K})(298\ \text{K})}{(5.49\ \text{L-atm/mol})(0.72\ \text{cm/s})}$$
$$= 919.42\ \text{s/cm}$$

Therefore, $K_L = 1.088 \times 10^{-3}$ cm/s. This is almost identical to k_L, indicating that the dominant resistance is provided by the liquid phase. Hence, since the resistance to transport in the air phase is essentially negligible, any increase in wind velocity would have inconsequential effect on volatilization of the contaminant. From Equation 2.28, the first-order rate constant for volatilization can be estimated by $k = K_L/d = (1.088 \times 10^{-3}\ \text{cm/s})/(200\ \text{cm}) = 5.44 \times 10^{-6}\ \text{s}^{-1}$.

The dispersion coefficient characterizing contaminant transport in the stream may be estimated via the McQuivey and Keefer correlation (Equation 2.8):

$$D_z = 0.058 \frac{(0.3\ \text{m/s})(3\ \text{m})(2\ \text{m})}{(0.0004)(3\ \text{m})}$$
$$= 87\ \text{m}^2/\text{s} = 8.7 \times 10^5\ \text{cm}^2/\text{s}$$

Note that this dispersion is eleven orders of magnitude greater than the molecular diffusivity of benzene in water (D_L)—this indicates that turbulent mixing is substantially greater than dispersion at a molecular level.

The actual concentration at a distance z meters downstream is computed using the Wehner and Wilhelm equation (Equation 2.15), and the computation proceeds as follows:

$$\text{Pe} = (0.3\ \text{m/s})(z)/(87\ \text{m}^2/\text{s})$$
$$= 0.00345\ z$$

$$a = [1 + 4(5.4 \times 10^{-6}\ \text{s}^{-1})(87\ \text{m}^2/\text{s})/(0.3\ \text{m/s})^2]^{1/2}$$
$$= 1.01$$

Therefore,

$$C/C_0 = \frac{4.04 \exp(1.725 \times 10^{-3}\ z)}{4.0401 \exp(1.7423 \times 10^{-3}\ z) - 10^{-4} \exp(-1.7423 \times 10^{-3}\ z)}$$

which can be tabulated for different values of distance:

z (meters)	C/C_0
100	0.998264
200	0.996534
500	0.991367
1000	0.982825
10000	0.841117
50000	0.421041
100000	0.17728
200000	0.031429
300000	0.005572

Clearly, there is some volatilization of benzene into the atmosphere. However, even at 100,000 meters (100 km), almost 20% of the initial concentration is present. For more volatile compounds (i.e., compounds with higher values of K_H), or streams with shallower depths or more rapid velocities (and hence higher values of k), more rapid loss of volatile material would occur.

ATMOSPHERIC DISPERSION

As the previous example illustrates, volatile chemical contaminants in the water or the terrestrial environment may be transported into the atmosphere. In addition, of course, there may be deliberate emissions of such compounds in processes such as incineration. In order to evaluate potential impacts, the concentration in the atmosphere, at some point downwind of a source of airborne contamination, must be estimated.

A starting point for such computations is knowledge of the emission rate (mass/time) for a given contaminant from a source. In the previous case of a river, for example, the flux (mass/area-time) is given as the product of the mass transfer coefficient (K_L) and the concentration. By integrating the flux over the length of interest of the river, the emission rate can be derived.

In the case of a location such as a dump site, where contaminants may either be admixed with soil and exposed at the surface, or where they may be covered by a soil or clay cap, the flux may be estimated from consideration of the various resistances which can be combined into an overall mass transport coefficient, K_s, as in Equation 2.39. The vapor phase in contact with the admixed waste can be considered to be at equilibrium with the admixed waste (providing that the concentration in the admixed waste of the volatile contaminant is known). In Equation 2.39, C_G is the concentration of volatile material in the atmosphere above the waste site.

$$J = K_S(C_S - C_G) \tag{2.39}$$

In the presence of a "cap," there are two resistances in series for transport to the bulk atmosphere which must be considered. The first resistance is the transport of volatile material through the vapor-filled pore space in the soil which may overlie the contaminated area. This can be characterized by a mass transport coefficient (analogous to k_G or k_L), which will be designated as k_S. The second resistance is the gaseous film resistance, which is identical to that discussed previously and designated by the mass transport coefficient k_G. These may be summed as follows:

$$\frac{1}{K_S} = \frac{1}{k_G} + \frac{1}{k_S} \tag{2.40}$$

To use Equation 2.40, k_G can be estimated using Equation 2.31. The soil resistance mass transport coefficient k_S is estimated by:

$$k_S = \frac{\epsilon D_G}{h\tau} \tag{2.41}$$

where ϵ is the porosity of the overlying soil cap, D_G is the molecular diffusion coefficient of the volatile material through the air, h is the thickness of the soil cap (=

0, and hence $k_S = \infty$ in the absence of a soil cap), and τ is the tortuosity of the soil cap (defined as the ratio of the actual path length traveled during diffusion to the physical thickness of the soil layer).

This approach considers that the transport of volatile materials to the atmosphere is governed by diffusive flow of the volatile contaminant, and that the bulk velocity of gas through the soil cap is zero. However, in certain circumstances, there may be a significant bulk velocity which could increase (or decrease) the flux of materials to the atmosphere.

In the presence of a velocity through the soil cap (positive if in the outward direction), the flux of material is determined from a modification of Equation 2.39 (Thibodeaux et al., 1982):

$$J = K_S(C_S - C_G)(\phi e^{\phi})/(e^{\phi} - 1) \tag{2.42}$$

where $\phi = hv/D_S$.

v is the superficial velocity of gas through the soil cap (defined as volumetric flow rate divided by total area normal to flow). One possibility is that the velocity may be caused by barometric pressure differences, and, in this case, using a quasi-steady-state assumption, can be estimated as:

$$v = \kappa_G(p - \pi)/\mu h \tag{2.43}$$

with p and π being the external and internal barometric pressures (in atmospheres), respectively, and κ_G being the permeability of the soil cap to gas. Note that a negative pressure differential will result in a negative velocity and thus a reduction in flux to the atmosphere. A second possibility is that the bulk gas flow results from the generation of methane by biological decomposition within the landfill itself. If r_G is the rate of gas generation (mass/volume-time), h_W is the thickness of the waste deposit in which gas generation at that rate is occurring, and ρ_G is the average density of this gas as it travels through the pores, then

$$v = h_W r_G/\rho_G \tag{2.44}$$

If both mechanisms of bulk flow are operative, then it is necessary to dynamically model the variation of internal pressure with time.

Estimation of the flux to the atmosphere requires knowledge of the gaseous concentration, that is, C_S, of volatile contaminant in equilibrium with the soil or landfill. While it is possible that specific chemical interactions with the bulk waste, or with other materials, may substantially alter (either in a positive or negative sense) the component vapor pressure, a reasonable estimate may be obtained using a modification of Raoult's law (Shen, 1982). For pure materials, the vapor pressure (p in mm Hg) versus temperature (T in Kelvin) can generally be given by Antoine's equation 2.45 where A and B are parameters depending only upon the volatile component and which are generally available:

$$\log_{10} p^{\text{vap}} = -0.2185\,(A/T) + B \tag{2.45}$$

The vapor phase concentration in equilibrium with a *pure* volatile organic contaminant can be estimated using the ideal gas law:

$$C'_S = p^{vap}M/RT \tag{2.46}$$

with M being the compound mole weight, and R being the ideal gas constant (e.g., 62.3 mm Hg-L/K-mol). To correct for the reduction in vapor pressure from the pure component to the mixture, the following is used:

$$C_S = C'_s x_v \tag{2.47}$$

where x_v is the weight fraction of volatile material in the mixture of waste and soil from which volatilization is occurring.

Regardless of the source, the transport of contaminants to the atmosphere (and, in particular, the value of C_G above a source of contamination) will be partially governed by the dispersion processes in the atmosphere. In the simple case of a point source of contamination (e.g., a stack from an incinerator), if the flux is continuous and steady, the average concentration of material in the atmosphere at any point in space can be given by a modified form of Equation 2.13:

$$C = \frac{W}{2\pi v_x \sigma_y \sigma_z}\left\{\exp\left[-\frac{y^2}{2\sigma_y^2} - \frac{(z-h_{\text{eff}})^2}{2\sigma_z^2}\right] + \exp\left[\frac{y^2}{2\sigma_y^2} - \frac{(z+h_{\text{eff}})^2}{2\sigma_z^2}\right]\right\} \tag{2.48}$$

In this equation, W is the emission rate (mass/time) of contaminant from the source, x is the downwind direction, z is the vertical direction (above ground), and y is the transverse horizontal direction from the wind, which is at a velocity v_z. The source is located at $x = 0$, $y = 0$, $z = h$. The term h_{eff} is the effective height of the source, which accounts for a tendency of gasses which are heated or which have a high velocity to rise vertically prior to their dispersion. This may be estimated by:

$$h_{\text{eff}} = h + (v_s d/v_x)[1.5 + 0.00268\ P(\Delta T)d/T_s] \tag{2.49}$$

In this equation, h and h_{eff} are both in meters, v_s—the vertical exit velocity from the source—and v_x are both in meters/second, d—the diameter of the source—is in meters, P is the atmospheric pressure in millibars, and ΔT is the source temperature (T_s) minus the ambient air temperature with temperature measured in Celsius.

The σ's in Equation 2.48 have units of length and account for horizontal and vertical dispersion in a similar manner to the D's in Equation 2.13. Figures 2-5 and 2-6 in conjunction with Table 2-2 present correlations for these dispersion parameters as a function of distance downwind from the source and meteorological conditions.

In actuality, a landfill or a contaminated pond or lagoon really is an area source, which, only far away, may be approximated as a point source. To account for the finite size of the source at close distances, a modification of the above approach is required.

To account for the area source, Equation 2.48 is still used. However, the position of a virtual (i.e., imaginary) point source that would result in the same spread of contaminant is located. This is done by locating on Figure 2-5 the distance that would result in a value of σ_y equal to the maximum length of the areal source normal to the wind direction divided by 4.3. The numerical value of this distance is added to the physical distance from the areal source in computing the value of x used to determine σ_y and σ_z for use in Equation 2.48. For example, assume a circular area

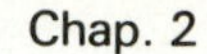

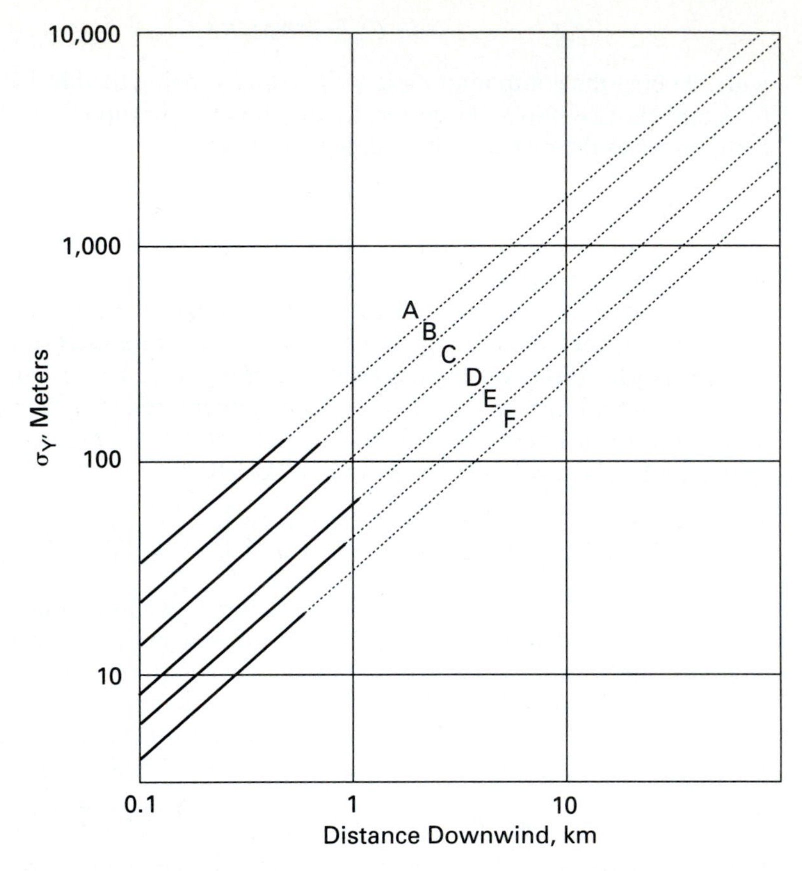

Figure 2-5 Horizontal dispersion coefficient as a function of downwind distance from the source. (Turner, 1970.)

source of diameter 1,000 meters. This divided by 4.3 equals 233 meters. If the wind speed is 4 m/sec, during daylight with strong sunlight, the stability class is *B* (Table 2-2). From Figure 2-1, a value of σ_y of 233 meters is achieved at a distance of about

TABLE 2-2 PASQUILL STABILITY CATEGORIES (From Turner, 1970.)

Surface wind speed (m/s)	Insolation			Night, mainly overcast or ≥ 4/8 low cloud	≤ 3/8 low cloud
	Strong	Moderate	Slight		
2	A	A–B	B	—	—
2–3	A–B	B	C	E	F
3–5	B	B–C	C	D	E
5–6	C	C–D	D	D	D
6	C	D	D	D	D

* A, extremely unstable; B, moderately unstable; C, slightly unstable; D, neutral; E, slightly stable; F, moderately stable.

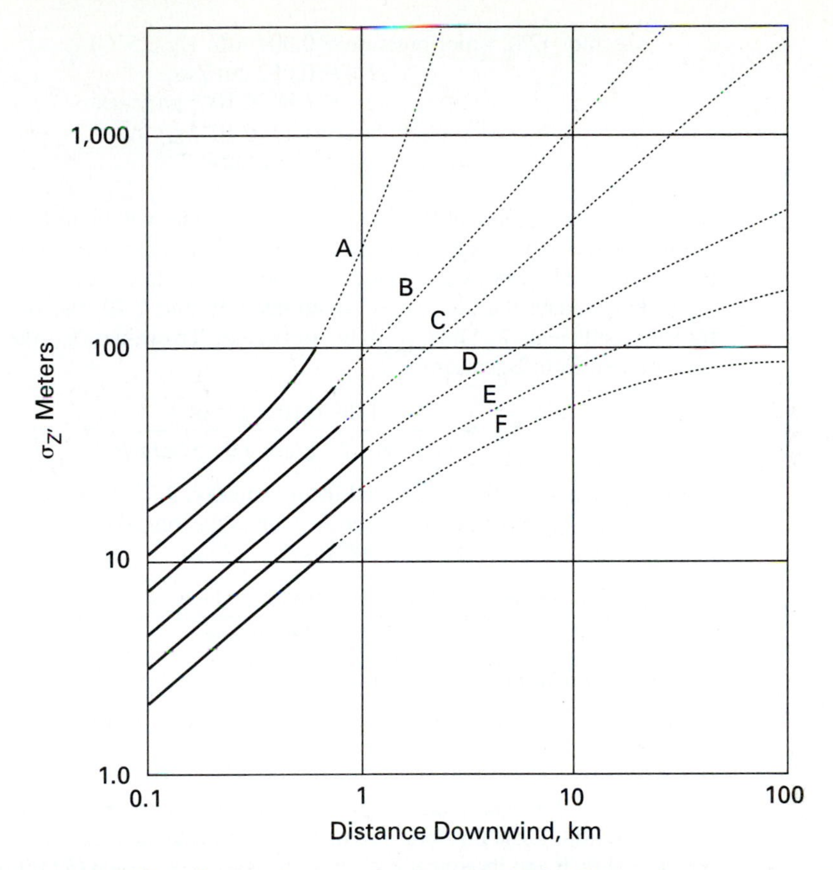

Figure 2-6 Vertical dispersion coefficient as a function of downwind distance from the source. (Turner, 1970.)

1,500 meters. Hence, if one wanted to compute the concentration 500 meters downwind of a source, the x value used to determine the pollutant dispersivities (σ's) would be 2,000 ($= 500 + 1{,}500$) meters.

By coupling the appropriate relationship for flux to the atmosphere with the dispersion equation, one can determine the ambient air concentration above the site, the flux from the site, and the concentration at all points downwind.

Example 2-3

Shen (1982) describes the Caputo dump in Glens Falls, New York. This site consisted of heavily contaminated soil laden with PCB's (specifically the commercial mixture Arochlor 1242). The site has an irregularly shaped area of 35,000 m^2 with 300 meters as its longest dimension and 180 meters as its shortest dimension. The soil concentration of Arochlor 1242 is 5,000 ppm. It is necessary to estimate the flux of PCB's into the atmosphere from the site, the concentration 100 meters downwind and the thickness of soil cover which would be needed to reduce the atmospheric flux by 99%. The wind speed is equal to 400 cm/s (9 mph) and is normal to the longest dimension of the site. The sky is overcast (stability class D).

The physical data relevant to the transport process are as follows:

Arochlor 1242 vapor pressure = 0.004 mm Hg (25°C)

$$D_G = 0.052 \text{ cm}^2/\text{sec}$$
$$\mu_G = 1.86 \times 10^{-4} \text{ g/cm-sec}$$
$$\rho_G = 1.17 \times 10^{-3} \text{ g/cm}^3$$
$$M = 328 \text{ g/mol}$$

First, an estimate of the flux to the atmosphere will be made assuming that the ambient air concentration (C_G) is much less than the vapor concentration at the source (C_s). The validity of this assumption will be checked later.

To evaluate the flux using Equations 2.39 and 2.40, the two individual mass transfer coefficients, k_G and k_s must be computed. To evaluate k_G, the Schmidt number is computed from Equation 2.32:

$$\text{Sc}_G = \frac{1.86 \times 10^{-4} \text{ g/cm-s}}{(1.17 \times 10^{-3} \text{ g/cm}^3)(0.052 \text{ cm}^2/\text{s})} = 3.06$$

The Reynolds number is computed from Equation 2.33; however it is first needed to determine the fetch (L). The geometric mean of the longest and shortest dimension will be used—that is, $L = (180 \text{ m} \times 300 \text{ m})^{1/2} = 232$ m. Therefore,

$$\text{Re} = \frac{(400 \text{ cm/s})(232 \text{ m} \times 100 \text{ cm/m})(1.17 \times 10^{-3} \text{ g/cm}^3)}{1.86 \times 10^{-4} \text{ g/cm-sec}} = 5.84 \times 10^7$$

Now, using Equation 2.31, k_G can be computed:

$$k_G = \frac{(0.036)(0.052 \text{ cm}^2/\text{s})(3.06)^{1/3}(5.84 \times 10^7)^{0.8}}{232 \text{ m} \times 100 \text{ cm/m}} = 0.19 \text{ cm/s}$$

In the absence of a soil cover, there is no second resistance, and hence $K_G = k_G$.

To compute the flux, one also needs to compute the value of C_G. Since the vapor pressure of PCB and its concentration in the waste are known (5,000 ppm equals 0.005 weight fraction), the equilibrium vapor concentration is computed using Equations 2.46 and 2.47:

$$\begin{aligned} C'_G &= (0.004 \text{ mm Hg})(328 \text{ g/mol})/[(62.3 \text{ mm Hg-L/K-mol})(298 \text{ K})] \\ &= 7.07 \times 10^{-5} \text{ g/L} \end{aligned}$$

$$\begin{aligned} C_G &= (7.07 \times 10^{-5} \text{ g/L})(0.005) \\ &= 3.54 \times 10^{-7} \text{ g/L} \\ &= 3.54 \times 10^{-4} \text{ g/m}^3 \\ &= 354\ \mu\text{g/m}^3 \end{aligned}$$

Substituting both the mass transfer coefficient and the vapor concentration into Equation 2.39, one obtains the flux:

$$\begin{aligned} J &= (3.54 \times 10^{-4} \text{ g/m}^3)(0.19 \text{ cm/s})(0.01 \text{ m/cm}) \\ &= 6.73 \times 10^{-7} \text{ g/m}^2\text{-s} \end{aligned}$$

The mass per unit time emitted from the site (W) is the product of the flux and the area; hence

$$\begin{aligned} W &= (6.73 \times 10^{-7} \text{ g/m}^2\text{-s})(35{,}000 \text{ m}^2) \\ &= 0.0236 \text{ g/s} \\ &= 2.035 \text{ kg/d} \\ &= 743 \text{ kg/yr} \end{aligned}$$

which is a substantial rate.

The concentrations in the atmosphere can be computed using the dispersion equation. At a stability class D, the distance at which σ_y equals 70 meters (≈ 300 m/4.3) is about 1,000 meters. At a distance 100 m away, the "corrected" value of x is 1,100 meters, and the corresponding values of σ_y and σ_z are, respectively, 70 and 30 m. Therefore, at the ground level ($h_{\text{eff}} = z = 0$) and the plume centerline ($y = 0$), the concentration is found by:

$$C_G = \frac{0.0236 \text{ g/s}}{2\pi(4 \text{ m/s})(70 \text{ m})(30 \text{ m})}[\exp(0) + \exp(0)]$$
$$= 8.9 \times 10^{-7} \text{ g/m}^3$$
$$= 0.89 \ \mu\text{g/m}^3$$

Observe that since this number is substantially less than the computed value for C_s, it was correct to neglect C_G in computing the flux.

The result may be compared with monitoring data presented by Shen (1982) in which the range of PCB concentrations at the site was reported as 0.07-540 μg/m^3. The geometric mean of 27 determinations was 2 μg/m^3 with a spread factor of 3.75. Hence the model appears reasonably consistent with the measured data.

The thickness of a soil barrier (cap) needed to reduce gaseous emissions by 99% may be computed. Since the flux is proportional to the mass transfer coefficient, it is desired to reduce the mass transfer coefficient from 0.19 cm/s to 0.0019 cm/s. From Equation 2.40, the necessary value of k_s can be computed:

$$\frac{1}{0.0019 \text{ cm/s}} = \frac{1}{0.19 \text{ cm/s}} + \frac{1}{k_s}$$

$$526.3 \text{ s/cm} = 5.263 \text{ s/cm} + \frac{1}{k_s}$$

$$k_s = 0.00192 \text{ cm/s}$$

From Equation 2.41, the thickness of a soil cap can be computed if the properties of the soil (ϵ, τ) and the diffusivity in vapor phase are known (this applies to the diffusive transport mechanism only—in the presence of velocities in the soil layer, it becomes necessary to know the hydraulic permeability, K). Typical values might be 0.2 for porosity and 1.8 for tortuosity; hence

$$0.00192 \text{ cm/s} = \frac{(0.2)(0.052 \text{ cm}^2/\text{sec})}{h(1.8)}$$

$$h = 3 \text{ cm}$$

In general, however, caps substantially thicker than this would be used to prevent cracking, and to provide increased resistance to any gas migration that might occur via either the barometric or biogenic mechanisms.

GROUNDWATER TRANSPORT

Essentially, the analysis of the movement of chemicals in groundwater requires solution of two coupled problems. The first problem is similar to those previously posed in this chapter in terms of contaminant dispersion. For solution, information about dispersion coefficients (as well as possible sorption or degradation processes) in the groundwater environment is necessary.

However, the second problem is the estimation of fluid velocities as functions of position. In the cases of dispersion that were previously discussed, the velocities were known and were constant with space. In groundwater, they may not be known from direct measurement and may in fact be a function of both space and time. Thus the velocity distribution must be estimated before estimating the contaminant dispersion.

The determination of velocities in groundwater requires different methods depending upon the physical system. One classification involves time dependency of velocities. Steady simply means that they are not functions of time. The second type of differentiation is whether the soil is saturated with water, that is, all the pores are filled with water, or unsaturated, in which some air space remains. Under unsaturated conditions, the transport of water involves the relationship among vapor pressure, soil adsorption of water, and hydraulic conductivity, thus complicating the analysis (McKee and Bumb, 1988).

Although unsaturated and unsteady flows are of interest, and may be of some practical importance, the analysis under such conditions generally involves numerical solutions of the hydraulic governing equations. The interested reader should consult Bear (1979) or Milly (1982) for further information.

For the case of steady, saturated flow, the Laplace equation governs the flow of water through the soil:

$$\frac{\partial^2 h}{\partial x^2} + \frac{\partial^2 h}{\partial y^2} + \frac{\partial^2 h}{\partial z^2} = 0 \tag{2.50}$$

where h (called the "head") represents the pressure at any location. h has units of length—representing the column of water that would exert a pressure equal to that experienced.

The Laplace equation is solved after substitution of the relationship between hydraulic head and velocity. This is given by Darcy's law, which states that velocity is linearly related in magnitude (and in direction opposite) to the gradient in head. If v_i is the approach velocity (Darcy velocity) in direction i (either x, y, or z), assuming that the medium porosity is unity, then Darcy's law can be written as:

$$v_i = -\kappa_i \frac{\partial h}{\partial i} \tag{2.51}$$

κ_i is called the hydraulic conductivity or permeability in direction i and has units of length/time. Depending upon the nature of the soil, κ_i may range from 10^{-10} to 10^{-2} cm/s. If the values of κ_i are identical (i.e., $\kappa_x = \kappa_y = \kappa_z$) then the system is described as being isotropic. Many systems involving clay minerals are not isotropic, with the hydraulic conductivity in the horizontal direction being greater than that in the vertical direction. If the κ_i's are independent of position (without necessarily being independent of direction) then the system is described as homogenous. The presence of lenses or inclusions in subsurface formations often produces wide variability in the hydraulic conductivity.

The Darcy velocity is related to the actual interstitial velocity of fluid relative to the stationary medium (v_i^e) by the following:

$$v_i = v_i^e \epsilon / \tau \tag{2.52}$$

The flow (Q_i, volume/time) passing in the ith direction perpendicular to a cross-sectional area A_i is given by:

$$Q_i = v_i A_i \tag{2.53}$$

Now given the geometrical and physical site characteristics necessary to specify boundary conditions, Equations 2.50 and 2.51 can at least be solved numerically to determine the values of the components of the velocity vector as functions of position. Appropriate boundary conditions specify the position and magnitude of any withdrawal or pumping wells, the existence of any impermeable boundaries. Numerical solution techniques involve the use of either finite element or finite difference methods, for which a variety of computer codes are available (Pinder and Gray, 1977).

For the most simple case of one-dimensional (e.g., in the vertical direction), steady, homogenous, isotropic flow, as might exist for water perched above a clay liner in a landfill, the solution is straightforward:

$$Q = -\kappa A(\Delta h/\Delta z) \tag{2.54}$$

where $\Delta h/\Delta z$ is the head loss per unit length (also often referred to as the hydraulic gradient) in the vertical direction. Therefore, given a measurement of the hydraulic gradient, the volumetric flow per unit area can be determined provided that the hydraulic conductivity has been measured or estimated.

For porous material, such as deposits of clay or sand, the velocity of flow is sufficiently small to be characterized as laminar. In other words, if d_p is the average (generally harmonic average) grain size of the deposit, then the Reynolds number defined by Equation 2.55 is under 2,000 (and, in virtually all realistic applications, is under 1.0).

$$\text{Re}_p = \frac{v_i^e d_p \rho}{\mu} \tag{2.55}$$

where

μ = fluid viscosity
ρ = fluid density

In such a system, the hydraulic permeability can be estimated by using the Kozeny–Carmen equation:

$$\kappa = \left(\frac{\rho g}{\mu}\right)\left(\frac{\epsilon^3}{(1-\epsilon)^2}\right)\left(\frac{d_p^2}{180}\right) \tag{2.56}$$

where g = gravitational acceleration (980 cm/s_2).

This approach is valid in media in which the flow can be considered to be through a porous layer. However, in certain situations, such as where the flow is through consolidated rock, or with material with very low porosities (below 0.1), the dominant transport of fluid may arise from passage through fractures. In such cases, the above equation may retain validity, but the porosity and diameter describe characteristics of these large fractures, rather than characteristics of a granular media per se.

Given estimates for the velocity, one can now apply them to analyzing the transport of chemical constituents using one of the dispersion equations. Dispersion characteristics may be estimated either by field investigation, or by use of one of several correlations. Bruch and Street (1967) related the longitudinal dispersion to the Reynolds number:

$$D_z = 1.8\left(\frac{\mu}{\rho}\right)(\mathrm{Re}_p)^{1.205} \tag{2.57}$$

At very low Reynolds numbers, the dispersion will be no less than the molecular diffusivity of the solute (corrected for porosity and tortuosity effects). The alternative correlation of Fried (1975) may be used in this situation:

$$D_z = D_L\left(\frac{0.67}{\tau} + 0.5\sqrt{\frac{v_i^e d_p}{D_L}}\right) \tag{2.58}$$

This equation is probably acceptable for relatively homogenous deposits, but where heterogeneity is present in the hydraulic conductivity, additional dispersion may be introduced by virtue of the multiple-transport paths (with a distribution of flow times) that contaminants may take between two points.

Gelhar and Axness (1983) have analyzed the effect of random spatial variations in the hydraulic conductivity of a medium on the dispersion of a dissolved material, and have provided a method for correcting D_z for such effects. The equation below is based on assuming that the distribution of hydraulic conductivity is isotropic and log-normal with a dimensionless standard deviation of σ. The spatial autocorrelation length, which is a measure of the scale over which the log hydraulic conductivity varies, is Λ. This model was found to provide good fit to field data in a large-scale tracer experiment (Sudicky, 1986). The corrected diffusivity, D'_z, accounts for both the local eddy transport and the effect of variation in hydraulic conductivity.

$$\begin{aligned} D'_z = \frac{6\sigma^2\Lambda}{6+\sigma^6}\Bigg\{1 - \upsilon\left[\frac{1}{1+\upsilon} + \frac{5}{3}\right] + \upsilon^2\left[-3 + 4\ln(1 + 1/\upsilon)\right] \\ + \upsilon^3\left[3 + \frac{2}{1+\upsilon}\right] + \upsilon^4\left[\frac{1}{1+\upsilon} - 4\ln(1 + 1/\upsilon)\right]\Bigg\} \end{aligned} \tag{2.59}$$

where $\upsilon = D_z/(v_z\Lambda)$.

If velocity is a function of time and/or position (or if the velocity vector has more than one component) it may be necessary to resort to a numerical model of the transport process (Wood et al., 1984).

The analysis of groundwater dispersion will be illustrated by the following example.

Example 2-4

Mercer et al. (1983) describe a portion of the Love Canal contaminant investigation. Data that they provide is for the site between the Niagara River and the edge of the dumpsite. The intervening distance is about 200 meters. After the contaminant plume has passed through this distance, contamination of the river occurs. The measured hydraulic permeability from field observations is 3×10^{-4} m/sec, and transport is essentially only one dimensional in the horizontal direction. The measured value for the hydraulic gradient is 1.52×10^{-4}. The intervening formation consists of saturated limestone with a density of 2.5 g/cm³. Porosity is 0.02. The tortuosity is not given, but it will be assumed to be equal to unity. While no estimates of solid partitioning are available, the authors speculate that a value of K_{sed} may range from 0–10 mL/g. From this information, it is desired to estimate the breakthrough of a contaminant wave at the river. The molecular diffusivity of a typical contaminant can be taken to be 5×10^{-6} cm²/s.

While no data are available on the variability of hydraulic conductivity in this system, based on the work of Sudicky (1986), it will be assumed that the standard deviation of the log-K distribution, σ, is 1, and that the correlation length, Λ, is 0.5 meter.

First, the effect of retardation can be estimated. Retardation may be computed as:

$$R = K_{sed}\,(2.5\ \text{g/cm}^3)(1\ \text{cm}^3/\text{mL})/0.02$$

For the range of K_{sed}, 0–10 mL/g, the retardation ranges between 0 and 1,250, which can be quite considerable. Hence, it would be very desirable to obtain more information to determine whether the limestone formation could substantially retard the specific contaminants in the leachate plume.

Now, the system hydraulics are considered. Based on the hydraulic gradient and permeability, the Darcy velocity is calculated:

$$v = (3.0 \times 10^{-4}\ \text{m/s})(1.52 \times 10^{-4})(100\ \text{cm/m}) = 4.56 \times 10^{-6}\ \text{cm/s}$$

The interstitial velocity is computed as:

$$v_e = \frac{4.56 \times 10^{-6}\ \text{cm/s}}{0.02} = 2.28 \times 10^{-4}\ \text{cm/s}$$

V_e is equal to V divided by epsilon. Divide that value by 0.02 to get 2.28×10^{-6} m/sec.

The travel time for a contaminant in the absence of dispersion can be computed by dividing the distance by the interstitial velocity, that is,

$$\begin{aligned} \text{travel time} &= (200\ \text{m})\ (100\ \text{cm/m})/2.28 \times 10^{-4}\ \text{cm/s} \\ &= 8.77 \times 10^{7}\ \text{seconds} \\ &= 1015\ \text{days} \end{aligned}$$

Now, the effect of dispersion of contaminants on the time required to reach the river can be estimated. In order to do so, an estimate of the dispersion itself is needed. However, in order to use the correlations presented, an estimate for particle diameter (which is not given) is needed. This can be estimated by applying the Kozeny–Carmen equation to the hydraulic permeability and computing the equivalent diameter that would describe the system.

$$3.0 \times 10^{-2}\ \text{cm/sec} = \left[\frac{(1\ \text{g/cm}^3)(980\ \text{cm/s}^2)}{10^{-2}\ \text{g/cm-s}}\right]\left[\frac{0.02^3}{(1-0.02)^2}\right]\frac{d_p^2}{180}$$

$$d_p^2 = 6.615$$

$$d_p = 2.57\ \text{cm}$$

This suggests that the dominant process may be transport through relatively large channels. Hence, the dispersion correlations with the Reynolds number must be used with a degree of skepticism. However, in the absence of field data needed for the more sophisticated estimation procedures, this method is all that is available.

To apply the Bruch and Street correlation, the particle Reynolds number is computed as:

$$Re_p = \frac{(2.28 \times 10^{-4}\ \text{cm/s})(2.57\ \text{cm})(1\ \text{g/cm}^3)}{10^{-2}\ \text{g/cm-s}} = 5.68 \times 10^{-2}$$

Therefore,

$$D_z = 1.8\left(\frac{0.01\ \text{g/cm}^2}{1\ \text{g/cm}^3}\right)(5.68 \times 10^{-2})^{1.205}$$
$$= 5.68 \times 10^{-4}\ \text{cm}^2/\text{s}$$

Using the Fried correlation, another estimate can be obtained:

$$D_z = 5 \times 10^{-6}\ \text{cm}^2/\text{s} \quad \left\{\frac{0.67}{1} + 0.5\left[\frac{(2.28 \times 10^{-4}\ \text{cm/s})(2.57\ \text{cm})}{5 \times 10^{-6}\ \text{cm}^2/\text{s}}\right]\right\}$$

$$D_z = 7.52 \times 10^{-4}\ \text{cm}^2/\text{s}$$

If it is desired to correct this estimate for the effect of hydraulic conductivity variations, the Gelhar and Axness equation can be used. The parameter υ is calculated using the value of D_z obtained from the Fried correlation (the larger of the two values is used here for conservation in predicting the breakthrough wave):

$$\upsilon = (7.52 \times 10^{-4}\ \text{cm}^2/\text{s})/(4.56 \times 10^{-6}\ \text{cm/s}) = 164.9\ \text{cm}$$

The relevant values can be substituted into Equation 2.46 to yield $D'_z = 0.01081\ \text{cm}^2/\text{s}$ —considerably larger than the dispersion due to eddy diffusivity alone.

Using Equation 2.10, at various values of time, the relative concentration (C/C_0) can be computed by:

$$\frac{C}{C_0} = \frac{1}{2}\left\{\text{erfc}\left[\frac{2 \times 10^4\ \text{cm} - (2.28 \times 10^{-4}\ \text{cm/s})(t)}{2(Dt)^{1/2}}\right] + \exp\left(\frac{2 \times 10^4\ \text{cm} \times 2.28 \times 10^{-4}\ \text{cm/s}}{D}\right) \times \text{erfc}\left[\frac{2 \times 10^4\ \text{cm} + (2.28 \times 10^{-4}\ \text{cm/s})(t)}{2(Dt)^{1/2}}\right]\right\}$$

In this equation, the time must be input as seconds. For comparison, values of the relative concentration will be tabulated at various times for two choices of dispersion parameter (D)—either D_z estimated from the Fried correlation ($7.52 \times 10^{-4}\ \text{cm}^2/\text{s}$) or the value estimated using the theory of Gelhar and Axness ($0.01081\ \text{cm}^2/\text{s}$):

	C/C_0	
Time (days)	$D = 7.52 \times 10^{-4}$ cm²/s	$D = 0.01081$ cm²/s
525	1.1E-298	1.19E-22
600	2.2E-188	6.94E-15
675	2.4E-113	1.44E-09
750	5.71E-63	5.85E-06
825	1.59E-30	0.001426
900	1.77E-11	0.042905
975	0.013333	0.289879
1050	0.968473	0.699571
1125	1	0.936549
1200	1	0.993165

Suppose that the concentration of a particular pollutant in the leachate was 10^5 mg/L or 10 g/L, which might occur for certain solvents, for example. If a critical concentration to reach the river was 1 μg/L (C/C_0 of 10^{-7}), then the breakthrough is estimated as occurring between 900 and 975 days, ignoring hydraulic conductivity variations, or 675–700 days considering such variations. In either case, this is more rapid than the estimate of 1,015 days based solely upon the transit time of water in the aquifer. This illustrates the substantial contribution that dispersion can make to the flux of a contaminant.

PROBLEMS

2.1. You are a monitoring engineer stationed on a moderate-sized river with a velocity of 0.2 m/s, a width of 2 m and a depth of 1.5 m. The hydraulic slope is 0.0001 m/m. You maintain a set of continuous analyzers which sample the water to monitor for the presence of toxic materials. The sensitivity of your monitor for PCB's is 1 μg/L. The Midnight Disposal Company is suspected of dumping waste transformer oil at a point 20 miles upstream of your monitoring location. Assuming that volatilization can occur, estimate the maximum *continuous* release that can occur upstream without your monitor's being able to detect it. (Assume average wind speed of 10 mph at a 45° angle to river flow, AROCHLOR 1242, H = 0.56 L-atm/mol, molecular weight = 254, diffusivity = 0.052 cm²/s (air), 5.23×10^{-6} cm²/s (water).

2.2. For the same river in problem 2.1, assume that on a given day, your monitor detects 8 μg/L of arsenic at 1 PM, at 3 PM it detects 10 μg/L, and at 5 PM it detects 8 μg/L. Assuming that arsenic is nonvolatile and nonreactive, and that this concentration resulted from a single instantaneous point discharge, estimate the time at which it occurred, the amount of mass released, and the distance upstream at which the discharge was released.

2.3. A petroleum storage tank ruptures and rapidly releases its contents into a neighboring river. The release consists of nonvolatile, nonadsorbable materials and amounts to 100 grams/square centimeter river cross section. At a monitoring station 20 km downstream, 28 hours after the spill, the concentration of material is measured as 149 mg/L in the river. There is a water treatment plant that uses the river as a source located 100 km downstream of the spill site. Based on the information you are given, estimate when the chemical concentration at the intake would first go above 0.01 mg/L after the spill occurs, and for what duration this elevated concentration would persist.

2.4. A dumpsite contains 500 ppm (w/w) of pentachlorophenol. The site is 100 meters square and the soil is contaminated to a depth of 5 meters and exposed to the atmosphere. The mean wind speed above the site is 10 mph. Estimate the flux to the atmosphere. If a clay cap is to be used to cover the site to reduce this flux by 99 percent, how thick should it be (assume particle diameter of 0.001 cm and porosity of 0.2).

Possibly useful data
Vapor pressure of pure PCP: 4.3×10^{-16} atm
Diffusivity of PCP in air: 0.059 cm^2/s
Molecular weight: 264 g/mol
Density of air: 1.2 g/L
Viscosity of air: 1.9×10^{-4} g/cm-s

2.5. (Modified from Freeze and Cherry). A hazardous waste landfill is to be sited on a five m thick deposit of dense clay that overlies an aquifer that provides drinking water to a small town. Assume that the leachate will accumulate at the landfill base and overlie the clay layer. The water table of the leachate in the landfill will be at 251.3 m above mean sea level (MSL). In the aquifer, the pressure head is measured at 250.5 m above MSL. The hydraulic conductivity of the clay is 2×10^{-11} m/s, and the porosity is 0.19. Estimate how long it will take for nonreactive contaminants to move through the clay into the aquifer. Consider whether dispersion significantly increases the breakthrough time (for this purpose, consider when the concentration passing through the clay barrier is 10% of that in the leachate).

2.6. A truck spill has just resulted in a small area of contaminated soil containing 1,000 ppm of dieldrin (a carcinogenic pesticide). The spill is roughly circular of 30 foot diameter. Estimate the atmospheric concentration that results 1,000 meters downwind if the wind speed is 15 mph, and the stability class is D. The molecular weight of dieldrin is 381 g/mol, and its diffusivitity in air is 0.04 cm^2/s.

2.7. The maximum desirable ambient air standard for contaminant X is 1 $\mu g/m^3$. An industry is considering siting an evaporation pond 500 meters from a residential area, which will receive hazardous liquids containing this volatile material. The lagoon is to be square with a side of 50 meters. Consider that typical evening conditions are 10 mph with clear skies. What is the maximum concentration of material in the lagoon that can be tolerated without exceeding the desirable air standard? (NOTE—Assume that the resistance to transport is solely in the gas phase—that is, there is no liquid mass transport resistance).

Physical properties
Henry's law constant: 3 L-atm/mol
Diffusivity in air: 0.02 cm^3/s
Molecular weight: 175

2.8. A plume of contaminated groundwater has resulted from surface industrial activities. The plume is migrating vertically through moderately pervious clay toward an aquifer

50 m away. As a remediation strategy, it is proposed that the side of the plume be injected with hydrogen peroxide in order to encourage biological growth and to effect some hydrolysis. This process will result in the first order disappearance of the polynuclear aromatic hydrocarbons (PAH's) in the plume with a rate constant of 0.03 d^{-1}. If the plume concentration at the source is 700 μg/L of PAH, given the following information, what is the maximum concentration of PAH which would be expected to reach the aquifer and how long will it take for 10, 50, and 90% of the maximum to be reached?

Data
Grain size in zone of transport: 35 μm
Vertical hydraulic gradient: 0.7 cm/cm
Porosity: 0.3

REFERENCES

ARCHER, S. R., BLACKWOOD, T. R., AND COLLINS, C. S. "Status Assessment of Toxic Chemicals: Polybrominated Biphenyls," U.S. EPA-600/2-79-210k (1979).

BALL, W. P., AND ROBERTS, P. V. "Long Term Sorption of Halogenated Organic Chemicals by Aquifer Material. 1. Equilibrium." *Environ. Sci. Technol.* 25(7): 1223-1235 (1991).

BALL, W. P., AND ROBERTS, P. V. "Long Term Sorption of Halogenated Organic Chemicals by Aquifer Material. 2. Intraparticle Diffusion." *Environ. Sci. Technol.* 25(7): 1237-1249 (1991).

BEAR, J. *Hydraulics of Groundwater.* McGraw Hill, New York (1979).

BIRD, R.B., STEWART, W. E., AND LIGHTFOOT, E. N. *Transport Phenomena.* Wiley, New York (1960).

CRANK, J. *The Mathematics of Diffusion,* 2nd edition. Oxford University Press, Oxford, UK (1975).

DRAGUS, J., AND HELLING, C. S. "Soil and Clay-Catalyzed Reactions. I. Physiochemical and Structural Relationships of Organic Chemicals Undergoing Free Radical Oxidation," in *Land Disposal of Hazardous Waste, Proceedings of the Eighth Annual Research Symposium.* U.S. EPA-600/9-82-002 (1982).

FREEZE, R. A. AND CHERRY, J. A. *Groundwater.* Prentice Hall, Englewood Cliffs, N. J. (1979).

FRIED, J. J. *Groundwater Pollution.* Elsevier, Amsterdam (1975).

GARABEDIAN, S. P., LEBLANC, D. R., GELHAR, L. W., AND CELIA, M. A. "Large Scale Natural Gradient Tracer Test in Sand and Gravel, Cape Cod, Massachusetts. 2. Analysis of Spatial Moments for a Nonreactive Tracer." *Water Resources Research* 27(5):911-924 (1991).

GARBARINI, D. R., AND LION, L. W. "Influence of the Nature of Soil Organics on the Sorption of Toluene and Trichloroethylene," *Environ. Sci. Technol.* 20(12): 1263-1273 (1986).

GAUTSCH, W. "Error Functions and Fresnel Integrals." In *Handbook of Mathematical Functions.* National Bureau of Standards, Washington (1964).

GELHAR, L. W., AND AXNESS, C. L. "Three Dimensional Stochastic Analysis of Macrodispersion in Aquifers," *Water Resources Research* 19:161-80 (1983).

GRADY, C. P. L. "Biodegradation of Hazardous Wastes by Conventional Biological Treatment," *Hazardous Waste and Hazardous Materials* 3: 333–365 (1986).

HANNA, S. R., BRIGGS, G. A., AND HOSKER, R. P., JR. *Handbook on Atmospheric Diffusion.* U.S. Department of Energy, DOE/TIC-11223 (1982).

KARICKHOFF, S. W., BROWN, D. S., AND SCOTT, T. A. "Sorption of Hydrophobic Pollutants on Natural Sediments." *Water Res.* 13: 241–248 (1979).

KOBAYASHI, H., AND RITTMAN, B. E. "Microbial Removal of Hazardous Organic Compounds." *Environmental Science and Technology* 16: 170A–182A (1982).

LAPIDUS, L., AND AMUNDSON, N. R. "Mathematics of Adsorption in Beds. VI. The Effect of Longitudinal Diffusion in Ion Exchange and Chromatographic Columns." *Journal of Physical Chemistry* 56: 984–88 (1952).

LEBLANC, D. R., GARABEDIAN, S. P., HESS, K. M., GELHAR, L. W., QUADRI, R. D., STOLLENWERK, K. G, AND WOOD, W. W. "Large Scale Natural Gradient Tracer Test in Sand and Gravel, Cape Cod, Massachusetts. 1. Experimental Design and Observed Tracer Movement." *Water Resources Research* 27(5): 895–910 (1991).

LEO, A. J., AND HANSCH, C. "Partition Coefficients and Their Uses." *Chemical Reviews* 71: 525–621 (1971).

LYMAN, W. J., REEHL, W. F., AND ROSENBLATT, D. H. *Handbook of Chemical Property Estimation Methods: Environmental Behavior of Organic Compounds.* McGraw-Hill, New York (1982).

MCKEE, C. R., AND BUMB, A. C. "Saturated and Unsaturated Flow of Groundwater: Part 2. Unsaturated Flow." *Hazardous Materials Control* 1(3): 14–21 (1988).

MCQUIVEY, R. S., AND KEEFER, T. N. "Simple Method for Predicting Dispersion in Streams." *ASCE Journal of the Environmental Engineering Division* 100: 997 (1974).

MERCER, J. W. et al. "Modeling Groundwater Flow at Love Canal, New York." *ASCE Journal of the Environmental Engineering Division* 109: 924–42 (1983).

MILLY, P. C. D. "Moisture and Heat Transport in Hysteretic, Inhomogeneous Porous Media: A Matrix Head-Based Formulation and a Numerical Model." *Water Resources Research* 18(3): 489–98 (1982).

NAUMAN, E. B., AND BUFFHAM, B. A. *Mixing in Continuous Flow Systems.* John Wiley, New York (1983).

O'CONNOR, D. J., AND DOBBINS, W. E. "The Mechanism of Reaeration in Natural Streams." *Journal of the Sanitary Engineering Division, ASCE* 82(SA6): 1115 (1956).

OGATA, A. "Theory of Dispersion in a Granular Medium." *U.S. Geological Survey Professional Paper* 411-1 (1970).

PAVLOSTATHIS, S. G., AND JAGLAL, L. "Desorptive Behavior of Trichloroethylene in Contaminated Soil." *Environ. Sci. Technol.* 25(2): 274–279 (1991).

PINDER, G. F., AND GRAY, W. G. *Finite Element Simulation in Surface and Subsurface Hydrology.* Academic Press, New York (1977).

PIONTEK, K. R. "A History of Human Exposure to 2,3,7,8-TCDD," mimeo, University of Missouri (1983).

RUTHERFORD, D. W., CHIOU, C. T., AND KILE, D. E. "Influence of Soil Organic Matter Composition on the Partition of Organic Compounds." *Environ. Sci. Technol.* 26(2): 336–340 (1992).

SCHWARZENBACH, R. P., AND WESTALL, J. "Transport of Nonpolar Organic Compounds from Surface Water to Groundwater: Laboratory Sorption Studies." *Environ. Sci. Technol.* 15(11): 1360–1367 (1981).

SHEN, T. T. "Air Quality Assessment for Land Disposal of Industrial Wastes." *Environmental Management* 6(4): 297–305 (1982).

STEELE, J. G. "Measurement of the Longitudinal Dispersion Coefficient Near a Junction in a Tidal River." *Water Resources Research* 27(5): 839–844 (1991).

SUDICKY, E. A. "A Natural Gradient Experiment on Solute Transport in a Sand Aquifer: Spatial Variability of Hydraulic Conductivity and its Role in the Dispersion Process." *Water Resources Research* 22, 2069–82 (1986).

THIBODEAUX, L. J., *Chemodynamics.* John Wiley and Sons, New York (1979).

THIBODEAUX, L. J., SPRINGER, C., AND RILEY, L. M. "Models of Mechanisms for the Vapor Phase Emission of Hazardous Chemicals from Landfills." *Journal of Hazardous Materials* 7: 63–74 (1982).

TUMA, J. J. *Engineering Mathematics Handbook.* McGraw-Hill, New York (1987).

WEHNER, J. F., AND WILHELM, R. H. "Boundary Conditions of Flow Reactor." *Chemical Engineering Science* 6: 89–93 (1956).

WOOD, E. F., et al. *Groundwater Contamination from Hazardous Wastes.* Prentice Hall, Englewood Cliffs, N.J. (1984).

3

Overview of the Waste Management Problem

The objective of this chapter is to outline the nature of the hazardous waste problem and to serve as a road map in the use of this book.

ONGOING VERSUS "OLD" SITES

Broadly speaking, there are two types of activities involving hazardous wastes which are of concern. One activity involves the evaluation, treatment, and remediation of past inappropriate waste disposal practices. The types of sites involved in this activity may include old dumps and landfills and abandoned industrial facilities. Frequently those responsible for the prior practices may be quite difficult to trace, and therefore a central activity becomes identification of (and apportioning responsibility between) "potentially responsible parties." The management of these sites is governed to a major extent under the federal "superfund" law.

The second type of activity involving hazardous waste is the ongoing management, treatment and disposal of the residuals from current operations. In these cases, the activity is governed to a major extent by the Resource Conservation and Recovery Act, and its amendments.

From a technical point of view, solutions to the two activities share many common features. In order to consider the appropriate technical solution to a hazardous waste problem, it is necessary to determine the types and quantities of materials present in the waste ("source"), the location of individuals or ecosystems that could be adversely impacted by these materials ("receptor"), and the transformation and attenuation processes that may change the nature or amount of the materials arriving at the receptor. A major objective of the technical response to a hazardous waste problem of either type is to take action such that the concentrations of materials at receptors do not exceed some critical level which may result in undesirable effects.

The technical actions that may be taken in either case may include treatment of the material by one of a number of processes of a physical, chemical, biological, or thermal nature. In the case of the existing "superfund" type of problem, treatment may occur *in situ* or above ground, after some means (excavation or pumping) has been used to remove the contamination from its physical location. In the case of the ongoing "RCRA" type of problem, additional options might include waste minimization to avoid, minimize, or reduce the risk from the waste being generated.

The focus of this book will be on the ongoing "RCRA" type of problem. As a result, subsequent chapters will include material on contaminant transport, treatment processes, and waste minimization. Ultimate disposal by landfilling is not included in this book, since the appropriate environmental geotechnical solution to a particular waste disposal problem is a very broad topic in itself, and beyond the scope of intended coverage here. Similarly, the processes of waste removal or pumping as preliminaries to treatment are not included since these are to a significant degree a focus of either the geotechnical engineering or the geohydrology disciplines.

SOURCE–RECEPTOR PARADIGM

To determine the required technical action that needs to be taken to solve a hazardous waste problem, it is necessary to characterize the waste or waste stream, and identify possible transport mechanisms that may result in migration of undesirable contaminants to a receptor. Then, the required action or degree of treatment may be determined by investigating the "acceptable" levels prior to attainment of an adverse response. In Chapter 2, these transport mechanisms were discussed and identified.

MATCHING APPLICABLE WASTES WITH APPLICABLE PROCESSES

After the degree of reduction or removal of contaminants from a particular waste that must be achieved is identified, it is necessary to devise an action strategy to produce the desired result. In general, one should use a hierarchical approach to this problem (Figure 3–1). It would be most desirable to avoid the generation of the waste, or at least avoid the introduction of the most undesirable constituents into the waste. This may be done either by a modification of the manufacturing operations that resulted in the waste being generated (waste minimization/pollution prevention) or by recycling or reuse of the materials themselves. These strategies will be discussed in Chapter 8.

The materials that cannot be eliminated or removed from the waste must then be subject to treatment. In general, treatment processes are designed to achieve volume and/or toxicity reduction. The designer must then determine which series of treatment processes may be feasible to achieve the desired performance, and of these feasible processes, which is most desirable considering a variety of criteria including

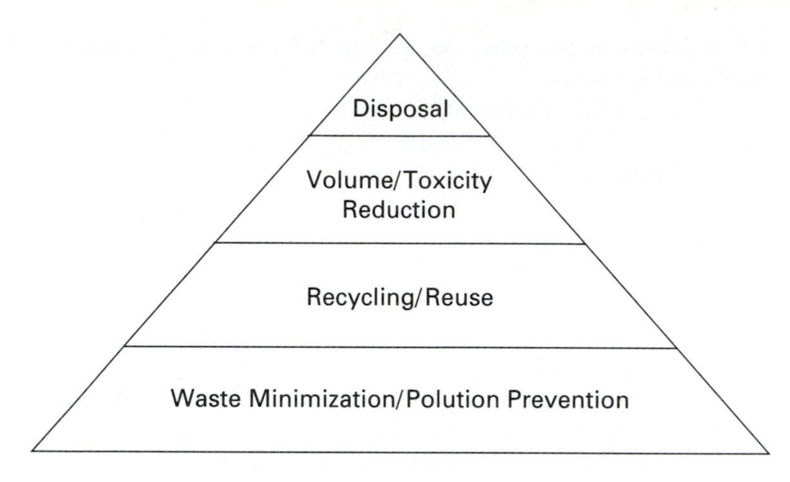

Figure 3-1 The waste management hierarchy.

cost and environmental compatibility. Chapters 4–7 consider the spectrum of treatment processes in detail.

In general, all treatment processes, including incineration and biological treatment, result in some residual the disposal of which must be considered. Under the USEPA "derived from" rules, such residuals may frequently themselves be hazardous wastes. Therefore, ultimate disposal of these materials is required. Almost exclusively (except for the underground injection of certain materials) disposal of the residuals from the waste management hierarchy occurs in secure landfills, or if the waste is no longer hazardous then it may be disposed in sanitary landfills. The design and operation of these land disposal facilities is beyond the scope of this text.

LEVELS OF DESIGN

To approach the problem of designing a series of treatment processes to achieve a particular specified performance objective (defined as above, or by regulatory stipulation), a sequence of steps is generally performed. The first step, termed conceptual design, amounts to a structured "brainstorming" activity, in which a combination of treatment steps may be considered. Based on literature data, the experience of the design engineer, or perhaps on performance of similar facilities, the ability of the process to achieve the desired goals is assessed. At this stage, various alternative sequences of processes may be considered; for example, in the treatment of waste containing chlorinated phenolics and heavy metals, one might consider chemical oxidation (for phenol removal) followed by metal precipitation or vice versa.

In the second stage of design, the plausible and presumed feasible process sequences are examined in greater detail. This stage may be termed the intermediate design level. The interactions between processes need to be examined. Continuing the above example, it may be found that the presence of certain heavy metals acts as a catalyst in chemical oxidation, and thus oxidation followed by precipitation might be more desirable. Or it might be determined that the metals may precipitate during the chemical oxidation stage resulting in an undesirable sludge production. This intermediate stage requires laboratory testing, although perhaps only at the laboratory or small pilot stage. Certain feasible process sequences may also be rejected at this stage on the basis of incompatibility due to specific design constraints. For example, the use of incineration may be precluded by the scale of the system (waste volume too small to justify such a capital intensive process).

For small problems, the intermediate stage results in a very small number of process sequences whose performance and economics are known to a sufficient level of reliability to allow direct development of plans and specifications, or bid documents. If the estimated facility cost is $50,000, for example, known to a reliability of ±30%, it would not be justifiable to conduct $20,000 of additional engineering work to substantially improve the certainty of the system cost. However, for larger systems, or where substantial noneconomic objectives exist, a more detailed engineering design investigation is required prior to the final design process.

In this third tier of design, pilot plant testing is conducted to determine and refine design parameters for the full scale facility. Often this testing is conducted over a long period of time (months) with the real waste stream to assess long term process reliability and maintenance factors. In this final design stage, the uncertainty of process costs might be reduced to under 10%, and factors such as amount of residuals generated, environmental acceptability (hazardous or nonhazardous nature), fouling and breakdown of process equipment, and cleaning requirements are determined.

The objective of this book is to allow the conceptual and intermediate design stages to be conducted. In addition, the engineer involved in a final design investigation needs to have understanding of the basic operational variables that govern the details of process performance. These are also set out in Chapters 4–7.

The final stage of a design problem, particularly if a network of treatment processes is being designed or developed, is overall system optimization. There may be issues of location and type of treatment facilities that are to be developed. The formal analysis and optimization of such systems is covered in Chapter 8.

4

Physical Waste Treatment Processes

AIM OF PRESENTATION

In this chapter, and the following two chapters, unit operations and processes that are useful in the recovery or treatment and disposal of hazardous wastes will be summarized. The intent of the presentation is to pinpoint key design variables that influence process performance and economics. While preliminary design equations may be provided for individual processes, the design of a hazardous waste processing system very often necessitates laboratory or pilot scale testing to determine characteristics of an individual waste stream. Hence, the design equations should be regarded as a method to summarize the effect of such waste characteristics on process performance, rather than as a means to produce a completed process design.

PROCESS CLASSIFICATION

Physical processes are those that work in the absence of any chemical or biological reaction other than a possible phase transfer. And so, while there might be a change of phase, there is no chemical or biological change in the hazardous material(s) per se. The classification of physical treatment processes can be based upon their dominant means of achieving separation of materials. Table 4-1 provides this classification and enumerates the specific physical treatment processes that will be described in this chapter.

Since no change in the chemical components is achieved in a physical process, there is always a concentration of hazardous components (or properties) in one output stream and a diminution of hazardous components (or properties) in another output stream in any physical process. Hence, in general, physical processes at most

TABLE 4-1 CLASSIFICATION OF PHYSICAL TREATMENT PROCESSES

Separation by Differential Action of a Force
Gravitational
Sedimentation
Centrifugal
Centrifugation
Magnetic
Magnetic Filtration
Electrical
Electrostatic Precipitation
Electrophoresis
Separation by Differential Passage through Fixed Media
Particulates
Granular Filtration
Vacuum Filtration
Filter Presses
Bag House Filters
Macromolecular
Ultrafiltration
Molecular
Reverse Osmosis
Dialysis
Electrodialysis
Separation by Differential Phase Partitioning
Liquid-Liquid
Solvent Extraction
Vapor-Liquid
Distillation
Air Stripping
Steam Stripping
Absorption
Solid-Liquid or Solid-Vapor
Adsorption

achieve a volume reduction (rather than an elimination) of a hazardous waste. Potentially, however, all of the streams exiting a physical separation process (in the case of a specifically listed waste) may remain hazardous by definition.

SEPARATION BY DIFFERENTIAL ACTION OF A FORCE

Physical processes that achieve separation by differential action of a force generally are used to separate large particles or macromolecules. Either gravitational, centrifugal, magnetic, or electrostatic forces can be used. The differential action of a force upon the particle relative to the fluid imparts a relative velocity between the bulk fluid and the particle that can be used for separation.

Gravitational Separation Processes

The simplest type of physical processes are those that rely upon gravitational forces to separate spent solids from a fluid (either gas or liquid) phase in which they are suspended. In the process of sedimentation, the solid particles are denser than the fluid and so they settle out under the action of gravity.

Sedimentation may be employed either as an initial operation for particle removal (to prevent fouling or excessive solids buildup in downstream process units), following chemical treatment in which soluble materials are precipitated, or following biological treatment in which a suspended microbial floc is generated. Sedimentation can also be employed for separation of different solids by means of their differential settling characteristics. For example, grits and soil particles typically have higher settling velocities than do inorganic chemical precipitates, especially hydroxides.

For freely suspended particles, at sufficiently low concentrations, their removal is governed by discrete or type I sedimentation behavior. For such systems, the removal characteristics of a sedimentation tank are governed by two primary variables: the settling velocity of the suspension (v_s) and the overflow velocity of the sedimentation tank (v_0). Settling velocity may be obtained from a laboratory settling column test. For real suspensions, there may be a distribution of settling velocities defined by a function $P(v_s)$ representing the fraction of solids with settling velocities less than or equal to v_s. The overflow velocity is defined as the volumetric influent flow divided by the area perpendicular to the vertical. For upflow sedimentation tanks, where the flow pattern of supernate is primarily vertical, the fraction of solids removed (f) may be computed as:

$$f = 1 - P(v_0) \tag{4.1}$$

In other words, all solids with settling velocities in excess of the overflow velocity will be removed. In the case of horizontal flow sedimentation tanks (or any sedimentation tank with a component of flow in the horizontal direction), there is some additional removal due to settling as the suspension flows over a collecting area and the removal may be computed from:

$$f = 1 - P(v_0) - (1/v_0)\int_0^{v_0} x\,dP(x) \tag{4.2}$$

However, real sedimentation tanks contain nonidealities. In particular, the longitudinal gradients of solids concentration will cause a dispersive flux of particles as settling occurs. This can be accounted for by ascribing a turbulent dispersion to the settling particles. The resulting equation for the particle removal fraction may be written as (Ostendorf and Botkin, 1987):

$$f = 1 - \lambda \int_0^{v_0} \mathrm{ierfc}\{[(v_s/v_0) - 1]/2\lambda\}\, dP(x) \tag{4.3}$$

where λ is a parameter which characterizes the clarifier dispersion, and ranges between strong dispersion (1) and absence of dispersion (0), and can be correlated to the clarifier depth. In Equation 4.3, ierfc is the integral error function. The limit, for large dispersions, can be given by (Cordoba-Molina et al., 1978):

$$f = 1 - \int_0^{v_0} \exp(-v_s/v_0)\, dP(x) \tag{4.4}$$

Equations 4.2 and 4.4 represent the upper and lower limits of removal in Type I clarifiers due to sediment diffusion, and thus provide a reasonable margin of safety in scale-up.

Example Problem 4-1

A nonflocculating slurry is to be partially clarified by plain sedimentation. Tests are conducted in which a settling column is filled with slurry and allowed to settle. At various times the settling suspension is measured for solids. The solids samples are collected at a depth 25 cm below the air-liquid interface. If the initial solids concentration is 5,000 mg/L, use the following data to determine the upper and lower limits to efficiency of removal of a clarifier if the design overflow velocity is 0.001 ft/s (0.03048 cm/s).

Time (sec)	Mg/L
100	4800
500	4300
1000	3700
2000	2000
2500	1100
3000	500
3500	250
4000	25

The first step is to convert the data to a distribution of sedimentation velocity (V_s) versus fraction of solids with settling velocity less than or equal to that value. The proportion of solids remaining at a given time precisely equals the proportion of solids with a settling velocity less than or equal to h/t—where h is the depth of measurement and t is the sampling time. Hence, from the above table, one can immediately compute the value of $P(V_s)$ versus V_s as in the following table:

V_s	$P(V_s)$	Ideal	Disperse
0.25	0.96		
0.05	0.86		
0.03048(*)	0.766304	0.010933	0.491172
0.025	0.74	0.010203	0.480542
0.0125	0.4	0.003828	0.292876
0.01	0.22	0.001803	0.168326
0.008333	0.1	0.000703	0.079461
0.007142	0.05	0.000316	0.040664
0.00625	0.005	0.000015	0.004536

Each row is obtained from the previous table, except for the starred row. In this row, the value of $P(V_0)$ is obtained via linear interpolation from the surrounding values. The final two columns (labeled ideal and disperse) represent values of the integrals in Equations 4.2 and 4.4 from which ideal and maximum dispersion removals can be computed. These integrals are computed using the trapezoidal rule, but other numerical integration techniques can be employed.

Therefore, for ideal sedimentation, the value of the integral in Equation 4.2 is 0.010933, and, from this, the fractional removal is:

$$f_{\text{ideal}} = 1 - 0.766304 + (1/0.03048)(0.010933)$$
$$= 0.592 \text{ or } 59.2\% \text{ removal}$$

Similarly, for completely disperse conditions, the result is:

$$f_{\text{disperse}} = 1 - 0.491172$$
$$= 0.508827 \text{ or } 50.9\% \text{ removal}$$

Thus, the effect of dispersion is to diminish removal by almost 10 percentage points.

For flocculent materials, the sedimentation velocity tends to increase with extent of sedimentation due to the agglomeration of particles into more rapidly settling aggregates. This phenomenon results in a time-dependent sedimentation velocity distribution of a suspension. If this is described by $P(t, h)$ which represents the fraction of particles remaining in a settling column as a function of time and distance down from the interface (h), then the removal can be described by:

$$f = 1 - (H/C_0) \int_0^H C dh \tag{4.5}$$

where H is the height of the settling zone in a clarifier, C_0 is the influent concentration of solids and C is the concentration of solids at a given depth (h) and time (t). By expressing C as an explicit function of h and t, this equation can be fit to experimental settling curves of Type II suspensions and used for design purposes. For example Berthouex and Stevens (1982) found the quadratic form of Equation 4.6 useful.

$$C = a_0 + a_{10}t + a_{01}h + a_{20}t^2 + a_{02}h^2 + a_{11}ht \tag{4.6}$$

The coefficients in this equation (a's) can be obtained from curve fitting to data from batch settling tests.

For highly concentrated suspensions, or when the primary purpose is the production of a highly concentrated sludge, the performance of a sedimentation tank may become limited by the ability of the developing sludge blanket to concentrate. It has been observed that, in such systems, the downward velocity (V_i) of the interface may be described as a function of the solids concentration (C_i) in one of the two forms:

$$V_i = V^*(C_i/C^*)^{-n} \tag{4.7}$$

$$V_i = V_0 \exp(-aC_i) \tag{4.8}$$

where V^*, C^*, n, V_0, and a are adjustable model parameters. In the first equation, C^* may be treated as an arbitrary constant so as to make the model dimensionally consistent. Application of these relationships to a thickener mass balance results in a relationship between the underflow velocity of sludge (U) and its concentration (C_u) which, for the above two models, can be found to be, respectively (Baskin and Suidan, 1985):

$$U/V^* = (n-1)[n/(n-1)]^n (C_u/C^*)^{-n} \tag{4.9}$$

$$U/V_0 = (\psi - 1)\exp(\psi) \tag{4.10}$$

$$\text{with } \psi = \frac{aC_u + [(aC_u)^2 - 4aC_u]^{\frac{1}{2}}}{2}$$

In practice, the overflow velocities of sedimentation tanks used for industrial and hazardous waste treatment (defined as the volumetric flow rate divided by surface area) are in the range of 12–60 meters per day depending upon the material to be treated. Highly concentrated suspensions and sludges generally require lower overflow rates. Typical residence times for such sedimentation tanks are several hours.

Example Problem 4-2

A lime sludge from the precipitation of plating rinsewater is to be thickened to a solids concentration of 2% (20,000 mg/L). Given the following data in which the initial settling velocity was measured as a function of time, if the sludge is produced continuously at a rate of 100 m^3/day and a concentration of 5,000 mg/L, what should the surface area of the thickener be to achieve the desired underflow solids concentration?

Experimental Data

Initial solids (mg/L)	Initial settling velocity (m/d)
12000	3940
10000	8050
9000	6550
8000	10260
7000	11030
6000	18210
5000	32850
4000	41940
2000	242320

Using the log–log model, a plot of ln (V_i) versus ln(C_i/C_*) yields a straight line with a correlation coefficient of 0.99. The comparison of observed and regressed values is shown in Figure 4–1. In constructing this plot, a value of 10,000 mg/L was used for C^*.

A similar analysis using the semi-log model shows a much poorer fit to the data. Therefore, these data may be described using the log–log model with the following constants:

V^* 43.8
n 2.25

Since the desired underflow solids concentration is known, the sludge withdrawal rate can be computed by:

$$U/V^* = (n-1)[n/(n-1)^n (C_u/C^*)^{-n}$$

$$U = 43.8(2.25)(2.25/1.25)^{2.25}(20{,}000/10{,}000)^{-2.25}$$

$$= 43.2 \text{ m/d}$$

Therefore, for thickening, the required area can be computed by equating the mass flows

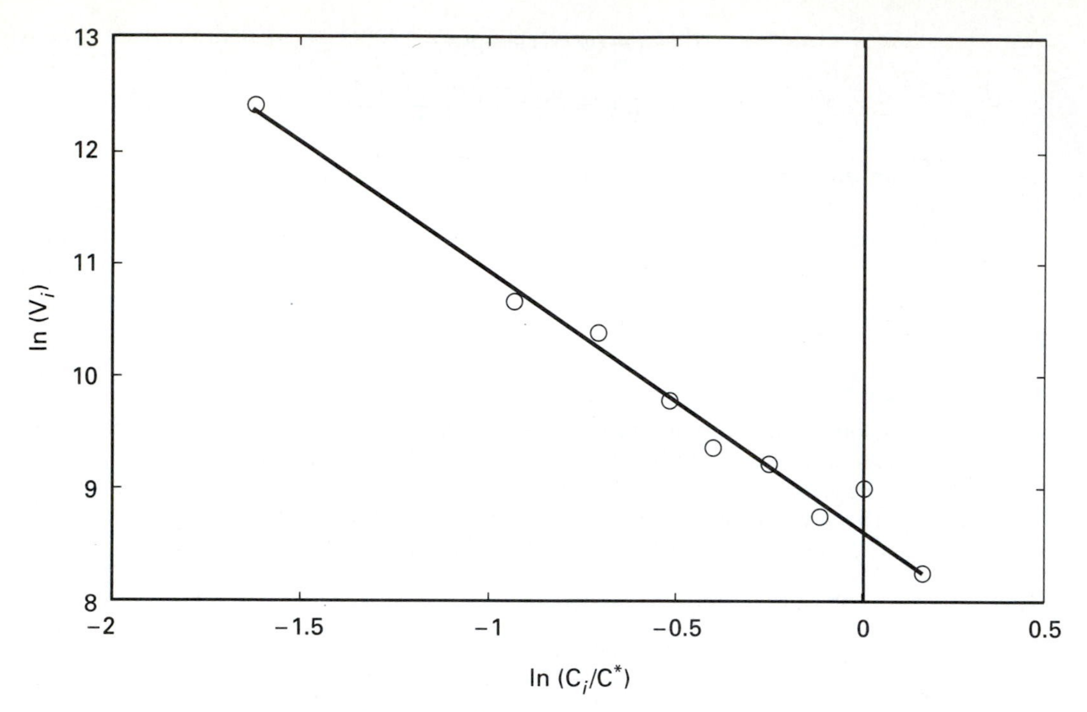

Figure 4-1 Initial settling velocity plot.

of sludge entering and exiting:

$$A_{\text{thick}} U C_u = Q C_0$$
$$A_{\text{thick}} = Q C_0 / U C_u$$
$$= 100 \text{ m}^3\text{/d } (5000 \text{ mg/L})/(43.2 \text{ m/d})(20{,}000 \text{ mg/L})$$
$$= 0.57 \text{ m}^2$$

If the initial setting velocity is computed for the feed sludge (5,000 mg/L), this is found to be equal to 970 m/d. Since this is far greater than the limiting solids flux, the area for thickening governs the design. Thus, a thickener area of 0.57 m^2 is required. This can be achieved in a circular thickener of 0.85 m in diameter.

The primary application of sedimentation is solids removal and/or recovery, particularly if the suspending fluids stream is relatively less hazardous or easier to manage. For example, dissolved metals may be precipitated to form a metal sludge. By sedimentation to remove the solids, the bulk of the metal from the original waste will have been concentrated in the sludge. Ideally, then, the supernate may be of either diminished toxicity, or may (assuming that it is not specifically listed and that cyanides or sulfides are not present) no longer be hazardous.

Sedimentation is a very well understood technology. It is easy to design over a borad range of flow rates. Generally, it involves a low capital cost and a low equipment cost but potentially a large land area. Thus, for treating large flows, a tremendous amount of land may be required.

If sedimentation is applied to a waste containing volatile materials, a portion of the volatiles may escape into the atmosphere and result in unacceptable levels of air pollution. If the volatile material is combustible (in the case of gasoline, for example) this might present a major flammability risk.

Centrifugation Processes

To reduce the land area required for particle separation and to minimize the risk of volatile emissions, a centrifugal separation can be employed. In principle, a centrifuge is nothing more than a sedimentation tank in which the driving force is centrifugal rather than gravitational. Since use of a centrifuge permits greater forces to be applied to a particle, larger settling velocities can be achieved (and, as a consequence, separation can be conducted in a smaller space).

There are several different types of centrifuges available—basket, disk, and screw centrifuges. Figure 4-2 depicts these types. These are generally continuous in operation with a constant feed and continuous withdrawal of both a clarified liquid and a thickened slurry.

The principal design variables for centrifugation are rotational speed (ω) and the axis of rotation, both of which will influence the force on the particles, the influent flow rate, and the residence time (or bowl volume). Vesilind (1979) has analyzed the performance of centrifuges for clarification and has formulated the following relationship for the equivalent area for a centrifuge:

$$\Sigma = \frac{V\omega^2}{g \ln\left(\frac{r_{max}}{r_{min}}\right)} \tag{4.11}$$

where g is the gravitational acceleration, and r_{max} and r_{min} are, respectively, the maximum and minimum radii from the axis of rotation. Using this principle, if the ratio of flow rates to equivalent areas is maintained constant (i.e., if the equivalent hydraulic loading is maintained constant), then similar degrees of clarification will be achieved. There is, however, no such relationship that has yet been proposed to estimate the performance of centrifuges at thickening sludges.

Operationally, for many centrifugal processes, it has been found to be advantageous to add chemical conditioners to improve the ability for particles to agglomerate and thus settle more rapidly in the centrifuge. In general, the ability of centrifuges to thicken slurries is much more difficult to predict. For biological waste solids produced by the activated sludge process treating domestic wastewaters, centrifugation can increase the solids from under 1% in the feed to over 5% (in some cases, over 10%) with over 90% solids recovery (U.S. EPA, 1980). The best combination of design variables should be determined in a pilot plant study. The dewaterability of a slurry may vary dramatically from waste to waste, and even from batch of waste to batch of waste. Very often with centrifugation, the limiting factor is the formation of a cake of solids at the bottom at the outside of the centrifuge, which provides a restriction to the flow of water. This factor is difficult to predict, and should be determined from pilot plant experiments. Chemical additives used to condition the suspension can also have a dramatic influence on this resistance, and alum, lime,

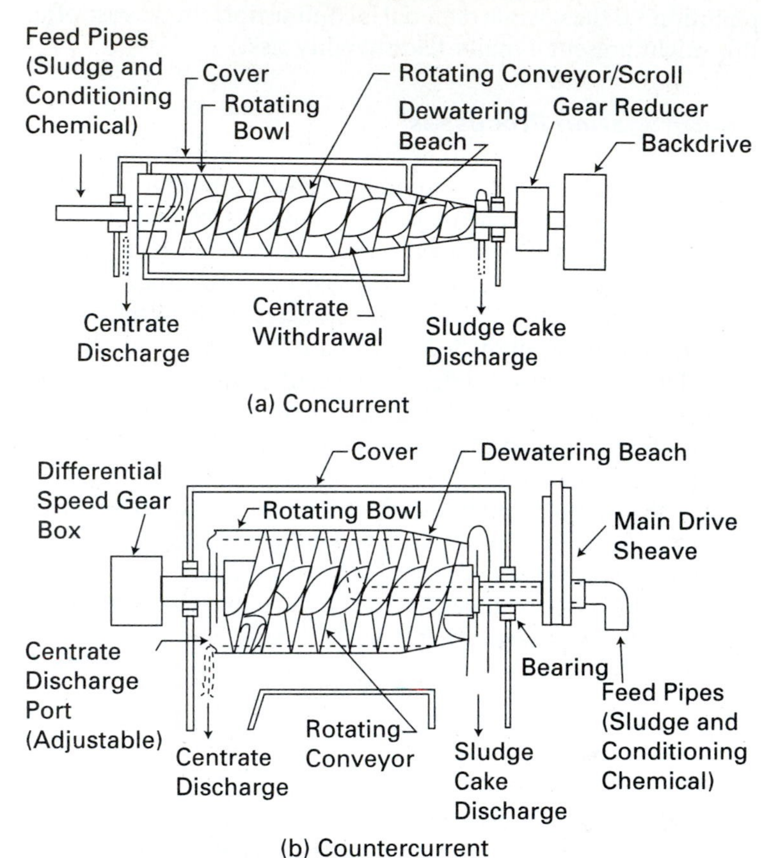

Figure 4-2 Types of centrifuges. (EPA Sludge Manual.)

ferric chloride, or polymers might be added to improve dewaterability in the centrifuge.

The emission of volatile organics from centrifugation can be better controlled with the centrifuge than in sedimentation tank since the former is a smaller physical device, and in general, can be placed under an air duct for collection of any evolved vapors. This allows the control (or at least the monitoring) of the emission of organic materials from the centrifuge.

Centrifuges may find similar applications as sedimenation. In addition, centrifugation can be applied to the handling of concentrated slurries in order to even further concentrate the solid materials in the slurry. With gravity sedimentation, it is very difficult to get above 1% solids (10,000 mg/L) in the underflow. However, the centrifugation of a well-conditioned material can produce a sludge containing 15–20% solids. This may be sufficiently concentrated such that the solid material no longer has free liquid present or so that little excess fuel is required for its combustion.

Imperforate Basket Centrifuge Schematic

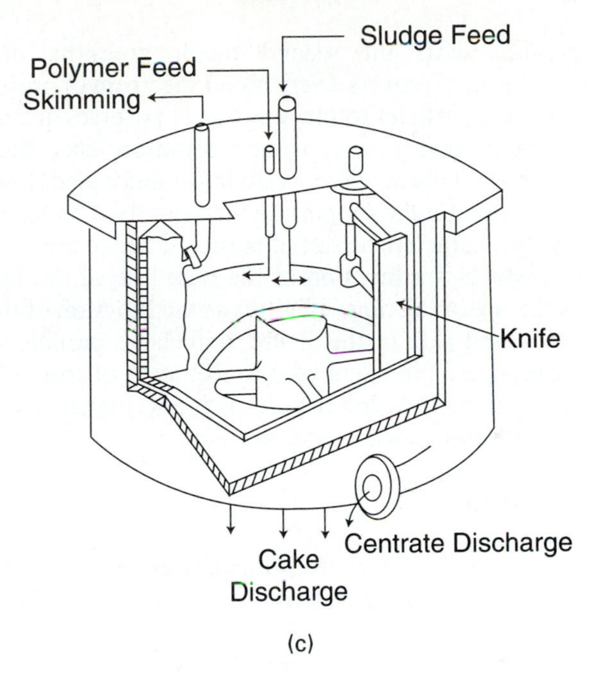

(c)

Disc Nozzle Centrifuge Schematic

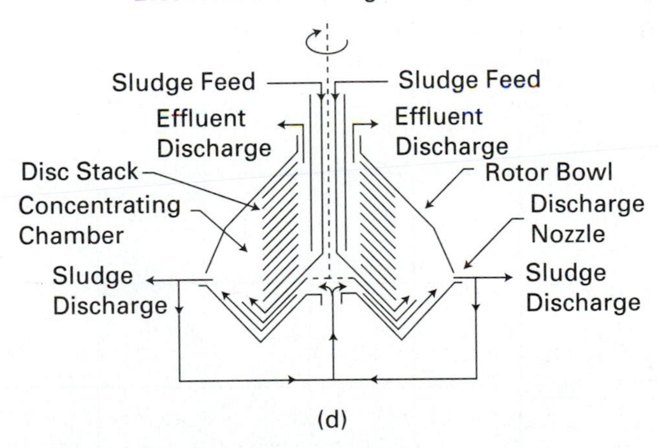

(d)

Figure 4-2 *(Continued)*

Centrifugation can be applied to nonaqueous materials. Solid separation from these materials presents a high potential for volatilization of organics. In addition, the settling velocities in oils and oily sludges tend to be much lower than aqueous materials since the viscosity of oil is much higher than water. Both of these factors provide significant advantages to the application of centrifugal separation (as compared to sedimentation).

SEPARATION BY DIFFERENTIAL PASSAGE THROUGH FIXED MEDIA

Filtration represents the second major grouping of solid-liquid separation processes. Filtration covers a very broad spectrum of processes, from the removal of large suspended particles to the removal of particles at the colloidal and molecular level. All the processes have, as their common facet, the passage of a suspension through a fixed medium (either granular or fabricated) and the removal of particles by their retention on the supporting medium itself. These processes can be roughly classed by the characteristic size of particles, which are separated, and the hydraulic resistance (which is a function of the pore size of the filter medium). Figure 4-3 describes the operating range of filters as the function of these two variables. At the upper right-hand side of this figure, with large particle sizes and relatively small pressure drops, are the filters used for removal of truly suspended particles (larger than 1 micrometer). At lower pore sizes and higher pressure drops are molecular filters (ultrafiltration and reverse osmosis).

Granular Filtration

In the case of granular filters, the principal design variable is the hydraulic loading, or volumetric flow rate per unit area. As loading increases, pressure drop in the filter

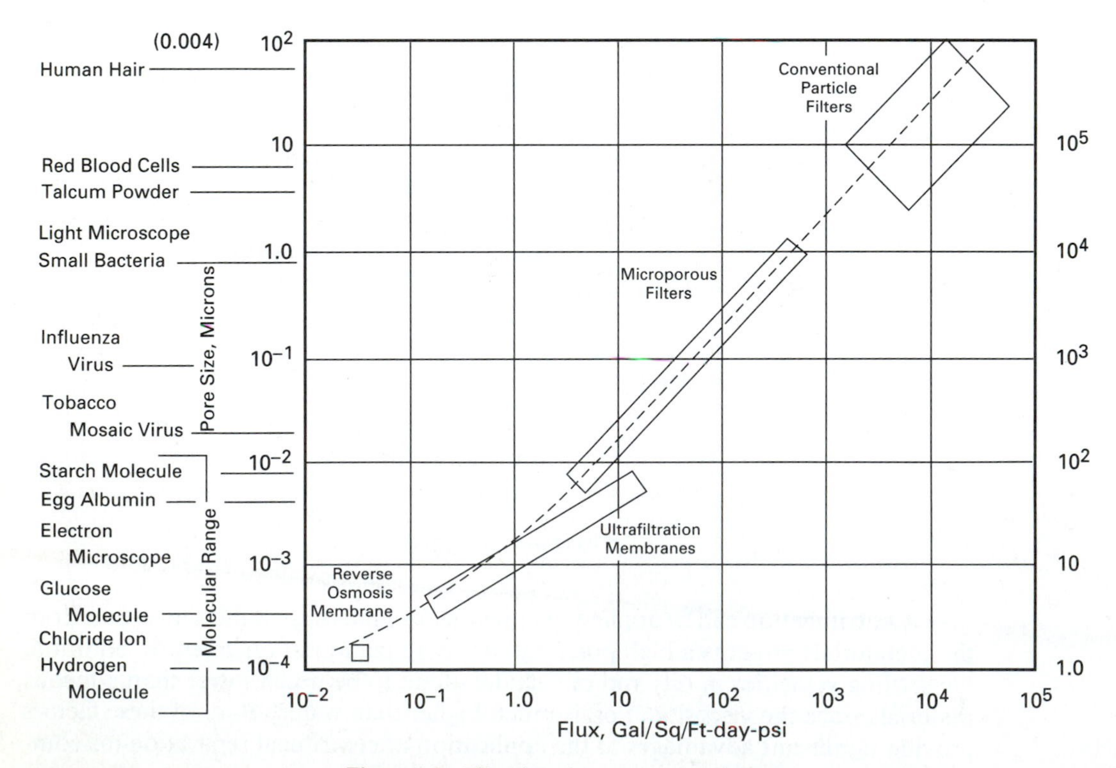

Figure 4-3 Filtration size versus pressure drop.

also increases, and the particle removal efficiency changes. Granular filtration is an inherently cyclical process; periodically the filter pressure drop builds up due to an accumulation of entrapped particles, or the particles may break through to the effluent. Then the filter must be cleaned by backwashing. At this time, the flow direction is reversed and the bed is fluidized (expanded in length) until the filter grain particles are no longer resting on each other. Both the relationship between removal efficiency and velocity and the characteristics of backwashing are functions of the physical properties of the filter medium.

In general, for granular filters, the hydraulic loading is in the range of 1 to 10 gpm/ft^2. There is a relationship between the hydraulic loading and the pressure drop through a filter which can be given by the Carmen–Kozeny equation:

$$h_t = \text{head loss through filter}$$
$$= f_p\,[(1-\epsilon)\tau^3/\epsilon^3][V_0^2/g][L/\Phi d_g] \qquad (4.12)$$

where

f_p = friction factor = $1.75 + 150(1-\epsilon)/\text{Re}$
Re = $\Phi d_g V_0 \rho/\mu$
ϵ = filter bed porosity
Φ = filter grain sphericity (1.0 is ideal sphere)
L = length of filter bed
τ = filter bed tortuosity (ratio of actual path length of flow to height of filter bed)

Also, in this equation, V_0, the approach velocity, is used to define the hydraulic loading. This is defined as the volumetric flow divided by the cross sectional area of the filter and physically represents the velocity of the water column towards the filter. Another velocity, of particular importance in defining removal characteristics of particles, is the interstitial velocity, U, which is the relative velocity of water moving within the filters with respect to the stationary filter medium. The two velocities are related by:

$$U = V_0\tau/\epsilon \qquad (4.13)$$

For type filter media used in treatment of liquids, the porosities are 0.3–0.5 and the tortuosities are 1.2–1.5. For graded sand media, the sphericity is close to 1.0, but for anthracite filter media, it can be as low as 0.7.

In actual filters which are backwashed, and which contain a graded size of a single type of filter medium (e.g., sand), after the first backwashing there will be a tendency for the larger (and hence, more rapidly settling filter grains) to be preferentially concentrated at the bottom of the filter. Conversely, the smaller grains will accumulate near the top of the filter. This tendency can be counteracted if various filter media (with different densities) are used so that the densest filter medium consists of the smallest particles. In such a scenario, the smaller particle sizes will now be on the bottom of the filter.

In any event, once the filter is stratified, the clean bed head loss must be predicted by applying the Carmen–Kozeny equation to each layer of the filter and summing the head losses over depth. The proportion of the filter bed depth which is accounted for by each layer is simply the mass fraction of filter medium in that weight fraction (providing that the porosity is independent of particle size). If this is done, it can be found that the dominant head loss is due to the smaller particles in the filter.

The backwashing of filters is commonly designed for once per day operation involving a velocity sufficient to fluidize the bed to 100–200% expansion. This involves a loading rate of 20 to 50 gpm/ft^2. The relationship between the degree of expansion of a filter bed and the velocity required can be given by relationships developed by Cleasby and Fan (1981).

During fluidization, there is a direct relationship between the bed length (L_e) and the bed porosity (ϵ_e), where the unsubscripted variables indicate static bed conditions:

$$L_e(1 - \epsilon_e) = L(1 - \epsilon) \tag{4.14}$$

Also, during fluidization, the head loss remains constant at the following value (and, in fact, the onset of a constant head loss versus velocity can be used operationally to indicate fluidization):

$$h_{\mathrm{Le}} = L(1 - \epsilon)(\rho_g - \rho)/\rho \tag{4.15}$$

If this head loss is substituted into the Carmen–Kozeny equation and the unknown velocity computed, this velocity (denoted as V_t) is the minimum velocity needed for fluidization to occur. However, it must be corrected for a slight additional effect of sphericity. If V_t is the theoretical minimum fluidization velocity thus computed, then the corrected minimum velocity for the onset of fluidization, V^* is given by:

$$V^* = 0.91 V_t \Phi^{-0.40} \tag{4.16}$$

Once fluidization is achieved, the relationship between the expanded bed porosity and the approach velocity is given by:

$$V_e / V^* = (\epsilon_e)^n \tag{4.17}$$

with

$$n = 4.45\ \mathrm{Re}^{-0.1} \Phi^{-(2.9237 \Phi^{0.844} \mathrm{Re}^{-0.363})} \tag{4.18}$$

$\mathrm{Re} = d_g V_s \rho / \mu$ (V_s is the discrete sedimentation velocity of a filter grain particle).

The maximum approach velocity which can be used during fluidization is given by the discrete settling velocity of the finest particles in the bed. Any excess velocity above and beyond this point will serve to remove filter medium.

The minimum backwash velocity is thus governed by the largest particles of filter medium—these require the greatest flow to achieve fluidization. The headloss during loading of a filter is governed primarily by the smallest particles of filter medium. Hence it is desirable to maintain as uniform a filter medium size as possible. The finer the filter medium, the greater the removal of small particles, regardless of how great the head loss.

Since the filter must be taken off line during backwashing, it is necessary to either have two filters or to have a feed water holding tank which can store the feed water during the backwashing operation.

Granular medium filters can produce a product with a low suspended solids concentration (<1 mg/L). However, they are not suitable for treatment of wastes with suspended solids above about 100 mg/L, due to excessive head loss buildup. Hence, they are often applied subsequent to sedimentation of a high-suspended solids material.

The removal of particles by a granular filter can be predicted from consideration of the fluid mechanics during filtration. Small particles are removed by a diffusional mechanism, while larger particles are removed by a combination of Van Der Waals, inertial and gravitational forces. There is a range of particle sizes in any granular filter which is removed the least. Consideration of the diverse mechanisms leads to the following relationship for the fraction of particles remaining after passage through one layer of filter medium (η)(O'Melia, 1985):

$$\eta = 4A_s^{1/3}\mathrm{Pe}^{-2/3} + 0.72A_s\mathrm{Lo}^{1/8}R^{15/8} + 0.0024A_sGr^{1.2}R^{-0.4} \quad (4.19)$$

with

$$A_s = (1-\alpha^5)/(1-1.5\alpha+1.5\alpha^5-\alpha^6)$$
ϵ = porosity of filter medium
$$\alpha = (1-\epsilon)^{1/3}$$

$$\mathrm{Pe} = 3d_gd_p\pi\mu U/kT$$

d_g, d_p = filter grain, particle diameters
U = interstitial velocity of fluid in the filters
k = Boltzmann constant
T = absolute temperature

$$\mathrm{Lo} = A/9\pi\mu d_p^2U$$

A = Hamaker constant which characterizes the filter grain-particle Van Der Waal's forces (10^{-13} erg is frequently assumed)

$R = d_p/d_g$
$Gr = 2d_p^2(\rho_p-\rho)g/9\mu U$
ρ_p = density of particles

This relationship assumes that every collision between a particle and a filter grain results in removal. However, if the particles are insufficiently destabilized (e.g., if they carry electrical charges of like signs) the removal efficiency will be substantially less. Hence, it is important to assure proper particle conditioning prior to filtration. Polyelectrolytes, or small concentrations of aluminum, iron, or calcium compounds may be used for this purpose.

The integration of a particle mass balance over the depth of a filter, assuming that the filter has uniform properties (grain size, porosity, cross section) over depth leads to the following relationship for the ratio of effluent to influent particle concentration (f):

$$\ln f = -(3/2)(1-\epsilon)\eta\eta_c(L/d_g) \quad (4.20)$$

where

L = length of filter column
η_c = dimensionless collision efficiency factor
(1 = ideal destabilization)

In actual filter operation, the effectiveness of removal of subsequent particle changes, along with the head loss characteristics of the filter bed, due to the collection of new particles. Recent work has shown that it is possible to consider this as a coupled set of partial differential equations characterizing the head-loss buildup and the modification of the collection efficiency of the surface of the filter (Tare and Venkobachar, 1985; Horner et al., 1986), however, the use of such approach in design has not been thoroughly explored.

Example Problem 4-3

A machine shop generates 40 liters/day of contaminated cutting oils containing metal filings. It is suggested that this oil can be recovered by filtering the metal filings through a bed of scrap iron. If it is desired to remove 90% of the particles with an initial head loss of less than 3 feet (91.44 cm), what properties (grain size, filter depth, and diameter) should the filter have? Use the following information:

$\epsilon = 0.6$
$\Phi = 0.8$
$\tau = 2.5$
$\rho_l = 0.9\ \text{g/cm}^3$
$\mu = 0.5\ \text{g/cm-s}$
$\rho_p = 3\ \text{g/cm}^3$
$\rho_g = 3\ \text{g/cm}^3$
$d_p = 10^{-3}\ \text{cm}$
$A = 10^{-13}\ \text{erg}$
$\eta_c = 0.1$
$T = 298\ \text{K}$

First, considering the head loss relationships, the following equations can be written:

$$\text{Re} = \Phi d_g V_0 \rho/\mu = 1.44 d_g V_0 (d_g \text{ in cm, } V_0 \text{ in cm/s})$$
$$f_p = 1.75 + 150(1-\epsilon)/\text{Re} = 1.75 + 41.7/d_g V_0$$
$$h_l = f_p[(1-\epsilon)\tau^3/\epsilon^3][V_0^2/g][L/\Phi d_g] = 91.44 \text{ cm}$$
$$\therefore \quad L = 2478 d_g^2/(1.75 V_0^2 d_g + 41.7 V_0) \qquad \text{(A)}$$

Now, from the relationships involving particle removal one can obtain:

$$\ln f = -(3/2)(1-\epsilon)\eta\eta_c(L/d_g)$$
$$\ln(0.1) = -(3/2)\eta(0.1)(L/d_g)$$
$$\eta L/d_g = 38.4 \qquad \text{(B)}$$
$$\alpha = (1-0.6)^{1/3} = 0.737$$
$$A_s = 12.96$$
$$U = 4.17 V_0$$
$$Pe = 4.78 \times 10^{11} d_g$$
$$L_0 = 1.6 \times 10^{-9}/V_0$$
$$R = 0.001/d_g$$

$$Gr = 2.19 \times 10^{-4}/V_0$$
$$\therefore \quad \eta = 1.54 \times 10^{-7} d_g^{-2/3} + 1.77 \times 10^{-6} V_0^{-1/8} d_g^{-15/8} + 2 \times 10^{-5} d_g^{0.4} V_0^{-1.2} \quad \text{(C)}$$

Equations (A) and (B) can be combined to produce the following relationship:

$$d_g = 45.7 V_0/(64.5\eta - 1.75\ V_0^2) \quad \text{(D)}$$

For an assumed value of V_0, equations (C) and (D) can be solved for d_g and η by iteration as follows:

- assume trial value for η
- compute d_g from (D)
- compute a revised value of η from (C)
- repeat until the desired degree of convergence is achieved, at which point L can be obtained from (A)

The following short table gives sample results.

V_0 (cm/s)	η	d_g (cm)	L (cm)
0.001	0.0225	0.0287	48.99
0.002	0.0138	0.0939	261.6
0.004	0.00914	0.283	1188.3

Based upon this, the 0.001 cm/s velocity is selected (due to the extreme heights associated with larger filter velocities). This velocity requires that the filter medium have a particle size of 0.29 mm, which is reasonable. The velocity is equivalent to a loading of:

$$0.001 \text{ cm/s } (3600 \text{ s/hr})(24 \text{ hr/d}) = 86.4 \text{ cm}^3/\text{cm}^2 - \text{d}$$

Hence, based on the production of waste, a filter of the following area would be required:

$$40 \text{ l/d } (1000 \text{ cm}^3/\text{l})/86.4 \text{ cm}^3/\text{cm}^2 - \text{d} = 463 \text{ cm}^2$$

This is equivalent to a circular filter of diameter 24.3 cm. As a margin of safety, and to allow for the buildup of headloss and breakthrough of solids with filter loading, a somewhat larger diameter and length of filter should be used.

Vacuum Filters

Vacuum filters consist of a cylinder around which a porous material is allowed to travel (Figure 4-4). A vacuum is maintained in the center of the drum, and as the porous material travels through the feed tray, solids are drawn onto the filter surface and water is expressed into the center of the drum. At the end of the cycle, the dewatered filter cake is scraped from the medium surface with a knife edge. The primary aim of vacuum filtration is solids concentration and dewatering. The variables that influence process performance, in addition to size and media type (typically woven fabric or wire or coil), are rotational speed, hydraulic flow-per-unit filter area, and degree of immersion of the medium into the feed tray.

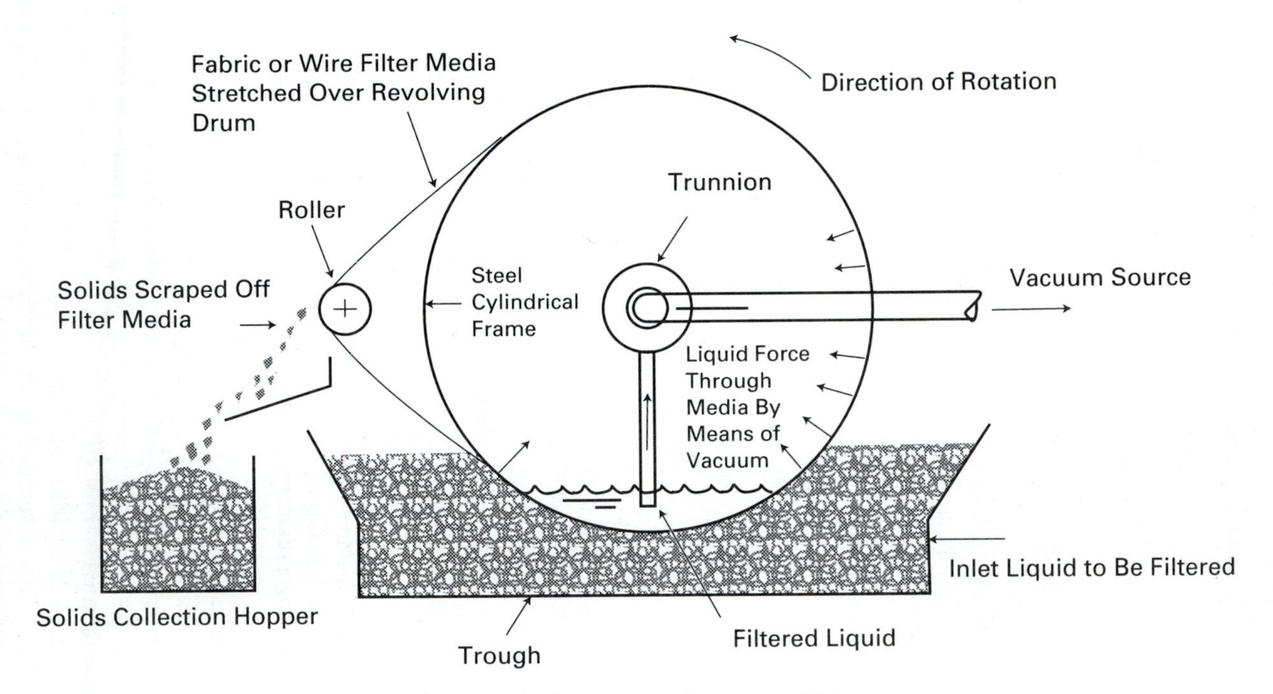

Figure 4-4 Schematic of a typical vacuum filter.

In vacuum filters, since the primary objective is the production of a highly dewatered cake, the primary resistance to filtration can be due to water travel through the pores of the solid sludge as it builds up over the surface of the filtered medium that forms on the drum. Therefore solids loading tends to be a much more important variable than hydraulic loading because the mass of solids per unit area will determine the headloss through the filter. Typical vacuum filters are sized on the basis of 2 to 10 lb/ft²/hr. Conditioning, as in the case of centrifugation, plays a considerable role in influencing the feasible solids loading rate.

The resistance of the solid sludge cake during vacuum filtration can be estimated by the laboratory measurement of specific resistance. In this technique, a laboratory filter funnel containing a paper filter is loaded with a sludge suspension and the rate of passage of water through the forming sludge cake is determined. The hydraulic resistance to filtration may be considered as the summation of the resistance of the developing filter cake and the filter medium itself. For incompressible filter cakes, the production of filtrate may be described by (Christensen and Dick, 1985):

$$\frac{dV}{dt} = \frac{P_T A^2}{\mu(rwV + RA)} \tag{4.21}$$

where

V = filtrate volume at time t
P_T = pressure differential
A = filter area
r = solids specific resistance (length/mass)
w = mass of cake deposited per volume filtrate produced
R = filter medium resistance (1/length)

By integration and rearrangement, this equation results in the following:

$$t/V = (\mu rw/2P_T A^2)V + \mu R/P_T A \tag{4.22}$$

hence, a plot of t/V versus V yields a line, from which the value of specific resistance can be determined. Raw sludges have a specific resistance of 10 – 100 Tm/kg, while well conditioned sludges have resistance of 0.1 – 1 Tm/kg. The specific resistance test can thus be used as a relatively rapid measurement of the adequacy of sludge conditioning for dewatering by filtration.

Pressure Filters

The third type of filtration process is the pressure filter. While the primary aim of pressure filters is to achieve solids concentration, as in the case of a vacuum filter, the mode of operation is discontinuous, or cyclic, like a granular filter. The filter press consists of a stack of parallel plates that are surrounded by a porous woven or mesh medium (Figure 4-5). Sludge is fed into the space between the plates. The inlet is then closed, and the plates are placed between a hydraulic jack, compressing the plates together, and forcing water to pass through the outside of the porous medium and into the collection channels. The material remaining inside the plates is compressed and dewatered. At the end of the cycle, the filter is opened and the solid cake is removed.

The dominant operating variables are cycle time (time under pressurization), the operating pressure and the feed solids concentration and condition. Typically the cycle time is on the order of hours, and the solids loading based on the cycle time is 0.2 to 2 $lb/ft^2/hour$. This solids loading is less than the vacuum filter and thus, in general, it would appear that a greater amount of filter press area is needed for a given sludge than vacuum filter area. However, the filter press can generally yield higher solid concentrations than the vacuum filter for the following reasons. In the vacuum filter, the available pressure drop is inherently limited. Even if an absolute vacuum were to be attained, the maximum pressure differential available (in the case of a vacuum filter) is one atmosphere. With filter presses in principle, the pressure is limited only by the strength of the hydraulic jack and the structural integrity of the cells. Thus, pressures far in excess of 1 atmosphere can be used to drive water through the pores of the developing filter cake. The performance of vacuum filters is generally limited to 20 or 25% cake solids. In a filter press, cake solids (for well conditioned sludges) of as high as 50% can be achieved.

Common Features of Pressure and Vacuum Filters

Vacuum or pressure filters (Figure 4-5) may both be used to treat biological or chemical sludges, including the solids from precipitation of inorganic wastewaters. In particular, the latter materials are quite amenable to filtration, generally attaining higher solids concentrations than biological sludges.

The major environmental impact of solids filtration is the disposal of the filtrate. In addition, volatile materials in sludges will be stripped off during vacuum filtration, but due to the volume system, the exhausted gases can be easily ducted and treated. And so the vacuum filter is a relatively easily controlled system from the point of view of controlling volatiles.

Macromolecular and Molecular Filtration

If the solid phase through which filtration is conducted has pores that are sufficiently small, the physical separation of molecules can be achieved. As indicated in Figure 4-3, this molecular selectivity also involves very low filtration rates or very high pressure drops. Generally molecular filtration is divided into two categories: ultrafiltration (UF) and reverse osmosis (RO) processes. The differentiating characteristic between RO and UF is the molecular weight cutoff of the membrane, and collaterally, the required pressure differentials to achieve a given membrane flux.

The cutoffs for UF membranes are generally from several thousand up to 10^6 daltons (i.e., molecules with molar weights between several thousand and one million grams/mole). The cutoff must be regarded as approximate, since the physical mechanism for retention of a molecule (termed rejection) involves both a size exclusion effect as well as electrostatic effects, and not all molecules of the same molecular weight will have similar sizes and electrostatic properties. All chemical components with a molecular weight greater than the cutoff will be retained by the membrane. All molecules with a molecular weight less than the cutoff will pass through the membrane. For example, UF membranes can be used to separate proteins and high

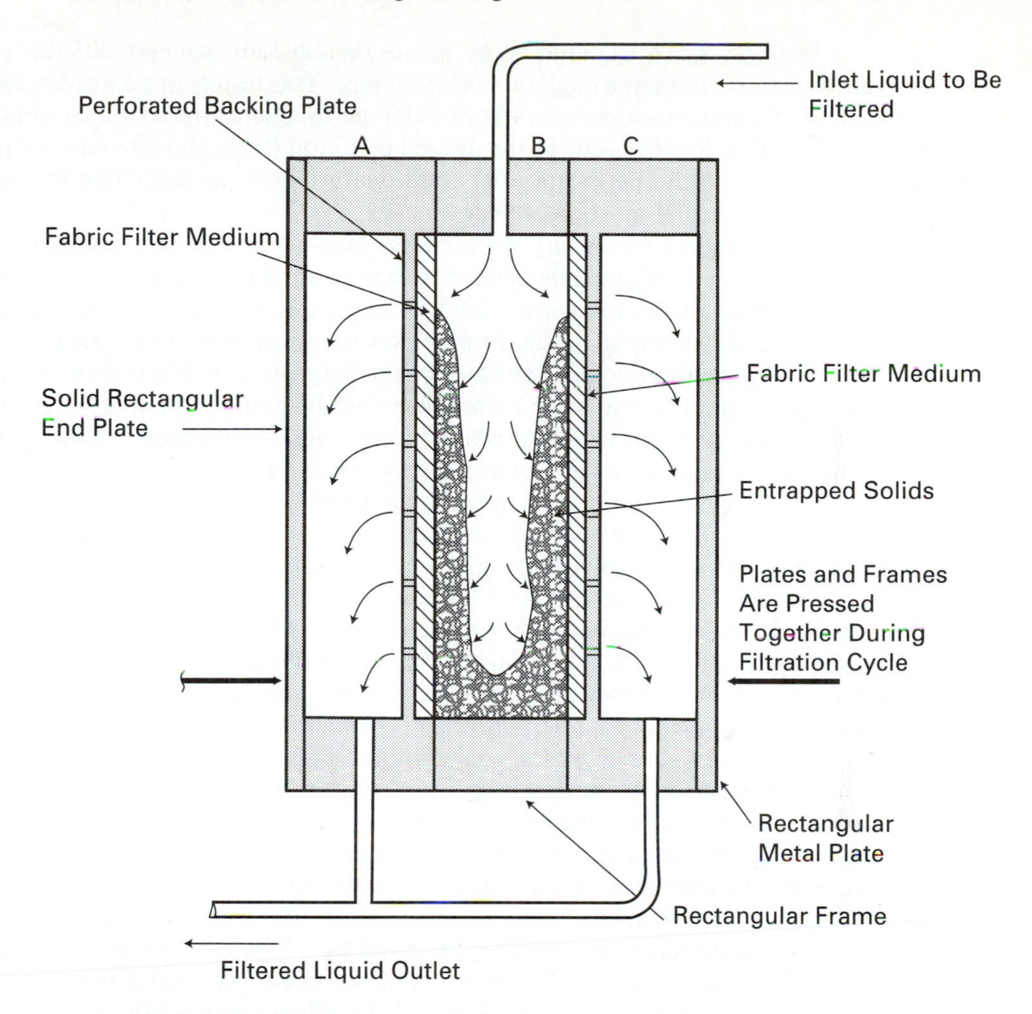

When the cavity formed between plates A and C is filled with solids, the plates are separated.
The solids are then removed and the medium is washed clean.
The plates are then pressed together and filtration resumed.

Figure 4-5 Schematic diagram of a pressure filter.

molecular weight carbohydrates from lower molecular weight dissolved materials, and certainly, inorganic materials.

RO membranes operate with molecular weight cutoffs that are much smaller. The typical RO membrane may have a cutoff of about 100–200 daltons. Therefore, the RO membrane will retain most organic materials, as well as many of the inorganic solutes. However, for low molecular weight materials, during RO, there is some dissolution of solute in the membrane material itself and bleed-through into the permeating liquid.

In the case of a RO membrane, a considerable ionic strength difference arises due to retention of some or all inorganic solutes. This results in a considerable back pressure from the osmotic pressure of water acting against the solute concentration gradient. For flow to occur in the desired direction (from the retentate side of the membrane into the permeate side) additional pressure to overcome the osmotic pressure differential must be provided.

RO membranes typically operate with trans-membrane pressures of up to 500 psi. UF membranes generally operate with pressure differentials only as high as 50 psi since most of the inorganics will pass right through the UF membrane, thus minimizing the potential for an osmotic pressure resistance. The pressure differential has significant economic implications since it costs a lot of money to maintain a pressure differential across the membrane. On the other hand, a lot of material is going to pass through the UF membrane. Very often, when reverse osmosis is used, upstream UF is provided as a pretreatment device for RO.

There are a number of operational problems with membrane processes. Probably the two major ones are the potential for chemical and biological fouling of the membrane, and, particularly with RO, membrane deterioration. Chemical fouling can occur by precipitation of solutes when their concentration on the retentate side of the membrane exceeds their solubility. This may occur for a variety of materials, most particularly calcium (calcium sulfate or carbonate), magnesium (magnesium ammonium phosphate), and iron (iron hydroxide) salts. Biological fouling can occur when microorganisms proliferate on the retentate side of the membrane due to the high local concentration of nutrients.

Membrane deterioration can occur by physical (e.g., pinhole leaks due to abrasion of particulates) or chemical/biological mechanisms. Chemical deterioration can occur with some membrane materials due to chemical oxidation or hydrolysis at extreme pH values. Biological deterioration can occur in the presence of biological fouling when the attached microorganisms "etch" the membrane surface. If a high degree of solute removal is desired, there will be a high concentration of dissolved materials at the retentate side of the membrane. Even a very small pinhole leak will then result in significant contamination of the effluent with solute. Under extreme situations, this deterioration may also manifest itself in a sudden decrease in hydraulic resistance across the membrane.

For cellulose acetate membranes, the rate of membrane chemical deterioration may occur by one of two mechanisms—hydrolysis or oxidation promoted by the addition of disinfectants such as chlorine. This process may be described by a rate constant and is temperature dependent. The deterioration typically has a half life on the order of 10^3-10^4 hours depending upon operating conditions (Ohya et al., 1981). The factors influencing membrane life include membrane material, pH (both in the presence and absence of halogen disinfectants) and very likely the presence of particulates (Glater et al., 1981). There is no substantial data base on the sensitivity of membrane materials to chemical constituents likely to be in hazardous and industrial wastewaters treated by membrane processes. However, the costs of planned membrane replacement may become significant factors in process economics.

In the presence of fouling, periodic membrane cleaning is required. For chemical fouling, depending upon the membrane material, a mild acid or alkali solution can be used. For precipitate fouling, a chelating agent such as EDTA or citric acid may also be used. For biological fouling, the mild acids and alkalies will be of assistance. For resistant membranes, dilute chlorine solutions can also be used to remove the built up slime or fouling layers. Some applications employ enzyme-detergent mixtures to fulfill this function.

The flux of water and a dissolved solute through a membrane (either reverse osmosis or ultrafiltration) may be described by a few simple relationships. Using the concept of steady state transport, the flux of water across the membrane (volume/area-time) can be written as:

$$J_l = -k_m(\tilde{v}_l/RT)(\Delta P - \Delta\pi) \quad (4.23)$$

where $\tilde{v}_l$ is the molar volume of water, ΔP is the transmembrane hydrostatic pressure differential, $\Delta\pi$ is the transmembrane osmotic pressure differential, and k_m is the mass transfer coefficient through the membrane. At osmotic pressures below 2 atmospheres, corresponding to concentrations below about 0.1 molar, $\Delta\pi$ can be estimated by:

$$\Delta\pi = (C_{2,m} - C_{2p})RT \quad (4.24)$$

where $C_{2,m}$ is the concentration of solute at the feed face of the membrane and $C_{2,p}$ is the solute concentration in the permeate. Note (as will be discussed below) that $C_{2,m}$ does not necessarily equal the bulk feed concentration due to concentration polarization effects. If the flow of water through the membrane is primarily via a solvation mechanism in the membrane matrix, then

$$k_m = D_{\text{water}}/\delta \quad (4.25)$$

with D_{water} being the diffusivity of water in the membrane matrix, and δ being the thickness of the membrane layer. Alternatively, if the flow of water is primarily via flow through fine capillaries of porosity ϵ, tortuosity τ, and radius r, then

$$k_m = \epsilon r^2 RT/(\tilde{v}_1 \mu\tau\delta) \quad (4.26)$$

In the presence of both mechanisms, then the overall mass transfer coefficient for water would be the sum of the preceding two equations.

The flux of a dissolved material (mass/area-time) through a membrane via a solvation mechanism may be given by the following:

$$J_2 = (D_{2m}K_2/\delta)(C_{2,m} - C_{2,p}) \quad (4.27)$$

with D_{2m} being the diffusivity of solute through the membrane material, and K_2 being the dimensionless partition equilibrium constant for the solute in the membrane (defined as the equilibrium concentration in the membrane divided by the equilibrium concentration in the liquid). If solvent transport can also occur by bulk motion through pores, then this may contribute an additional flux of solute. In this case, the relationship between the two concentrations becomes:

$$(C_{2,p}/C_{2,m}) = \frac{K_2}{(K_2 - \epsilon)\exp(J_1\tau\delta/D_{21}) + \epsilon} \quad (4.28)$$

with D_{21} being the diffusion coefficient of the solute in water. Furthermore, from a mass balance over solute, the following relationship must also hold:

$$C_{2,p} = J_2/J_1 \quad (4.29)$$

thus indicating the close connection between solute and solvent fluxes. The term rejection is generally applied to the ratio of the two concentrations on either side of the membrane, that is

$$R = 1 - (C_{2,p}/C_{2,m}) \tag{4.30}$$

However, the designer is interested in the rejection relative to the feed concentration of material. Due to the fact that not all solute is transported through the membrane, there is an accumulation of solute at the membrane surface relative to the bulk solution. A high degree of fluid turbulence reduces the extent of this concentration due to increasing eddy diffusion. The relationship between the concentrations in turbulent flow can be given by:

$$C_2/C_{2,m} = 1 - R + R\exp[-25(J_1/U)\mathrm{Re}^{1/4}\mathrm{Sc}^{2/3}] \tag{4.31}$$

with U being the velocity of fluid over (and parallel to) the surface of the membrane; Sc is the Schmidt number of the solute and Re is the bulk Reynolds number.

From these relationships, the tradeoffs among the design variables may be explored by the following procedure:

1. Assume a value for $\Delta P - \Delta\pi$ and, from Equation 4.23 calculate a value for the water flux (J_1).
2. From Equations 4.28 or 4.27 and 4.29 compute the ratio of $C_{2,p}/C_{2,m}$ and hence, from Equation 4.30 the rejection (R).
3. From the feed concentration, use Equation 4.31 to compute $C_{2,m}$, and thus also $C_{2,p}$.
4. From Equation 4.29 compute the osmotic pressure differential, and hence the hydrostatic pressure differential required to achieve the desired degree of separation.

Example Problem 4-4

Consider the potential application of a reverse osmosis system to treat a dilute rinsewater in order to recover the dissolved salts for reuse. Given the following data describing a hypothetical spiral wound membrane assembly, estimate the pressure differential and the recirculation flow (i.e., over the membrane surface) which would be necessary to produce a permeate with a solute concentration of < 1 mg/L($< 10^{-6}$ g/cm^3) from a feed of 10 mg/L (10^{-5} g/cm^3).

Diffusivity of water in membrane material	$= 10^{-4}$ cm^2/s
Solute molecular weight	$= 150$
Membrane thickness	$= 2.00 \times 10^{-5}$ cm
Membrane pore size	$= 5.00 \times 10^{-7}$ cm
Solute partition coefficient	$= 100$
Porosity	$= 0.9$
Tortuosity	$= 10$
D_{12}	$= 5 \times 10^{-6}$ cm^2/s
Viscosity	$= 0.01$ g/cm-s
Density	$= 1$ g/cm^3
Temperature	$= 288$ K
Liquid thickness (over membrane)	$= 0.01$ cm

From this data, the following parameters can be derived:

$$\tilde{v} = 18 \text{ cm}^3/\text{mol}$$

$$k_m = D_{\text{water}}/\delta + \epsilon r^2 RT/(\tilde{v}_1 \mu \tau \delta) = 5.0 \text{ cm/s}$$

$$\text{Sc} = 2{,}000$$

If one assumes a value for the quantity $(\Delta P - \Delta\pi)$(in atmospheres), then the water flux (in cm/s) can be obtained from Equation 4.23:

$$J_1 = -k_m(\tilde{v}_1/RT)(\Delta P - \Delta\pi) = (5)(18)(\Delta P - \Delta\pi)/(288)(82 \text{ cm}^3\text{-atm/mol-K})$$
$$= 3.811 \times 10^{-3}(\Delta P - \Delta\pi) \text{ cm/s}$$

Rejection is computed by use of Equations 4.28 and 4.29:

$$R = 1 - \frac{K_2}{(K_2 - \epsilon)\exp(J_1 \tau \delta / D_{21}) + \epsilon}$$
$$= 1 - 100/\{(100 - 0.9)\exp[J_1(10)(2 \times 10^{-5} \text{ cm})/(5 \times 10^{-6} \text{ cm}^2/\text{s})] + 0.9\}$$
$$= 1 - 100/[99.1 \exp(40 J_1) + 0.9]$$
$$= 1 - C_{2,p}/C_{2,m})$$

The degree of concentration polarization may be obtained from Equation 4.31.

$$C_2/C_{2,m} = 1 - R\{1 - \exp[-25\,(J_1/U)\,\text{Re}^{1/4}\text{Sc}^{2/3}]\}$$

where U is the velocity across the surface of the membrane, and Re is the Reynolds number with respect to the feed liquid, that is:

$$\text{Re} = U d_1 \rho / \mu$$

Since C_2 is known, one may determine the value of U necessary to achieve the desired value of product concentration ($C_{2,p}$) by trial and error.

Given the computed values of solute concentration across the membrane, the transmembrane osmotic pressure differential can be computed via Equation 4.24. Note, however, that a conversion from mass concentration to molar concentration is required. Finally, from the osmotic pressure, the mechanical pressure differential can be computed by addition. The results of this computation for several values of pressure are summarized below:

$\Delta P - \Delta\pi$ atm	J_1 cm/s	R	U cm/s	$C_{2,m}$ g/cm³	$\Delta\pi$ atm	P atm
70	2.66E-01	0.99997	626	0.042052	994	1064.
50	1.90E-01	0.9995	726	0.001995	47.2	97.2
40	1.52E-01	0.9977	840	0.000433	10.2	50.2
35	1.33E-01	0.9951	940	0.000203	4.8	39.8
30	1.14E-01	0.9895	1120	0.000094	2.2	32.2
25	9.52E-02	0.9776	1500	0.000044	1.02	26.02

Note that as the mechanical pressure differential decreases, the required velocity across the membrane tends to increase. Thus, there is an optimum, in terms of minimizing power required to maintain the pressure differential and to provide the high velocities required to minimize concentration polarization.

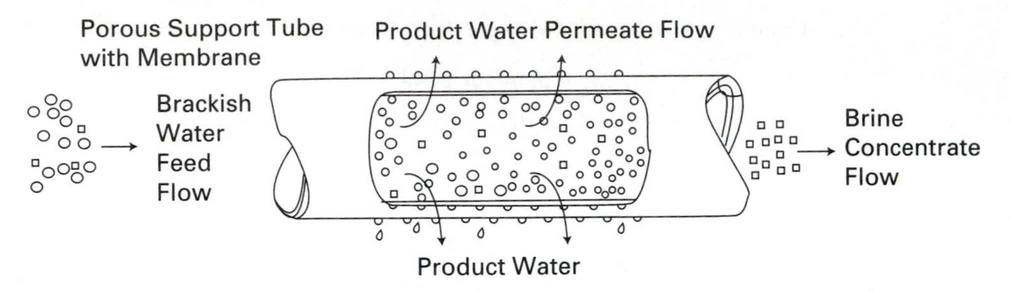

Figure 4-6 Schematic diagram of a tubular membrane separation process.

As in any filtration process, the volumetric flow of water is proportional to the area of membrane through which filtration occurs. To minimize the volume occupied by a membrane treatment process, filter units can be configured to maximize the area/volume ratio. Rather than having a single sheet of filter medium in a planar configuration, there are several more popular arrangements. Furthermore, in a flat-plane configuration, the buildup of a highly concentrated surface layer is encouraged, and thus the hydraulic resistance as well as the breakthrough of partially retained constituents is encouraged.

One configuration uses cylindrical or tubular membrane modules (Figure 4-6). In this system, there are caps at both ends of the cylinder and feed is introduced to the center of the membrane. The velocity across the face of the membrane can be set independently by using recirculation pumps for feed water and thereby reducing concentration polarization. Permeate passes through the membrane and is collected at the outside of the tube.

A second configuration, termed "hollow fiber," uses tubes of membrane of submillimeter size (Figure 4-7). Feed is introduced on the outside of the hollow fiber bundles and permeate is collected internally. While it is more difficult to achieve

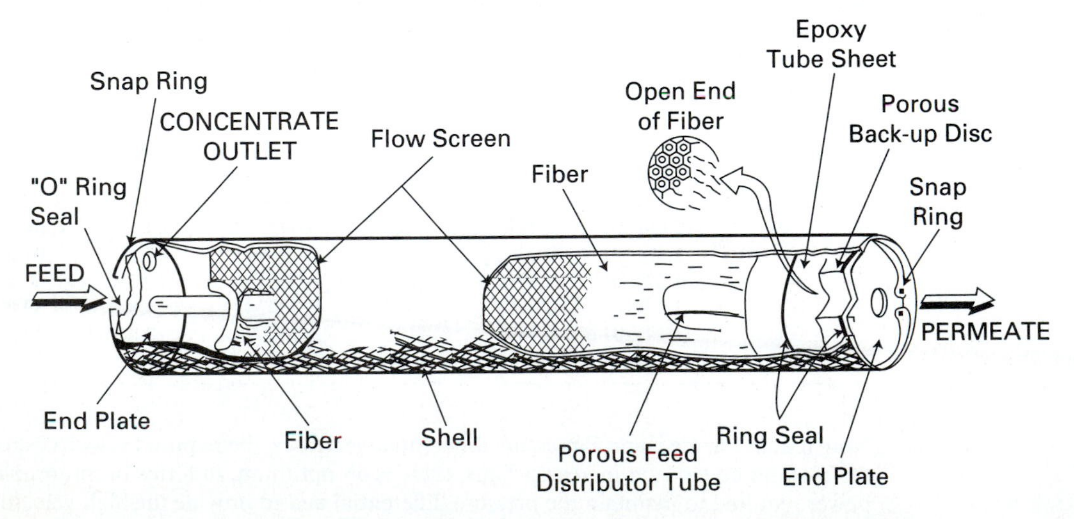

Figure 4-7 Schematic diagram of hollow fiber membrane separation process.

high velocities over the face of the membrane bundle, the extremely high surface area, which can be achieved in a small volume, acts as a compensating factor.

The third configuration, termed "spiral wound," consists of sheets of membrane material backed by spacer material and wound up jelly-roll fashion (Figure 4-8). The ends of the bundle are capped and flow is introduced across the membrane surface at a velocity that can be controlled in a similar manner to the tubular process. Permeate is collected in the spacer layer and flows to the central collecting tube where it may be discharged.

Hollow-fiber and spiral-wound membrane systems find wider application in treating fluids of lower solids (dissolved and suspended) content. Tubular membrane systems are more useful for more highly contaminated liquids and slurries (McArdle et al., 1988).

RO and UF find application whenever both recovery/reuse of the permeate fluid as well as removal or recovery of the dissolved constituents is attractive, for example, in the recovery of dilute chemicals from electroplating rinse water. Ultrafiltration can be used whenever there is a colloidal or high molecular component to the waste as a means of recovery or pretreatment prior to RO. UF also has found application in the removal of many high molecular weight dyes and pigments from such facilities as textile manufacturing and printing.

Membrane treatment units have also been employed as mobile above-ground treatment for contaminated groundwaters. RO systems have been found capable of removing up to 95% of the toxic organic carbon from such waters—however, such removals are functions of the hydraulic loading and percent water recovery (Taylor et al., 1987). Use of low pressure reverse osmosis for the removal of hazardous organics from coal liquefaction wastewaters has been found to be feasible, at least in laboratory scale (Siler and Bhattacharyya, 1985).

Recently, special types of membrane systems have also been developed that, rather than having a liquid on both sides, have a liquid on the feed side and a gas (at a negative pressure) on the permeate side. Their mode of operation is different than RO in that the membrane acts by a selective dissolution process. Preliminary studies indicate that they may be economically employed for recovery of volatile solvents from gaseous emissions (Peinemann et al., 1986). They may also find application in the removal of toxic gases, such as hydrogen sulfide. However, very little work on

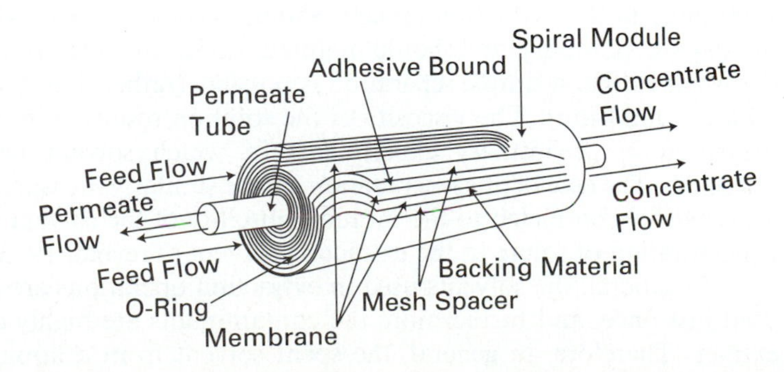

Figure 4-8 Schematic diagram of spiral wound membrane separation process.

their actual application, or the decline in flux and selectivity with exposure to actual process streams, has occurred.

SEPARATION BY DIFFERENTIAL PHASE PARTITIONING

Separation by differential phase partitioning involves the contact of the waste to be treated with a second phase. Certain contaminants, for which the process is designed, will preferentially accumulate in the second phase. Separation of the phases, for example by physical decantation, results in net contaminant removal from the waste.

Liquid–Liquid Separation

The most straightforward process for the treatment of a liquid stream might be its mixing with an immiscible solvent. This process is called solvent extraction. One can characterize the performance of solvent extraction by the relative concentrations of a pollutant in the waste (C_w) and the solvent ($C_{solvent}$) phases. If the extraction is at equilibrium, the concentration ratios assume a dimensionless constant value, termed the partition coefficient (K_p), which can be regarded as an equilibrium constant:

$$\frac{C_{solvent}}{C_w} = K_p \tag{4.32}$$

For relatively dilute waste streams containing few organic materials, this partition coefficient can be estimated from thermodynamic first principles. However, in many practical applications, the equilibrium properties must be determined experimentally.

Certain empirical rules exist for selection of a solvent for given types of solutes. These are summarized in Table 4-2. Solutes and solvents are classed into nine categories, and the relative mutual affinity as compared to the solute water affinity is noted. Those with increased affinity represent candidate solvent systems for evaluation.

In addition to having adequate solubility properties, the candidate solvent, when used in the extraction process, should have a viscosity which is amenable to mixing and pumping, and should maintain sufficiently high interfacial tension with the waste so that a simple separation is possible (rather than forming a difficult-to-separate emulsion). The viscosity of the solution resulting from extraction can be measured in preliminary experiments in which solvent performance data is obtained. The ease of phase separation can be similarly measured—however, this is also inversely corrected to the mutual solubility of the solvent in water and to the concentration of solute in the extracting solvent (Treybal, 1963).

In general, the solvents used in extraction operations are too expensive to be used just once, and furthermore the contaminants are highly concentrated in the extract. Therefore, in general, the spent solvent from a liquid–liquid extraction operation needs to be treated further to reclaim the solvent for reuse and to reduce

TABLE 4-2 CLASSIFICATION OF SOLVENT-SOLUTE INTERACTIONS FOR SOLVENT EXTRACTION FROM AQUEOUS WASTE MATERIALS. (Modified from Schweitzer, 1979, by permission of McGraw-Hill.)

Solute Class	Description	Solvent Class								
		1	2	3	4	5	6	7	8	9
1.	Acid, aromatic OH	0	+	+	+	+	0	−	−	−
2.	Paraffin OH, water, imides, amides with active H	+	0	−	−	−	−	−	−	−
3.	Ketone, aromatic nitrate tertiary amine pyridine, sulfone	+	−	−	0	−	+	−	−	−
4.	Ester, aldehyde, carbonate, phosphate, nitrite, nitrate, amide without active H, intra-molecular bonded	+	−	−	0	−	+	−	−	−
5.	Ether, oxide, sulfide, sulfoxide, primary or secondary amine or imine	+	−	−	−	0	+	0	−	−
6.	Multihalo paraffin with active H	0	−	+	+	+	0	0	−	0
7.	Aromatic, halogen aromatic, olefin	−	−	0	−	0	0	0	0	0
8.	Paraffin	−	−	−	−	−	−	0	0	0
9.	Monohalo paraffin or olefin	−	−	−	−	−	0	0	−	0

+: stronger mutual affinity, likely candidate solvents

−: weaker mutual affinity, unlikely candidate solvents

0: little interaction

further the volume in which the contaminant is contained. Some solvent repurification sequences include the use of distillation or adsorption, both of which will be discussed later.

Examples of solvents which fall into each of the solvent classes in Table 4-2 are:

Class 1— Phenol
Class 2— Pentanol
Class 3— Methylisobutylketone
Class 4— Ethyl acetate
Class 5— Ethyl ether
Class 6— Methylene chloride
Class 7— Toluene
Class 8— Hexane
Class 9— Butyl chloride

In the design of liquid–liquid extraction systems, the two principle design variables are the solvent: waste ratio and the nature and intensity of interfacial contacting. The simplest type of extraction unit is a batch fed tank which is charged with known volumes of waste (V_w) and solvent (V_s) and mixed until equilibrium is assumed. Providing that there is no volume change in either phase on mixing, and that the initial concentration of extractable contaminant in the waste is C_0, then a mass balance at equilibrium requires:

$$C_0 V_w = C_w V_w + C_{\text{solvent}} V_s \tag{4.33}$$

However, Equation 4.32 relates the concentrations at equilibrium to a partition coefficient (which might itself be a function of composition of the solvent at equilibrium). Therefore,

$$C_0 V_w = C_w V_w + K_p C_w V_s$$

or

$$\frac{C_w}{C_0} = \frac{1}{1 + K_p V_s / V_w} \tag{4.34}$$

It is possible, in practice, to conduct extractions in which a cascade of such vessels is used. The optimal performance is generally obtained when there is a countercurrent flow of waste and solvent. For this system, providing that the partition coefficient is constant with composition, equations developed by Kremser (Treybal, 1963) can be used.

$$N = \frac{\ln[(1/U)(1 - 1/R) + 1/R]}{\ln(R)} \tag{4.35}$$

with

$$U = C_{\text{effluent}}/C_0$$

$$R = K_p Q_{\text{solvent}}/Q_w$$

where N is the number of equilibrium stages that are required to achieve the separation. Equation 4.35 applies strictly when the solvent and the waste are immiscible and when the concentration of the partitioning material in the feed solvent is negligible, but it can be modified if this is not true. When the partition coefficient is not constant, the geometric mean of the partition coefficient at the feed and the effluent compositions may produce a suitable approximation (Treybal, 1963).

Use of the Kremser equation reveals a tradeoff between the solvent : waste ratio and the physical size (number of stages) required to obtain a given degree of removal of a material from the waste. Choice of the proper combination of these design variables requires consideration of the detailed economics of scale and of cost to manage (i.e., recycle) various amounts of solvent.

In addition to a countercurrent cascade, a number of other physical types of contact systems can be used. These include packed tower contactors, plate columns, agitated columns, and centrifugal extractors (Figure 4-9). Performance of these types of contactors can also be described in terms of the number of theoretical

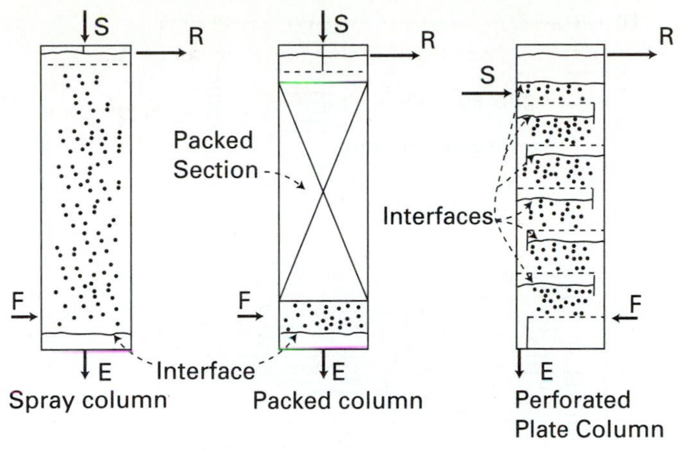

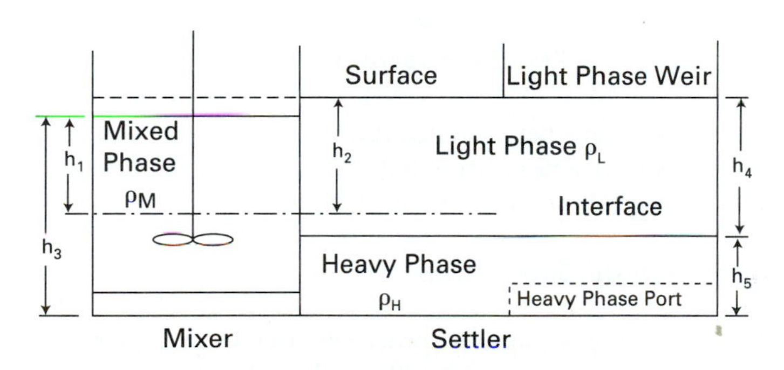

Figure 4-9 Types of extractors. (Reprinted from Hanson, 1968, by permission of McGraw-Hill.)

equilibrium stages to which they are equivalent. Generally, for most extractors, under 15 equilibrium stages are provided. Figure 4-10 presents a conceptual diagram of factors which influence the type of extractor which should be used in a given application.

The primary applications of solvent extraction in waste treatment are dephenolizations from petrochemical and coal-related industries. However, conceptual process designs for the removal of priority pollutants from aqueous wastewaters have been completed (Joshi et al., 1984). Diethylene glycol monomethyl ether has been found to be useful as a solvent for the removal of PCB's from mineral oils (Cowles and McNaughton, 1985).

Recent work has focused on the use of supercritical fluids, fluids maintained at the critical temperature and pressure, as solvents. With these systems, the solvent phase can be recovered from the entrained contaminants by releasing pressure and allowing the solvent to vaporize (generally leaving the contaminant as a sludge or oil). Use of a supercritical CO_2 for soil cleanup and regeneration of granular acti-

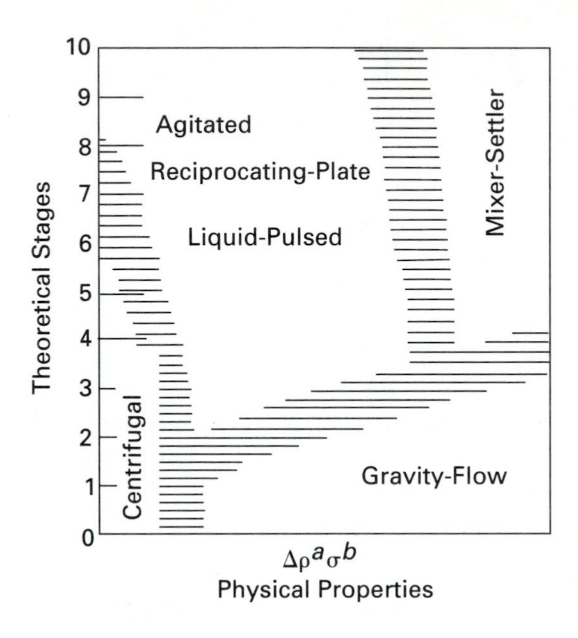

Figure 4-10 Factors influencing selection of extractor type. (Reproduced from Schweitzer, *Handbook of Separation Techniques for Chemical Engineers,* 1979 by permission of McGraw-Hill.)

vated carbon have been studied (Groves et al., 1985). Also under investigation are supercritical extraction using halomethanes for the removal of oils in a variety of materials (de Fillippi and Chung, 1985).

Vapor–Liquid Separation

In vapor–liquid separation processes, the second phase into which partitioning occurs is a vapor or gas, rather than a liquid. These processes include air stripping, steam stripping, and distillation.

Air stripping. In air stripping, air is passed through or over a contaminated liquid. Any volatile constituents in the liquid will be preferentially removed into the gas phase and will leave the system as a vapor. The most common type of stripping device that has been used is a packed tower (Figure 4-11). Various types of column packing materials are in use—plastics, ceramics, and stainless steel modules of cylindrical, ring, or saddle shape being the most common. In the top is fed contaminated liquids exiting at the bottom. Water is generally introduced in the top of the tower. Air is introduced into either the bottom (countercurrent flow), the top (cocurrent flow), or the side (cross flow) of the tower. The flows allow good interfacial contact between the phases and the volatile material, provided that its concentration in solution is above the saturation concentration based on the vapor phase composition will strip out and be removed from the fluid. The best driving force for mass transfer is presented by the countercurrent stripper in which the cleanest vapor stream is contacted with the most dilute liquid.

For vapor–liquid equilibria, Henry's law can generally be used to characterize the partitioning of volatile compounds. Above mole fractions of about 0.1 (in either phase) corrections for nonideality must be incorporated. If X and Y are, respectively,

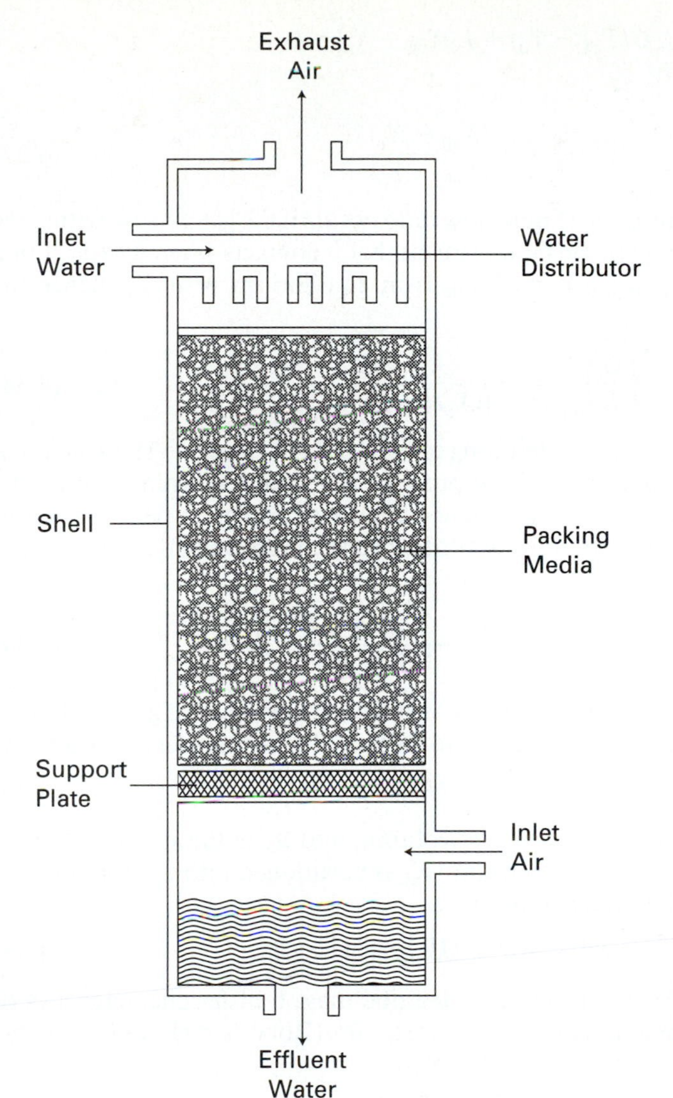

Figure 4-11 Schematic diagram of a packed tower stripper. (Reprinted from Eckenfelder, *Industrial Pollution Control,* 2nd ed., 1989, by permission of McGraw-Hill.)

the mole fractions of volatile material in the solution and the gas phases, H is the Henry's law constant (units of pressure), and P_T is the total pressure in the gas phase, then the following equilibrium relationship may be written:

$$Y = XH/P_T \tag{4.36}$$

Analogous to solvent extraction, the major design variables for strippers are the relative flows of gas and liquid and the column height. The gas to liquid ratio is set by Henry's law and the desired influent and effluent concentrations. If G and L are the molar flow rates of gas and liquid (assumed constant over the stripper), respectively, and the subscripts $_o$ and $_{\text{eff}}$ denote influent and effluent, an overall mass balance over any stripper at steady state requires:

$$G(Y_{\text{eff}} - Y_o) + L(X_{\text{eff}} - X_o) = 0$$

or, rearranging:

$$\frac{G}{L} = -\frac{(X_{\text{eff}} - X_o)}{(Y_{\text{eff}} - Y_o)} \tag{4.37}$$

For a stripper, the minimum gas : liquid flow ratio occurs if the gas stream exiting the stripper is in equilibrium with the liquid stream that it contacts when it exits. For a countercurrent stripper, this occurs if Y_{eff} is in equilibrium with X_o, hence the following:

$$\left(\frac{G}{L}\right)_{\text{min}} = -\frac{(X_{\text{eff}} - X_o)}{[(X_o H/P_T) - Y_o]} \tag{4.38}$$

In this situation, however, an infinitely long tower would be needed. The kinetics of mass transfer limit the efficiency of the process. For countercurrent towers, the tradeoff between the gas : liquid flow rate and the tower size may be expressed in terms of the number of transfer units according to the following (Kavanaugh and Trussell, 1980):

$$N = \frac{\ln[(C/C_o)(1 - 1/S) + 1/S]}{1 - 1/S} \tag{4.39}$$

with $S = Q_G H/P_T Q_L$, and the Q's denoting volumetric flow rates of gas and liquid.

The height of a transfer unit is related to an overall mass transfer coefficient by:

$$\text{HTU} = Q_L/(K_L A a_w) \tag{4.40}$$

where A is the cross-sectional area of the column, and a_w is the interfacial area/volume of column in service. Furthermore, K_L is partitioned into a gas and liquid resistance via the two-film model as follows:

$$K_L^{-1} = k_L^{-1} + (Hk_G/P_T)^{-1} \tag{4.41}$$

The Onda correlation has been found to describe mass transfer characteristics of air-water strippers to a precision of approximately 20% (Roberts et al., 1985). In this approach, k_L, k_G, and a_w are predicted from:

$$a_w/a = 1 - \exp\left\{-1.45(\sigma_c/\sigma_L)^{0.75}\left[\frac{L_M}{a_t \mu_L}\right]^{0.1}\left[\frac{L_M^2 a_t}{\rho_L^2 g}\right]^{-0.05}\left[\frac{L_M^2}{\rho_L \sigma_L a_t}\right]^{0.2}\right\} \tag{4.42}$$

$$k_L(\rho_L/\mu_L g)^{1/3} = 0.0051(L_M/a_w \mu_L)^{2/3}(\mu_L/\rho_L D_L)^{1/2}(a_t d_p)^{0.4} \tag{4.43}$$

$$k_G/a_t D_G = 5.23(G_M/a_t \mu_G)^{0.7}(\mu_G/\rho_G D_G)^{1/3}(a_t d_p)^{-2} \tag{4.44}$$

with subscripts on D, μ, and ρ indicating gas or liquid, and

a_t = nominal (dry) packing surface area/packed volume
L_M = liquid mass flow/area
G_M = gas mass flow/area
d_p = nominal packing diameter

σ_c = critical surface tension (property of packing material)
σ_L = liquid surface tension

With packing diameters less than 1.5 cm, the constant in Equation 4.44 should be changed to 2.0.

The hydraulics of flow of both the gas and liquid phases need to be considered during design of a stripper. For a given stripping tower, there is a maximum amount of liquid that can be handled at a given gas flow prior to the point where flooding occurs. At flooding, there is no longer a continuous gas phase available for transfer, and the gas phase pressure drop abruptly increases. Below flooding, the pressure drop of gas through the wetted column is adequately described by the Leva correlation, which is an empirical modification of the orifice equation (Leva, 1954):

$$\Delta P = C_2 10^{C_3 U_l} \rho_g U_g^2 \tag{4.45}$$

where ΔP is the pressure drop in inches of water per foot of column, U_l and U_g are, respectively, the liquid and gas superficial velocities in ft/sec, ρ_g is the gas density in lb/ft^3, and C_2 and C_3 are constants specific to the packing. Table 4-3 provides values for the Leva equation constants for various types of packing materials. It has been found that, to a reasonable degree, the point at which flooding occurs is defined by a given value of gas phase pressure drop which is a function of packing type—these critical values are also tabulated.

The major applications of air stripping have been the treatment of contaminated groundwater that is being used as a drinking water supply. Most commonly, the contaminants are gasoline or solvents, both of which are highly volatile. In these situations, the gas may be vented directly to the atmosphere. However, in many localities, it is necessary to treat the vapor emissions to minimize release of hydrocarbons to the atmosphere. This may be accomplished by adsorption. Depending upon the water quality, however, it may be more economical to remove the volatile contaminants directly by liquid phase adsorption, rather than by stripping to the gas phase with subsequent gas phase adsorption. One study (Amy et al., 1987) suggests

TABLE 4-3 LEVA EQUATION COEFFICIENTS AND PRESSURE DROPS AT FLOODING. (Adapted from Tables 18-9 and 18-10, Perry and Green, *Perry's Handbook of Chemical Engineers,* 5th ed., 1973, by permission of McGraw-Hill.

Type of packing	Nominal size (in.)	C_2	C_3	ΔP at flooding (in./ft)
Raschig rings	$\frac{1}{4}$			4
	$\frac{1}{2}$	3.5	0.0577	3.5
	$\frac{3}{4}$	0.82	0.0361	
	1	0.8	0.0348	4
	$1\frac{1}{2}$	0.3	0.0320	2.5
	2	0.28	0.0238	2.5
Berl saddles	$\frac{1}{2}$	1.5	0.0272	2.5
	$\frac{3}{4}$	0.6	0.0236	
	1	0.4	0.0236	2.5
	$1\frac{1}{2}$	0.2	0.0181	2.2

that stripping prior to liquid phase adsorption may be economically justifiable for increasing the breakthrough time of adsorption systems.

Possible drawbacks of air stripping processes other than the potential for air emissions relate to the effects of colder weather. The air exiting the stripping tower will be close to saturation with water. If the water temperature was higher than the ambient air temperature, it is likely that when the air is discharged, local fogging or the formation of condensation will occur. In very cold climates, this might lead to local icing. Furthermore, under cooler temperatures, the Henry's law constants for most volatile materials tend to decrease. This results in a poorer removal efficiency by stripping (or the necessity for using higher gas flow rates).

Example Problem 4-5

Groundwater is contaminated with toluene at a concentration of approximately 10 mg/L, and it is desired to reduce this to 0.1 mg/L using air stripping. The stripper is to be designed to treat a flow of 100 m^3/d and is to operate at 50% of the liquid flooding pressure drop and twice the minimum air : liquid flow ratio. The stripper is to be packed with raschig rings of nominal 2-inch (5.08 cm) size. Given the following data, determine the required size (area, depth) and air flow of the stripper.

temperature = 298 K | $a_t = 0.919$/cm
$\sigma_c = 80$ erg/cm^2 | $\sigma_L = 71$ erg/cm^2
$H = 0.27$ atm
$D_L = 8 \times 10^{-6}$ cm^2/s | $D_G = 0.088$ cm^2/s
$\mu_L = 0.01$ g/cm-s | $\mu_G = 1.8 \times 10^{-4}$ g/cm-s
$\rho_L = 1$ g/cm^3 | $\rho_G = 1.2 \times 10^{-3}$ g/cm^3
molecular weight of toluene = 92

Leva constants (from table)

$C_2 = 0.28$
$C_3 = 0.0238$
ΔP at flooding = 2.5 inches/foot

First, from the Leva equation, if U_L and U_G are, respectively, the velocities in ft/s of liquid and gas, based on the problem conditions, the following constraint must apply:

$$1.25 = 0.28\,(10^{0.0238 U_L})(0.075) U_g^2$$

The 0.075 in this equation is the gas density converted to units of lb/ft^3.

If G and L are, respectively, the molar flows of gas and liquid, then the minimum gas : liquid flow ratio is set by:

$$(G/L)_{\min} = (X_{\text{eff}} - X_0)/[(X_0 H/P_T) - Y_0]$$

However, the feed gas can be assumed to be clean of toluene ($Y_0 = 0$), and from the problem conditions, the mole fractions in the influent and effluent liquid are as follows:

$$X_0 = 10^{-2}\text{ g/L }(1/92\text{ mol/g})/[1000\text{ g/L }(1/18\text{ mol/g})]$$
$$= 1.956 \times 10^{-6}$$

and similarly

$$X = 1.956 \times 10^{-8}$$

The factor (1/18) mol/g in these equations is the molecular weight of water, which dominates the molar composition of the liquid. As a result, the minimum flow ratio is:

$$(G/L)_{\text{min}} = 3.67$$

Therefore, from the problem conditions, it is required that $(G/L) = 7.34$.

The liquid flow is given as 100 m³/d. This is also equal to 1157 cm³/s or 64.3 mol/s. Therefore, the required molar flow rate of air is 472 mol/s. Since the air density is given, and assuming a mean molecular weight of air of 30, the volumetric flow rate of air is 1.18×10^7 cm³/s. Providing that the cross sectional area in cm² (A) was known, the linear velocities through the column in ft/s could be determined by:

$$\begin{aligned} U_L &= (1157 \text{ cm}^3/\text{s})/[A\ (30.48 \text{ cm/ft})] \\ &= 37.99/A \end{aligned}$$

$$\begin{aligned} U_G &= (1.18 \times 10^7 \text{ cm}^3/\text{s})/[A\ (30.48 \text{ cm/ft})] \\ &= 387{,}140/A \end{aligned}$$

The above two equations can be used to transform the Leva equation to a single equation for A, which can be solved to yield the following results:

$$A = 50{,}163 \text{ cm}^2$$

$$U_G = 7.715 \text{ ft/s} \qquad = 235 \text{ cm/s}$$

$$U_L = 0.000757 \text{ ft/s} = 0.0231 \text{ cm/s}$$

Now, given the gas and liquid flows, the number of transfer units can be determined. First, the stripping factor is found by:

$$S = HU_G/P_TU_L = 0.27(235)/(1)(0.0231) = 2746$$

Therefore,

$$\text{NTU} = (1 - 1/2746)\log[0.01(1 - 1/2746) + 1/2746] = 4.57$$

The height of a mass transfer unit (HTU) is computed from the Onda relationship. First, based on the data, the following dimensionless quantities are computed:

$$\begin{aligned} \text{Re}_L &= L_m/a_t\mu_L = (0.0231 \text{ g/cm}^2\text{-s})/(0.919/\text{cm})(0.01 \text{ g/cm-s}) = 2.515 \\ \text{Re}_G &= G_m/a_t\mu_G = (0.282 \text{ g/cm}^2\text{-s})/(0.919/\text{cm})(1.8 \times 10^{-4} \text{ g/cm-s}) = 1705 \\ \text{Sc}_L &= \mu_L/\rho L D_L = (0.01 \text{ g/cm-s})/(1 \text{ g/cm}^3)(0.8 \times 10^{-5} \text{ cm}^2/\text{s}) = 1250 \\ \text{Sc}_G &= \mu_G/\rho_G D_G = (1.8 \times 10^{-4} \text{ g/cm-s})/(1.2 \times 10^3 \text{ g/cm}^3)(0.088 \text{ cm}^2/\text{s}) \\ &= 1.705 \\ \text{Fr (Froude Number)} &= L_m^2 a_t/\rho_L^2 g \\ &= (0.0231 \text{ g/cm}^2\text{-s})^2(0.919/\text{cm})/(1 \text{ g/cm}^3)^2(980 \text{ cm/s}^2) \\ &= 5 \times 10^{-7} \\ \text{We (Weber Number)} &= L_m^2/\rho_L\sigma_L a_t \\ &= (0.0231 \text{ g/cm}^2\text{-s})^2/(1 \text{ g/cm}^3)(71 \text{ erg/cm}^2)(0.919/\text{cm}) \\ &= 8.2 \times 10^{-6} \end{aligned}$$

Therefore, the interfacial area is given by

$$a_w = 0.919\{1 - \exp[-1.45\,(80\ \text{erg/cm}^2/71\ \text{erg/cm}^2)^{0.75}\,(2.515)^{0.1}\,(5 \times 10^{-7})^{-0.05}\,(8.2 \times 10^{-6})^{0.2}]\}$$
$$= 0.268/\text{cm}$$

The liquid Reynolds number is now corrected for the revised interfacial area:

$$\text{Re}_{L'} = L_m/a_w\mu_L = (0.0231\ \text{g/cm}^2\text{-s})/(0.268/\text{cm})(0.01\ \text{g/cm-s}) = 8.617$$

From the above, the liquid and gas film mass transfer coefficients are obtained:

$$k_L = 0.0051\,(\mu_L g/\rho_L)^{1/3}\,(\text{Re}_{L'})^{2/3}\,(\text{Sc}_L)^{1/2}$$
$$= 0.0051[(980)(0.01)/1]^{1/3}\,(8.617)^{2/3}(1250)^{1/2} = 3.003\ \text{cm/s}$$

$$k_G = 5.23(a_t D_G)(\text{Re}_G)^{0.7}(\text{Sc}_G)^{1/3}(a_t d_p)^{-2}$$
$$5.23(0.919/\text{cm})(0.088\ \text{cm}^2/\text{s})(1705)^{0.7}(1.705)^{1/3}[(0.919/\text{cm})(5.08\ \text{cm})]^{-2} = 0.423\ \text{cm/s}$$

The overall mass transport coefficient is determined by

$$K_L = \left(\frac{1}{k_L} + \frac{P_T}{k_G H}\right)^{-1}$$
$$= [1/3.003 + 1/(0.423)(0.27)]^{-1}$$
$$= 0.11\ \text{cm/s}$$

The height of a transfer unit may then be determined from

$$\text{HTU} = L_m/\rho_L K_L a_w = 0.0231\ \text{gm/cm}^2\text{-s}/(1\ \text{g/cm}^3)(0.11\ \text{cm/s})(0.268/\text{cm})$$
$$= 0.784\ \text{cm}$$

Therefore, the stripper height is equal to HTU $\times$ NTU = 3.58 cm.

At this point, since the stripper height is computed to be less than the depth of one layer of packing, the validity of the computation is in some doubt, since the method assumes countercurrent plug flow. The proposed design conditions may therefore not be suitable, and the designer may consider one of the following alternatives:

using a smaller packing

operating at a different G/L ratio or a different fraction of the flooding pressure drop

Distillation. In the process of distillation, a mixture is heated, allowing the vapor phase above the mixture to become enriched in more highly volatile materials. By condensing this material, a liquid enriched in the highly volatile fractions can be obtained. By conducting this process repeatedly, the degree of purity of the volatile fraction may be increased and the residue can be virtually exhausted of volatiles. Distillation units may be batch, in which a single charge of material is processed, with a continual withdrawal of volatile material in the overheads, or continuous, with constant introduction of new feed and constant withdrawal of both volatile overheads and nonvolatile bottoms. Batch units appear to be most suitable for hazardous waste applications since the operating conditions and schedules can be readily adjusted to account for variability in the waste to be treated. However, for unique, large scale applications, with anticipated constant flow and composition of waste materials, continuous units should be considered. Distillation, particularly in the continuous mode, is a common process employed in the petrochemical industry for separation of petroleum fractions. Figure 4-12 shows schematics of typical batch and continuous distillation systems.

For distillation to occur, it is necessary to volatilize one or more components. Thus, energy must be supplied in an amount at least equal to the enthalpy of vaporization of the amount of material to be distilled plus the sensible heat required to raise the mixture to the boiling point of the volatile material. It may, however, be possible to recover a portion of this energy from the cooling water discharged from the condenser.

The equilibrium during distillation can be characterized by a relationship between the vapor phase mole fraction of a material (Y) and its liquid phase mole fraction (X) in a manner similar to Henry's law. For a two component ideal mixture,

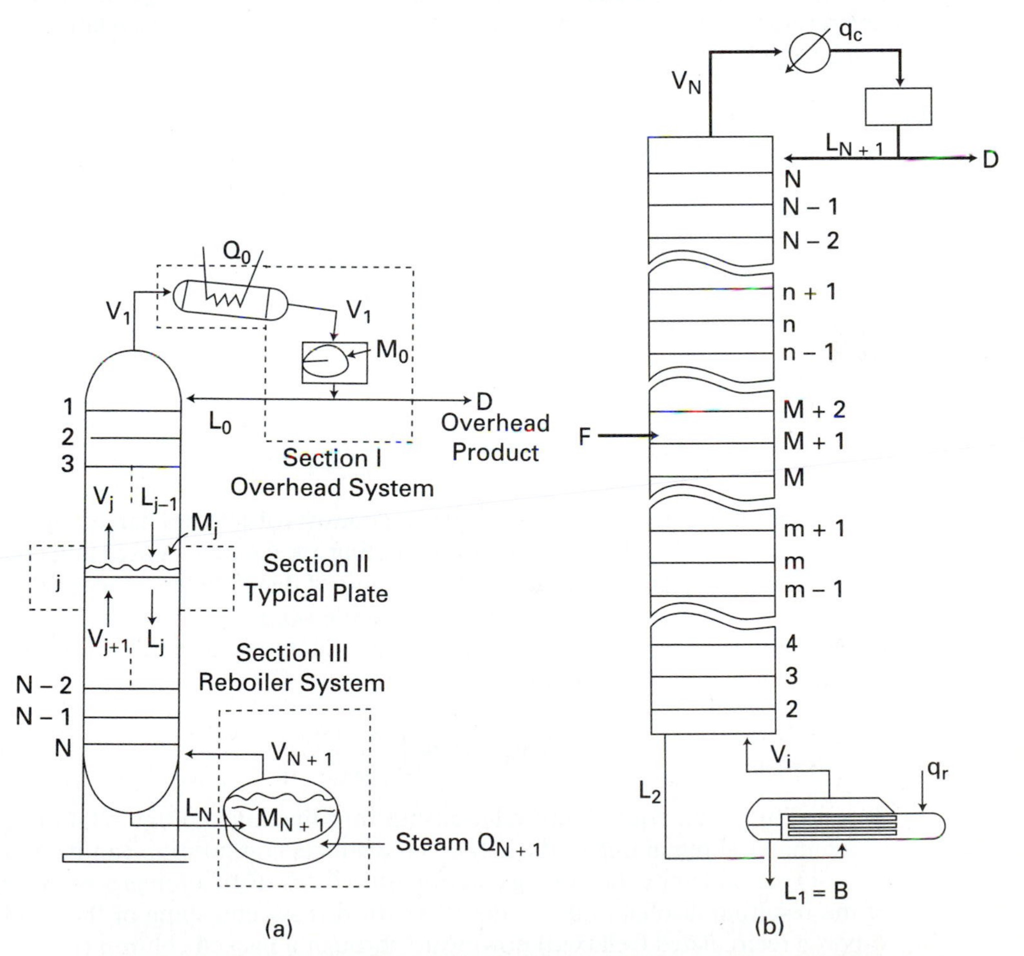

Figure 4-12 Schematic diagrams of (a) batch and (b) continuous distillation systems. (From Perry and Green, *Perry's Chemical Engineer's Handbook,* 6th ed., 1984, by permission of McGraw-Hill.)

this leads to two relations:

$$K_1 = y_1/x_1$$
$$K_2 = y_2/x_2$$

where component 1 is the most highly volatile. The equilibrium constants (K's) are functions of temperature, and (in the presence of nonidealities) composition. The relative volatility (α) is defined as the ratio of the equilibrium constant for the high volatile to the low volatile material:

$$\alpha = \frac{K_1}{K_2} = \frac{y_1 x_2}{y_2 x_1} \tag{4.46}$$

In the case of a nonideal mixture of two components, where the nonidealities are confined to the liquid phase, if the activity coefficients in the liquid mixture at the temperature of distillation are known (call these γ_1 and γ_2), then the relative volatility can be given by:

$$\alpha = \gamma_1 P_1 / \gamma_2 P_2$$

where P_1 and P_2 are the vapor pressures of the pure liquid components. Tabulations of activity coefficients are available for numerous binary liquid mixtures.

In the presence of only two components, the sum of the mole fractions in both phases is unity, and therefore,

$$\alpha = \frac{y_1(1 - x_1)}{(1 - y_1)x_1} \tag{4.47}$$

or, rearranging,

$$y_1 = \frac{\alpha x_1}{x_1(\alpha - 1) - 1} \tag{4.48}$$

The simplest type of distillation process consists of a kettle charged with a waste mixture, which is heated. The vapors emanating are then condensed and collected, and the process continues until a given amount of distillate has been produced, or until a given removal of material from the kettle occurs. If the initial charge is S_0 moles containing a mole fraction x_1^0 of volatile material, the relationship between material remaining and composition is given by:

$$\ln(S_0/S) = \int_{x_1}^{x_1^0} \frac{dx_1}{y_1 - x_1} \tag{4.49}$$

By substituting an equilibrium relationship into this integral, the relationship between material remaining in the kettle and distillate composition can be obtained.

By introducing a fractioning column into a batch distillation process, the purity of the resulting distillate can be increased. In this system, some of the condensed vapor is recirculated (refluxed) downward through a packed column (in a manner similar to a stripper). The heat released by the downcoming condensate is partially transferred to the newly forming vapor. The volatile fractions in the upcoming vapor and the downcoming liquid have extended opportunity to contact, and the degree of separation increases.

Two common modes of fractionation in batch distillation are practiced—constant reflux and constant composition at variable reflux. In the constant reflux method, the molar ratio of the downcoming liquid and the vapor (L/V) is maintained constant. In the variable reflux method, the reflux ratio is set to maintain a constant composition of the distillate. As the distillate declines in volatile material, reflux increases.

The exact computation of distillation performance in fractionation systems is quite complex in the general case. With the following simplifying assumptions, approximate calculations can be performed:

1. The molar flow rate through the fractionator is constant with both time and position. The constancy with position is only strictly true if the heating is identical for vaporization of all the components.
2. There are no heat losses through the fractionator column.
3. The amount of material in the fractionator is negligible with respect to the amount of material in the kettle.
4. The vapor-liquid partitioning in the fractionator may be considered a number of equilibrium stages in series. These assumptions lead to the McCabe–Thiele method for batch distillation systems. First, the equilibrium relationship ("equilibrium curve") is plotted (y_1 vs. x_1 as, for example, given by Equation 4.48). The origin and the point (1, 1) are connected by a straight line ("diagonal").

For the constant reflux operation, the mass balance on the system leads to Equation 4.49. However, with the fractionator, the distillate composition and the kettle composition are no longer related via the equilibrium relationship, but must be obtained graphically. A series of parallel lines ("operating lines") of slope equal to the reflux molar ratio (L/V) are drawn toward the left from the diagonal originating at the mole fraction of the distillate (y_1). The greater the number of operating lines, the more precise will be the integration of Equation 4.49. From the point of intersection with the diagonal, a horizontal to the equilibrium curve is constructed. At the intersection with the equilibrium curve, a vertical to the operating line is constructed. The process is continued $N + 1$ times, where N is the number of stages in the fractionator. The final vertical is extended to the abscissa, and its intersection defines the value of x_1, which should be used in evaluating Equation 4.49. The computational procedure is illustrated in Figure 4-13.

For the variable reflux operation at constant overhead composition, the mass balance over the system leads to the following equation:

$$\Theta = \frac{S_0(y_1 - x_1)}{V} \int_{x_1}^{x_1^0} \frac{dx_1}{(y_1 - x_1)^2(1 - L/V)} \qquad (4.50)$$

where Θ is the time at which the vapor and liquid compositions are, respectively, y_1 and x_1. For this system, the vapor and liquid compositions are related through an

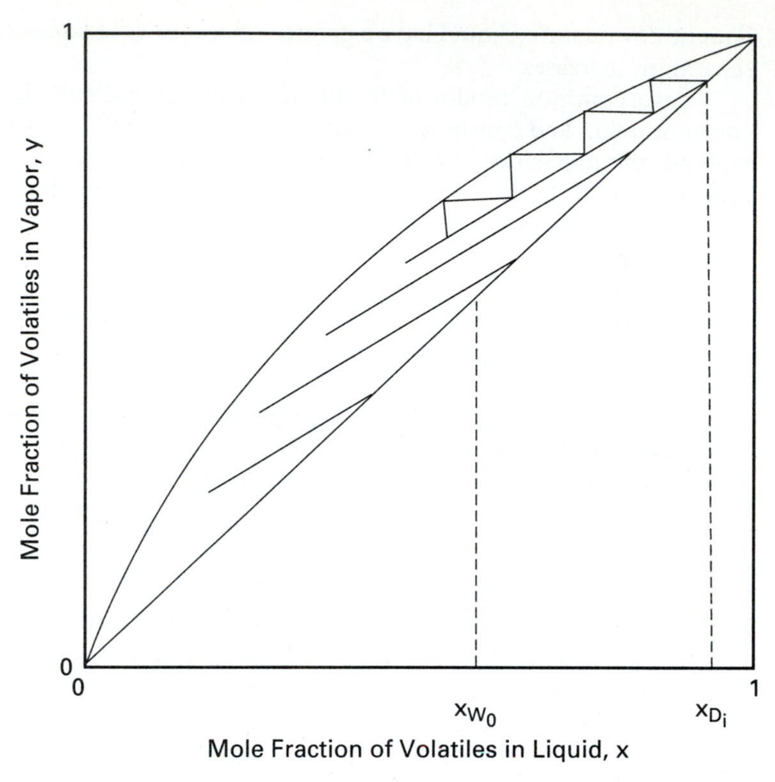

Figure 4-13 McCabe–Thiele diagram for constant reflux operation. (Reprinted from Ellerbe, 1988, by permission of Chemical Engineering.)

alternative McCabe–Thiele diagram. The equilibrium curve and the diagonal are drawn as previously. From the point on the diagonal at which the distillate composition is to be fixed, a series of operating lines of differing slopes (L/V ratio) are drawn. From each operating line, the number of equilibrium stages is "stepped off" as previously, and the value of x_1 corresponding to a given L/V value is obtained. This pair of numbers is used in the integration of Equation 4.18. The graphical construction is shown in Figure 4-14.

Distillation has a number of environmental implications. First, there is a potential for vapor emissions from the fractionator and condenser. With distillation systems, there is a considerable energy input. To cool the vapor, a cooling system, generally using water, is required. Hence, the release of heated and, if chemicals are used for descaling and antifouling, chemically polluted, water must be discharged. Also, in a batch distillation system, at the end of the process, the residue in the feed vessel may consist of a material which may have undergone polymerization and/or partial decomposition. This residue, or "still bottoms" may, in fact, be a specifically listed hazardous waste depending upon the feedstock.

In general, distillation has found wide application for solvent recovery. It is also widely used in the re-refining of motor and lubricating oils. In general, whenever a waste contains organic liquids of some volatility, distillation is a candidate process for use in resource recovery and treatment process trains (OTA, 1983).

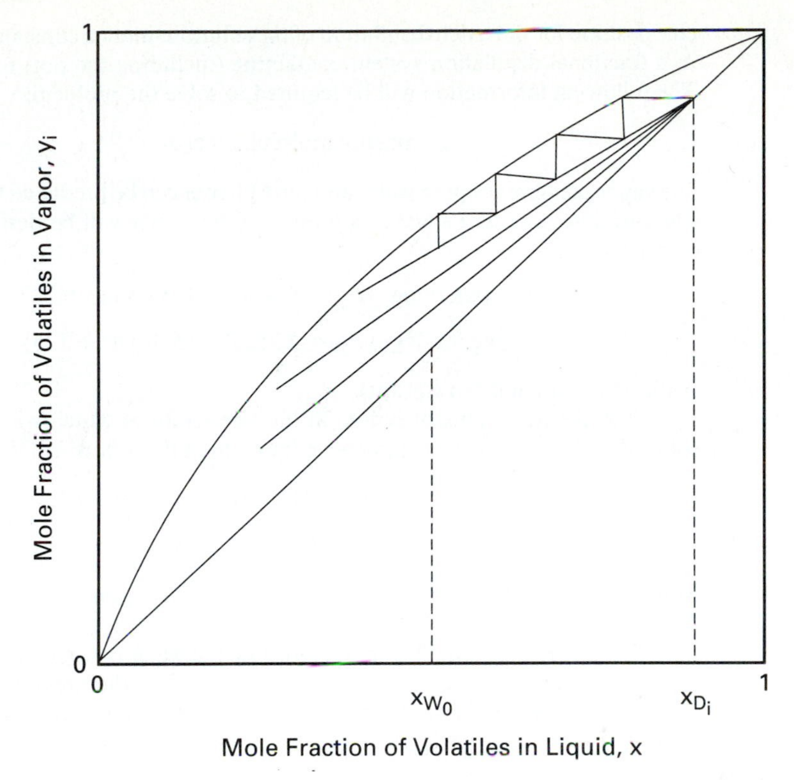

Figure 4-14 McCabe–Thiele diagram for constant overhead composition with variable reflux. (Reprinted from Ellerbe, 1988, by permission of Chemical Engineering.)

There are several pitfalls which one must be careful to avoid in the application of distillation. First, the elevated temperature required to volatilize materials at atmospheric pressure may lead to polymerization or other chemical transformations of the species which might be recoverable. Or, perhaps even worse, the still bottoms may polymerize to form a solid mass which must be physically removed from the kettle.

Second, a number of materials can form low boiling azeotropes. A low-boiling azeotrope is a nonideal mixture of two or more compounds with a boiling point lower than any of the pure compounds from which it is comprised. Hence, as the temperature is raised, the azeotrope will first be distilled, and the distillate will consist principally of the components in their azeotropic proportions. Only after all the azeotrope is distilled will the possibility exist for a pure component to be volatilized. There is really very little way to predict this behavior, although extensive tables of azeotropic systems exist (Holland, 1981).

Example Problem 4-6

A waste from polymer production consists predominantly of an aqueous solution of phenol along with polymer byproducts. Assuming that the polymer byproducts are negligible in terms of influencing distillation, if the initial waste consists of a 1% mole fraction of phenol, develop the curves of volume reduced, temperature, and distillate composition for the batch distillation of the solution under a constant reflux ratio of 0.1 in a fractional distillation system containing (including the pot) 6 theoretical stages. The following information will be required to solve the problem:

$$\text{phenol molecular weight} = 94$$

The vapor pressures of pure water and pure phenol can be predicted from the following relationships (note that since it is more volatile, water will be designated as component 1):

$$\text{water: } \log_{10}(P_1^\circ) = 5.0683 - 1657.46/(t + 227)$$

$$\text{phenol: } \log_{10}(P_2^\circ) = 4.2559 - 1518.1/(t + 175)$$

(with P° in atm and t in Celsius).

For the water-phenol system at the temperatures which will be of interest, the solution activity coefficients have been found to fit the following (Van Laar) equations:

$$\gamma_1 = A_{12}/[1 + A_{12}x_1/A_{21}x_2]^2$$

$$\gamma_2 = A_{21}/[1 + A_{21}x_2/A_{12}x_1]^2$$

with $A_{12} = 0.36$ and $A_{21} = 1.40$ and with the x's representing the liquid phase mole fractions.

To develop the equilibrium relationship (curve of y_1 versus x_1), we need to determine the composition of the vapor which boils off at the pressure of the distillation (P_t—one atmosphere) as a function of x_1. The following constraint is used to determine this:

$$y_1 + y_2 = 1 = \gamma_1 P_1/P + \gamma_2 P_2/P$$

These equations can be solved by trial and error for a given value of x_1, and, as a result the equilibrium curve (and the temperature at which the liquid of that composition boils) can be determined. Figure 4-15 plots both the equilibrium curve and the operating line for this system.

The construction of the McCabe–Thiele Diagram for this situation can be readily performed in tabular form. Integration is to proceed at various values of distillate composition. If the point (x_{di}, x_{di}) is the starting point, and if the equilibrium curve has the relationship:

$$y = f(x)$$

then the coordinates of the point at the end one stage (at the intersection of the vertical with the reflux line) are given by:

$$(f^{-1}(x_{di}),\ Rf^{-1}(x_{di}) + (1 - R)x_{di})$$

where R is the reflux ratio (0.1 in the present case). By a similar analysis, it can be shown that if the coordinates (x_{si}, y_{si}) are those at the end of stage "i" then the coordinates at the end of stage "$i + 1$" are as follows:

$$(f^{-1}(y_{si}),\ Rf^{-1}(y_{si}) + (1 - R)x_{di})$$

The determination of f^{-1} can be done using an inverse interpolation process from the equilibrium curve. The procedure is illustrated in the table, with the composition at each stage indicated.

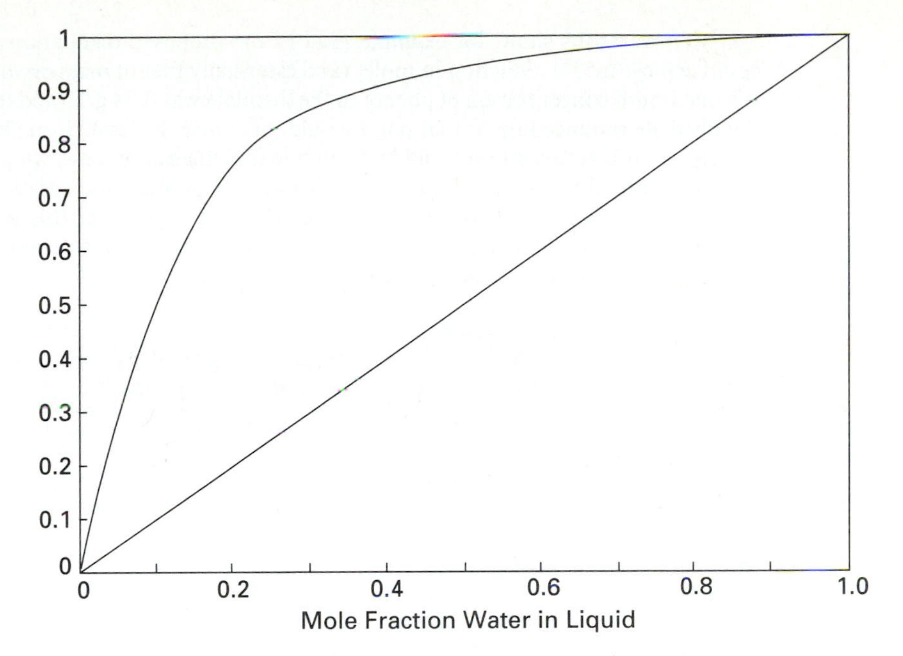

Figure 4-15 Equilibrium and operating lines.

	End of stage 1		End of stage 2		End of stage 3		End of stage 4		End of stage 5		End of stage 6	
x_{di}	x	y	x	y	x	y	x	y	x	y	x (x_{wi})	y
0.999999	0.999082	0.999907	0.998074	0.999806	0.996141	0.999613	0.986790	0.998678	0.973056	0.997304	0.952523	0.995251
0.99999	0.999073	0.999898	0.998065	0.999797	0.996132	0.999604	0.986781	0.998669	0.973047	0.997295	0.952515	0.995242
0.99995	0.999033	0.999858	0.997109	0.999665	0.993253	0.999280	0.986461	0.998601	0.972981	0.997253	0.946074	0.994562
0.99992	0.998087	0.999736	0.995156	0.999443	0.986622	0.998590	0.972970	0.997225	0.946047	0.994532	0.900895	0.990017
0.9998	0.996135	0.999433	0.986612	0.998481	0.972864	0.997106	0.945935	0.994413	0.900787	0.989898	0.826558	0.982475
0.9995	0.986678	0.998217	0.966205	0.996170	0.932255	0.992775	0.875879	0.987137	0.777534	0.977303	0.677245	0.967274
0.997	0.945834	0.991883	0.851664	0.982466	0.727331	0.970033	0.608875	0.958187	0.529821	0.950282	0.493672	0.946667
0.99	0.826643	0.973664	0.611154	0.952115	0.494625	0.940462	0.432346	0.934234	0.429483	0.933948	0.429351	0.933935
0.95	0.493526	0.904352	0.348840	0.889884	0.307438	0.885743	0.306007	0.885600	0.305958	0.885595	0.305956	0.885595

At the left side of the table, the distillate composition (x_d) is given (by x), at the end of stage 6, the pot composition (x_w) is given (also by x). From this table, the following integral can be evaluated at various arbitrary x_1 values. The maximum value of x_d that should be used to evaluate this integral is that which would result in a pot composition given by the actual starting composition (i.e., in this case, 0.99 mole fraction of volatile).

$$\ln(S_o/S) = \int_{x_1}^{x_1^0} \frac{dx_1}{y_1 - x_1}$$

This gives directly the relative number of moles remaining to be distilled. By difference, one can determine the amount of distillate produced. Based on the value of x_w, one can also determine the temperature at which that mixture boils. The following tabulates the fraction of moles remaining versus the temperature at which the pot is boiling and the distillate concentration (in mg/L) of phenol at that condition.

These results show, for example, that in the proposed distillation process, one could achieve a 53% reduction in moles (and essentially that in mass or volume), while the maximum concentration of phenol in the distillate was 0.34 g/L (and the average of the distillate produced up to that point would, of course, be less). It might be that the distillate of this composition could be treated in a biological process, while the volume reduction would allow more ready solidification, or perhaps incineration, of the remaining materials. Note also that the analysis indicates that, at this point, the pot temperature would be about 120°C—which might represent a convenient means of determining when to terminate the run.

Moles left	°C	Moles removed	g/L phenol in distillate
1.000	101.0	0.000	0.03
0.880	101.1	0.120	0.16
0.461	101.9	0.539	0.34
0.255	103.0	0.745	0.73
0.132	106.1	0.868	1.83
0.083	112.3	0.917	9.14
0.073	115.4	0.927	33.94
0.059	123.1	0.941	156.67

Steam distillation/steam stripping. The processes of steam distillation and steam stripping differ from their parent processes (i.e., distillation and stripping) only in the use of steam in the vapor phase. Generally, if the liquid phase is immiscible with water, the process is referred to as steam distillation, while if the liquid phase is miscible with water (or an aqueous fluid), the process is designated as steam stripping.

Steam distillation involves the addition of liquid water to the kettle in which a distillation of a material with boiling point above the boiling point of steam is to be conducted. The evolved vapor consists of a mixture of water vapor and vapor of the volatile component. For immiscible fluids, the vapor pressures of each of the components will be set solely by their fluid-phase compositions (and the temperature). Thus, the presence of the steam will result in the attainment of atmospheric pressure at lower temperatures. The resulting distillate, when condensed, will form immiscible phases which can then be separated by decantation or similar physical operations.

Physical equipment in which steam distillation is conducted is much the same as in either simple or fractional distillation. The efficacy of a single-stage steam-distillation process can be rapidly assessed by use of a Hausbrand diagram (Ellerbe, 1974) in which the vapor pressure of volatile material (in the mixture to be steam distilled) versus temperature is plotted versus 760-p_{H_2O} (where p_{H_2O} is the vapor pressure of water in mm Hg (Figure 4-16). The point at which the curves intersect indicates the temperature at which steam distillation will occur and the partial pressure of the organic material in the distillate. This partial pressure divided by the total pressure (i.e., 760 mm) gives the distillate mole fraction. For steam distillations in which fractionation columns are used to increase purity, more elaborate computations of the McCabe–Thiele type may be developed.

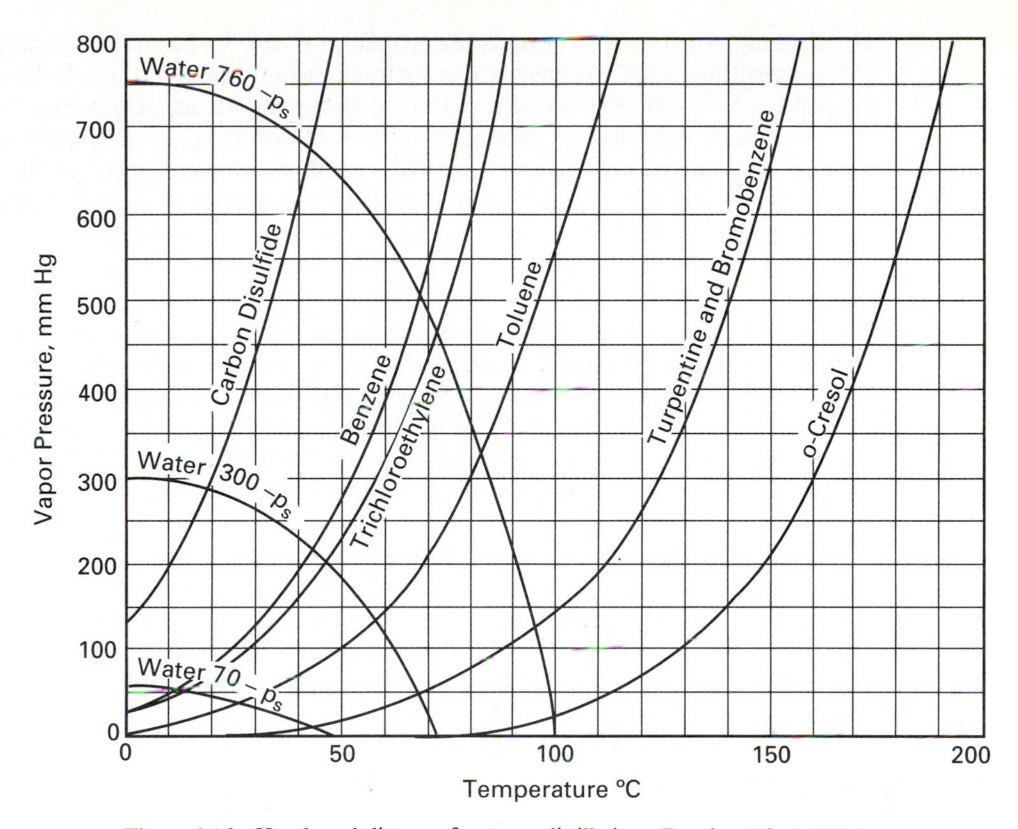

Figure 4-16 Hausbrand diagram for steam distillation. (Reprinted from Ellerbe, 1988, by permission of Chemical Engineering.)

In steam stripping, physical contactors (packed towers) similar to air strippers are employed. However, the increase in temperature that the steam provides will tend to increase the Henry's law constant of a volatile material in solution. Therefore, a higher removal efficiency will generally be attainable, or removal of less volatile materials may be achieved. Steam strippers can be designed using the same procedures as air strippers, provided that the increased temperature (and the properties of steam, rather than air) is taken into account in computing equilibria and transport constants (diffusion, viscosity, density).

Steam stripping has been used extensively in the removal of ammonia and sulfites from petroleum refinery wastes (U.S. EPA, 1980). It also has been used in the removal of volatile solvents and organo-sulfur compounds from paper plant wastewaters (McKance and Burke, 1980).

Solid – Liquid and Solid – Vapor Separation

In the process of adsorption, phase partitioning occurs between a bulk solution (either a liquid solution or a gaseous solution) by the selective accumulation of one or more components at the interface formed with an added solid phase. By physical separation of the solid phase, the net removal of the dissolved constituent results.

The driving force for adsorption is the reduction of surface tension (interfacial free energy) resulting when the adsorbed constituent accumulates at the interface. Two general types of materials are of importance commercially as adsorbents.

Activated carbons are produced by partial combustion of specific organic solids, generally coal or vegetable charcoals. Under such conditions, a high surface area in the interior of the porous solid that is produced remains, onto which adsorption can occur. Activated carbons may be produced in large particle size, approaching 1 mm, or in much smaller sizes. The larger sizes, termed granular, are substantially more expensive but allow the adsorbent to be used in a packed bed and regenerated. The smaller carbon sizes, termed powdered, are generally added as a slurry to the material (usually liquid) to be treated and discarded after one use.

The other type of adsorbent is polymeric resin of neutral surface charge. These are often more expensive but have properties that are more closely controlled. The most common of these are Rohm and Haas and XAD's series resins. Resin adsorbers are generally produced in larger particle sizes suitable for multiple applications in packed bed contactors.

For the packed-bed adsorption contactors, which are the most frequently employed, the major design variables are the size (height) of the column, or equivalently, the residence time in the column, and the time required (or volume throughput which has been applied) for breakthrough to occur. For column contactors, breakthrough is defined as the time at which the effluent concentration exceeds some predetermined tolerance limit (or standard), as in Figure 4-17. The time of exhaustion may be substantially later than the breakthrough time, and occurs when the concentration in the effluent equals the influent concentration. In systems in which

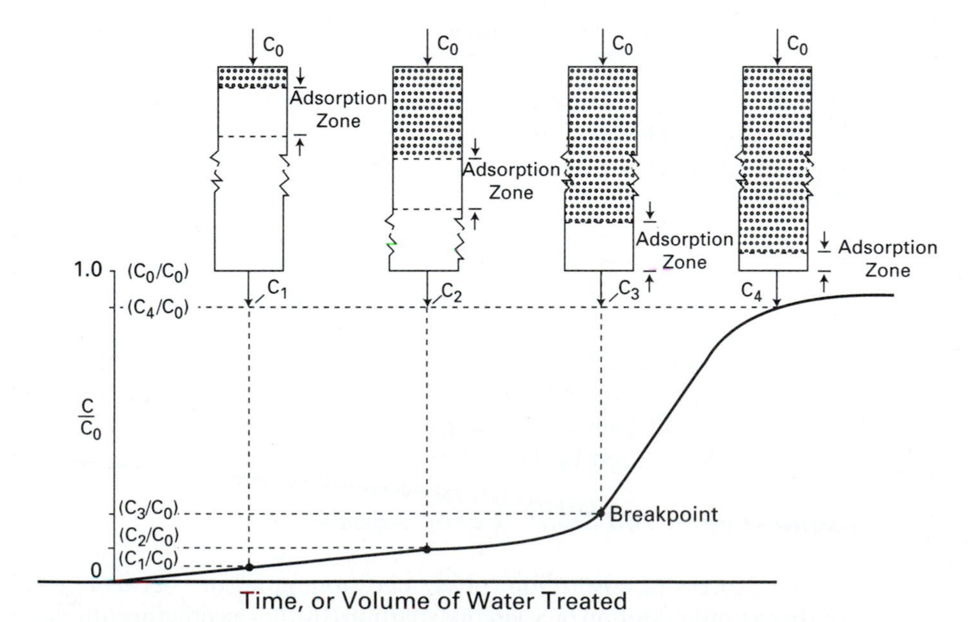

Figure 4-17 Typical breakthrough curve for a fixed-bed adsorber. (Weber, *Physicochemical Processes for Water Quality Control,* 1972; reprinted by permission of John Wiley & Sons, Inc., copyright 1972.)

some biological or chemical decomposition of contaminant occurs in the column, exhaustion may never occur. In addition, the headloss through the contactor is important in specifying system hydraulics.

The equilibrium relationship characterizing adsorption is an isotherm which relates the solution phase compositions and the solid phase concentrations of adsorbable materials. If C_i and q_i are, respectively, these concentrations (with q_i typically in mass contaminant sorbed/mass sorbent) then, in general terms, the following relationship (termed an "isotherm" since it is temperature dependent) exists:

$$q_{i,eq} = f_i(C_i, q_{j \infty i}) \tag{4.51}$$

For systems containing only a single adsorbable component, this results in a direct correspondence between solution phase concentration and solid phase concentration. Usually either a Langmuir (Equation 4.52) or a Freundlich (Equation 4.53) isotherm can be used to describe the system:

$$q_{i,eq} = b_i q_{i,\infty} C_i / (1 + b_i C_i) \tag{4.52}$$

$$q_{i,eq} = K_i C_i^{1/n_i} \tag{4.53}$$

where the parameters b_i, $q_{i,\infty}$, K_i, and n_i characterize the particular equilibrium (a function of the adsorbing component and the solid phase), and should be obtained from laboratory tests on a particular waste.

In the absence of kinetic limitations, the breakthrough will be extremely sharp, and the time for breakthrough will equal the time for exhaustion. The mass of carbon required for treatment (m) of a given volume of liquid passed through can be estimated from the equilibrium with the incoming solution (concentration $C_{i,0}$) as in Equation 4.54:

$$m q_{i,eq} = V C_{i,0} \tag{4.54}$$

with the substitution of $q_{i,eq}$ from the applicable equilibrium relationship, the minimum dose of carbon (mass/volume treated) as a function of influent waste composition can be established.

With multiple materials in solution that are capable of being adsorbed, characterization of the equilibrium relationship becomes more complicated. If the assumption is made that the adsorbed phase consists of an ideal surface solution of molecules, then the single component isotherms can be used without further investigation to characterize the surface. For the case when the single component isotherms are all of Freundlich form, this approach (termed Ideal Adsorbed Solution Theory, or IAST), leads to the following equations:

$$C_i = C_i^\circ \left[q_i \Big/ \left(\sum q_j \right) \right] \quad \text{(for each } i\text{)} \tag{4.55}$$

$$\sum [q_i / q_i^\circ] = 1 \tag{4.56}$$

$$n_1 q_1 = n_i q_i \quad \text{(for each } i \neq 1\text{)} \tag{4.57}$$

$$q_i^\circ = K_i (C_i^\circ)^{1/n_i} \quad \text{(for each } i\text{)} \tag{4.58}$$

In these equations, C_i and q_i are the solution and adsorbed concentrations at equilibrium, while q_i° and C_i° are hypothetical concentrations which would exist at equilibrium with one another in the absence of other components at the same surface spreading pressure as in the mixture. (The surface spreading pressure is a two-dimensional pressure, computed from the isotherm and expressing the intensity of adsorption at a surface). These equations can be solved for the equilibrium conditions and the hypothetical concentrations, providing that one additional condition per component is set, typically in the form of a mass balance or a solution concentration.

It has been found that, for adsorbable contaminants at the mg/L level present in the absence of substantial nonadsorbable (or poorly adsorbable, such as humic matter) organic matter, the IAST method performs adequately (Crittenden et al., 1985). Such situations would include contaminated groundwater cleanup, for example. However, in the treatment of more concentrated wastes, or when there is substantial humic material present, the equilibrium may either be nonideal (Price and Danner, 1988) thus necessitating more detailed laboratory characterization, or adsorption may be irreversibly altered by interference with humic materials (Summers and Roberts, 1987).

An alternative approach, when IAST fails to describe multicomponent adsorption equilibria is the use of an empirical multicomponent isotherm. For example, Sheindorf, et al. (1981) proposed the following extension of the Freundlich isotherm:

$$q_i = K_i C_i \left(C_i + \sum_{j \neq i} \alpha_{ij} C_j \right)^{n_i - 1} \tag{4.59}$$

where the α's are binary interaction parameters which must be experimentally determined, but which are related by the following type of constraints:

$$\alpha_{ij} = \alpha_{ji} \quad \text{and} \quad \alpha_{ij}\alpha_{jk}\alpha_{ki} = 1 \tag{4.60}$$

Thus, by the determination of single component isotherms (to obtain the K_i's and n_i's) and selected experiments to estimate the α's, the multicomponent equilibrium may be described. To date, however, there has not been an extensive test of this approach, particularly in situations where IAST is inadequate.

Equilibrium theory provides a lower limit to the dose of adsorbent required to treat a given waste. There may be substantial kinetic limitations which cause the effluent concentration to exceed the breakthrough threshold before the carbon bed is exhausted. These kinetic limitations include the following potential processes:

transport of material from the bulk solution to the surface of the adsorbent particle (fluid phase mass transfer);

transport of material from the outer extremities of the adsorbent particle into the interior where there is available capacity (intraparticle transport);

chemical reactions between the surface and the sorbing molecule which must be completed prior to sorption (chemical kinetic limitations).

In most applications, either fluid phase mass transfer or intraparticle transport (or both processes) are substantial limitations.

For systems in which the interaction between contaminants is negligible (and they each may be characterized by a single component isotherm), it is possible, knowing only the fluid-phase mass transfer coefficient and the diffusion coefficients characterizing intraparticle transport, to estimate the breakthrough time (Crittenden et al., 1987). These two parameters may be obtained in straightforward laboratory tests at the bench level.

For more complicated systems, it is necessary to conduct a series of pilot tests to obtain information on the breakthrough curve. These should be conducted at the same hydraulic loading which is to be used in full scale, but the bed depth can be varied. In general, the breakthrough wave in the pilot system will match, to a reasonable degree, the breakthrough in the full scale system if these two design variables, as well as the type of adsorbent, are kept constant. By analyzing the breakthrough time versus the column size, the optimal system may be designed.

The regeneration of adsorbents may be carried out using thermal regeneration, by steam stripping, or by the use of solvents. In most cases, however, the regeneration will be conducted off site by a commercial service. Only in large treatment situations does on-site regeneration, particularly using thermal techniques, appear economic.

The most widely used form of regeneration is thermal regeneration. In this process, the spent adsorbent is exposed to a controlled regime of temperature and gaseous atmosphere (oxygen, CO_2, and water vapor) in which pollutants are volatilized and combusted while, to the least extent possible, combusting the adsorbent itself. Thermal regeneration is performed in a furnace, typically a rotary hearth or a fluidized bed. Thermal regeneration is fairly expensive, and also, during thermal regeneration, there is some loss of adsorbent capacity, the amount of which depends on the physical characteristics as well as the chemical properties of the adsorbent. Typical losses of capacity might be anywhere from 5–20% per cycle.

Steam stripping for regeneration can be used whenever the materials adsorbed are totally volatile at lower temperatures, and, particularly, when it is desirable to achieve their recovery. Live steam, generally above saturation temperature, is passed through the adsorbent. The exhaust vapors, containing a mixture of steam and volatile contaminants, are condensed. Organics may then form an oil or a sludge which can either be used as a supplemental fuel, or in some other manner. Steam stripping is generally less costly than thermal regeneration, and also allows for a lower attrition of adsorption capacity between cycles. However, if nonvolatile contaminants are present in the waste, and if they are adsorbable, there will be progressive loss of capacity with steam regeneration.

With certain organics, it is possible to pass solvents through an adsorption bed and achieve a regeneration of capacity (Himmelstein et al., 1973; Fox, 1978). For example, sodium hydroxide can be used to recover acidic organics, such as phenols, which have been removed by adsorption. Ethanol and methanol have also been used as solvent regeneration systems for a variety of organics. Nonextractable adsorbed materials will progressively build up in the bed necessitating more severe regeneration processes.

Adsorption is widely used for removal of phenols and petroleum hydrocarbons from contaminated groundwaters. Adsorption also finds a large application in the treatment of gaseous constituents such as removal of solvent fumes. In particular, it can be used for the removal of volatile organics in air generated by air stripping processes.

Like all other physical processes, adsorption does not, in and of itself, result in pollutant destruction or detoxification (other than any destruction that might occur during regeneration). If you adsorb a listed hazardous waste, the spent adsorbent itself becomes a hazardous waste, and its transport, including to an off-site regeneration facility, must be manifested. Furthermore, as a hazardous waste, there is continual and perpetual liability on the part of the generator for consequences associated with the adsorbent regeneration and disposal.

If a waste contains high concentrations of suspended solids, they will increase the pressure drop through the column. Therefore, either an upstream filter or periodic backwashing of the adsorbent bed is needed to physically remove the solids. Any humic material in the waste will generally be highly adsorbed. The problem with this is that if they are present, they may be adsorbed preferentially, or may interfere dramatically with the adsorption of the more toxic materials. And so it is desirable in many instances to try to separate the toxics and the humics from these colored materials so that the adsorption process can concentrate on removing toxics rather than removing the humics.

Example Problem 4-7

A small lagoon contains 100 μg/L of trichloroethylene (TCE) and 200 μg/L of carbon tetrachloride. It is proposed to reduce the concentration of these compounds by a one time addition of activated carbon, followed by a dredging of the sediments (which would then contain settled carbon). Given the following isotherm data, compute the dose (mass of carbon per volume of lagoon) which would be necessary to reduce both dissolved compounds to below 10 μg/L (for each). Compute the dose assuming that there is no interaction among compounds, and assuming the ideal adsorbed solution theory (IAST).

	Freundlich isotherm parameters	
	K	$1/n$
Tichloroethylene (compound 1)	0.0363	0.592
Carbon tetrachloride (compound 2)	0.0068	0.469

The above are for units of g/g in q and mg/L in C.
(Isotherm data from Amy et al., 1987)

From mass balance considerations, it must be required that

$$C_1 = C_{T1} - 1{,}000\ mq_1$$

$$C_2 = C_{T2} - 1{,}000\ mq_2$$

By the problem conditions,

$$mq_1 \geq 0.09/1000, \text{ or } mq_1 \geq 9 \times 10^{-5}$$

$$mq_2 \geq 1.9 \times 10^{-4}$$

combining the inequalities,

$$m = \max(9 \times 10^{-5}/q_1,\ 1.9 \times 10^{-4}/q_2) \qquad \text{(A)}$$

If the interactions are ignored, and each species is assumed to adsorb independently, the

Freundlich isotherm can also be used and thus m obtained,

$$q_1 = 0.0363(0.01)^{0.592} = 2.376 \times 10^{-3}$$

$$q_2 = 0.0068(0.01)^{0.469} = 7.84 \times 10^{-4}$$

Therefore, the carbon dose is governed by the reduction of carbon tetrachloride below the desired level, the dose = 0.242 g/g.

Using the IAST model, the problem is to solve the following system of equations:

$$C_1 = C_1^\circ[q_1/(q_1 + q_2)]$$

$$C_2 = C_2^\circ[q_2/(q_1 + q_2)]$$

$$q_1/q_1^\circ + q_2/q_2^\circ = 1$$

$$(1/0.592)q_1^\circ = (1/0.469)q_2^\circ$$

$$q_1^\circ = 0.0363(C_1^\circ)^{0.592}$$

$$q_2^\circ = 0.0068(C_2^\circ)^{0.469}$$

$$C_1 = 0.1 - 1000\ mq_1$$

$$C_2 = 0.2 - 1000\ mq_2$$

The above system of eight equations in nine unknowns ($C_1, C_2, C_1^\circ, C_2^\circ, q_1, q_2, q_1^\circ, q_2^\circ, m$)

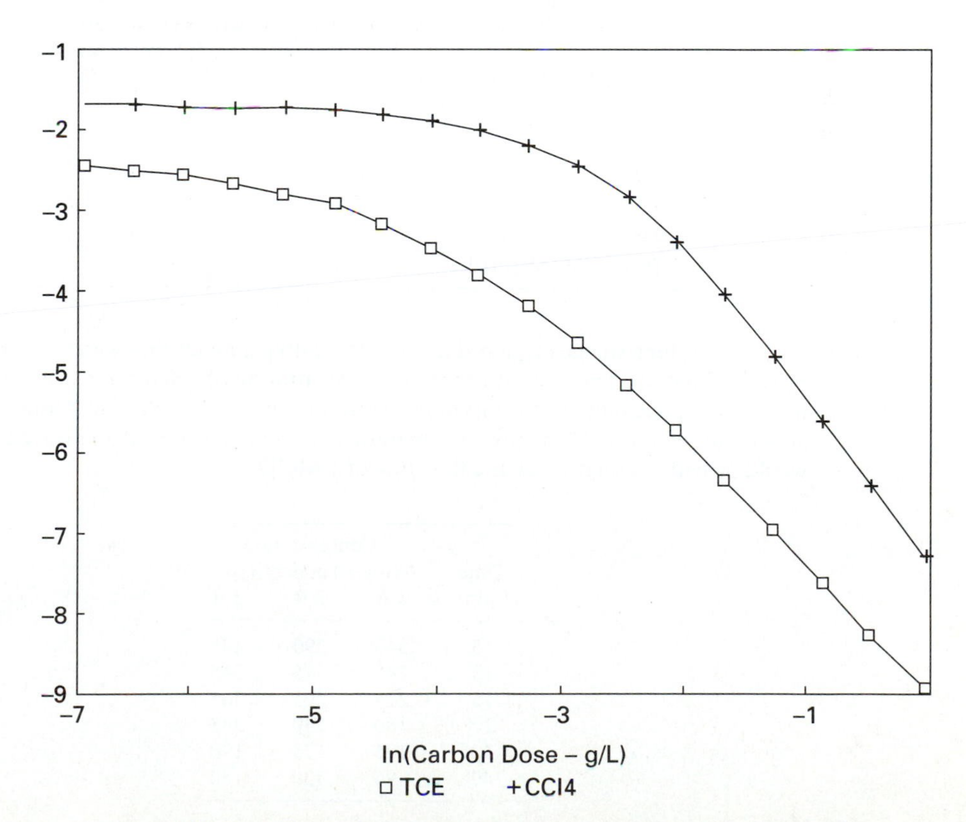

Figure 4-18 Logarithm of solution equilibrium concentration versus carbon dose.

can be solved for various carbon doses. This system is highly nonlinear but standard equation solvers are capable of developing the necessary solutions. By plotting the solution concentrations versus carbon dose, as shown in Figure 4-18, it is seen that, in order to assure that the concentration of each of the contaminants remain below 10 μg/L (i.e., ln(mg/L) = −4.6) it is necessary to reduce the TCE concentration far below this value. By graphical interpolation, it is determined that the ln(g/L) carbon dose is −1.27065, that is, the carbon dose is 0.28 g/L. This is about 15% above the carbon dosage determined by assuming that the compounds were independently adsorbed; hence the interactions are minor.

PROBLEMS

4.1. A waste cutting oil contains a suspension of two densities of particles—a light organic material (dust, grease) and heavy metal particles (from a grinding operation). Given the following data, and assuming that the suspension behaves as a Type I material (with fluid properties approximately those of water at 25°C), determine the largest overflow velocity that could be used in an ideal sedimentation tank to produce an effluent completely freed of the denser particles. Using this overflow velocity, estimate the size distribution of both classes of particles in the effluent.

	Mass fraction of particles at that size	
Diameter (cm)	Light fraction density 1.05 g/cm³	Heavy fraction density 2.1 g/cm³
0.1	0.1	0.3
0.05	0.2	0.25
0.025	0.3	0.2
0.0125	0.3	0.15
0.00625	0.1	0.1
TOTAL MASS (mg/L)	50	50

4.2. A flocculant lime sludge produced from precipitating a metal rinsewater is to be treated by sedimentation. Starting with an initial concentration of 500 mg/L placed in a laboratory column, readings of concentration versus time are taken at various depths below the air–liquid interface. Using this information, determine the area of an ideal basin which would provide 55% removal at a flow rate of 1 MGD.

Time	Concentration (mg/L) at depth of:		
(min)	1 ft	2 ft	3 ft
5	340	390	405
10	245	315	345
15	215	265	305
20	180	230	245
30	125	175	190
40	95	130	150

4.3. The following are data on initial settling velocities of a biological sludge exhibiting hindered settling characteristics. Using this information, determine the area of a thickener sufficient to treat 2,000 m^3/d of a sludge containing 1,000 mg/L of solids if the desired underflow solids concentration is to be 5,000 mg/L.

Initial solids concentration (mg/L)	Initial settling velocity (m/day)
500	120
1,000	95
2,000	59
2,500	32
3,000	13
3,500	8.6
4,000	6.0
5,000	3.6

4.4. It is claimed that a filter which consists of a bed of sand with the following properties can be designed to remove 95% of particles with a 90 micron radius and a density of 1.07 g/cm^3 if a superficial velocity of 10 gpm/ft^2 is used. Assuming fluid properties equal to that of water at 20°C, determine if this claim is reasonable. Specifically, is the headloss in a reasonable range?

sand diameter = 1 mm
sand density = 2.65 g/cm^3
filter porosity = 0.35
collision efficiency factor = 0.01

If this filter is to be backwashed, what velocity is required to achieve an expanded bed porosity of 0.75?

4.5. Gasoline is to be removed from an aquifer by air stripping. The flow to be treated is 1,000 m^3/d, and the material to be stripped can be approximated (in terms of physical properties) as 40 $\mu g/L$ of decane (molecular weight = 142), which must be reduced to 10 $\mu g/L$. The Henry's law constant is 4.93 atm-m^3/mol. The air flow is to be twice the minimum required for equilibrium, and the area is to keep the pressure drop at 0.005 lb/ft^2-ft. Using the following data, compute the area and depth of the required system:

Packing = 1/2 inch berl saddles — $D_L = 4 \times 10^{-6}$ $cm^2/2$
Porosity = 0.62 — $D_g = 0.06$ cm^2/s
Specific surface = 142 ft^2/ft^3
16,700 saddles/ft^3
Leva equation constants: $M = 0.0374$, $N = 0.0034$

4.6. 1,1,2,2-tetrachloroethane (TC) is used as a cleaning agent in a certain process. During use, it picks up dirt and moisture. It is proposed to use distillation to repurify this solvent. The material to be treated can be considered TC that is saturated with water (at 25°C, the solubility of water in TC is 1.0 mole percent). Given the following data, if a steam distillation process is to be used, and if sufficient water is added such that the "pot" remains saturated with water, determine the operating temperature of the process, and the vapor phase composition.

$$\log_{10}(P - mm) = 6.09073 - 959.602/(t + 149.778) \qquad \text{for TC (pure)}$$

Activity coefficients (TC = 1; Water = 2)

$$\ln\gamma_1 = x_2^2\left[\frac{0.491}{(x_1 + 0.276x_2)^2} + \frac{1.629}{(x_2 + 0.558x_1)^2}\right]$$

$$\ln\gamma_2 = x_1^2\left[\frac{0.909}{(x_2 + 0.558x_1)^2} + \frac{1.778}{(x_1 + 0.276x_2)^2}\right]$$

4.7. Consider the vapor phase produced from problem 6, which now contains a relatively pure mixture of TC and water. It is desired to recover TC with a moisture content < 0.01 mole fraction by batch distillation at a constant reflux ratio of 0.5, and in a system with 4 stages. This will be done by removal of the more volatile water vapor. Estimate the fraction of TC which can be recovered in the still pot and the temperature at which the distillation should be stopped.

4.8. A gas phase adsorption unit is to be used to remove volatile hydrocarbons from the off-gases from air stripping. The water vapor is initially removed by passage over a drying column and the dry gases contain 100 $\mu g/m^3$ each of toluene and methylcyclohexane. These are to be reduced to less than 1 $\mu g/m^3$ of each of these compounds by passage through activated carbon. If the adsorption process is characterized by equilibrium ideal adsorbed solution theory, and the isotherms are as given below, determine the mass of carbon required per cubic meter of air treated (isotherm data from Yu and Neretnieks, 1990).

toluene $Q = 72.4\ P^{0.2954}$

methylcyclohexane $Q = 31.2\ P^{0.2580}$

P = partial pressure (atm)

Q = surface coverage (mol/kg)

REFERENCES

AMY, G. L., NARBAITZ, R. M., AND COOPER, W. J. "Removing VOC's from Groundwater Containing Humic Substances by Means of Coupled Air Stripping and Adsorption." *Journal of the American Water Works Association.* August 1987, pp. 49–54.

BASKIN, D. E., AND SUIDAN, M. T. "Unified Analysis of Thickening." *Journal of Environmental Engineering* 111(1): 10–26 (1985).

BERKOWITZ, J. B. *Unit Operations for Treatment of Hazardous Industrial Wastes.* Noyes Data Co., Park Ridge, N.J. (1978).

BERTHOUEX, P. M., AND STEVENS, D. K. "Computer Analysis of Settling Test Data." *Journal of the Environmental Engineering Division, ASCE* 108(5): 1065–9 (1982).

CAMP, T. R. "Sedimentation and the Design of Settling Tanks." *Transactions, ASCE* 111 (1946).

CHRISTENSEN, G. L., AND DICK, R. I. "Specific Resistance Measurements: Nonparabolic Data." *Journal of Environmental Engineering* 111(3): 243–57 (1985).

CHRISTENSEN, G. L., AND DICK, R. I. "Specific Resistance Measurements: Methods and Procedures." *Journal of Environmental Engineering* 111(3): 258–70 (1985).

CLEASBY, J., AND FAN, L. *Journal of the Environmental Engineering Division, ASCE* 107, 455 (1981).

CORDOBA-MOLINA, J. F., HUDGINS, R. R., AND SILVESTON, P. L. "Settling in Continuous Settling Tanks." *Journal of the Environmental Engineering Division, ASCE* 104(6): 765–779 (1978).

COWLES, E., AND MCNAUGHTON, K. "Pollution Control Still a Major Issue for CE's." *Chemical Engineering* 92(17): 59–68 (1985).

CRITTENDEN, J. C., ET AL. "Prediction of Multicomponent Adsorption Equilibrium using Ideal Adsorbed Solution Theory." *Environmental Science and Technology* 19(11): 1037–43 (1985).

CRITTENDEN, J. C., HAND, D. W., ARORA, H., AND LYKINS, B. W., JR. "Design Considerations for GAC Treatment of Organic Chemicals." *Journal of the American Water Works Association* 79(1): 74–82 (1987).

ELLERBE, R. W. "Steam Distillation Basics." *Chemical Engineering.* July 18, 39–42. (1988).

FANG, C. S., AND KHOR, S.-L. "Reduction of Volatile Organic Compounds in Aqueous Solutions Through Air Stripping and Gas-Phase Carbon Adsorption." *Environmental Progress* 8(4): 270–278 (1989).

DE FILLIPPI, R. P., AND CHUNG, M. E. "Laboratory Evaluation of Critical Fluid Extractions for Environmental Applications," EPA/600/2–85/045 (1985).

FOX, C. R. "Plant Uses Prove Phenol Recovery with Resins." *Hydrocarbon Processing* 57(11): 269 (1978).

GLATER, J., MCCUTCHAN, J. W., MCCRAY, S. B., AND ZACHARIAH, M. R. "The Effects of Halogens on the Performance and Durability of Reverse-Osmosis Membranes." In *Synthetic Membranes: Volume I—Desalination.* A. F. Turbak [ed.], American Chemical Society Symposium Series 153 (1981).

GROVES, F. R., BRADY, B., AND KNOPF, F. C. *CRC Critical Reviews in Environmental Control* 15: 237–74 (1985).

HAND, D. W., CRITTENDEN, J. C., GEHIN, J. L., AND BENJAMIN W. LYKINS, J. "Design and Evaluation of an Air Stripping Tower for Removing VOC's from Groundwater." *J. AWWA* 78(9): 87–97 (1986).

HIMMELSTEIN, K. J., FOX, R. D., AND WINTER, T. H. "In Place Regeneration of Activated Carbon." *Chemical Engineering Progress* 69(11): 65 (1973).

HOLLAND, C. D., *Fundamentals of Multicomponent Distillation.* McGraw Hill, New York (1981).

HORNER, R. M. W., JARVIS, R. J., MACKIE, R. I. "Deep Bed Filtration: A New Look at the Basic Equations." *Water Research* 20(2): 215–20 (1986).

JOSHI, D. K., SENETAR, J. J., AND KING, C. J. *Industrial Engineering Chemistry Process Design and Development,* 23:748–54 (1984).

KAVANAUGH, M. C., AND TRUSSELL, R. R., "Design of Aeration Towers to Strip Volatile Contaminants from Drinking Water." *Journal of the American Water Works Association* 72: 684–92 (1980).

LEVA, M., "Flow Thru Irrigated Dumped Packings." *Chemical Engineering Progress Symposium Series,* 50(10): 51–9 (1954).

MALLEY, J. P., JR., ELIASON, P. A., AND WAGLER, J. L. "Point of Entry Treatment of

Petroleum Contaminated Water Supplies." *Water Environment Research* 65(2): 119–128 (1993).

McArdle, J. L., Arozarena, M. M., and Gallagher, W. E. "Treatment of Hazardous Waste Leachate: Unit Operations and Costs." U.S. EPA (1988).

McKance, K. E., and Burke, H. G. "Contaminated Condensate Stripping—an Industry Survey." *Pulp and Paper Canada* 81(11): 78–81 (1980).

Office of Technology Assessment, U.S. Congress. *Technologies and Management Strategies for Hazardous Waste Control.* Washington, D.C. (1983).

Ohya, H., Negishi, Y., Kamoto, K., Matsui, K., and Inoue, H. "Prediction Method for the Life of Reverse Osmosis Membranes." In *Synthetic Membranes: Volume I—Desalination.* A. F. Turbak [ed.], American Chemical Society Symposium Series 153 (1981).

O'Melia, C. R. "Particles, Pretreatment and Performance in Water Filtration." *Journal of Environmental Engineering* 111, 6: 874–90 (1985).

Ostendorf, D. W., and Botkin, B. C. "Sediment Diffusion in Primary Shallow Rectangular Clarifiers." *Journal of Environmental Engineering* 113(3): 595–611 (1987).

Peinemann, K.-V., Mohr, J. M., and Baker, R. W. "The Separation of Organic Vapors from Air." In *Recent Advances in Separation Techniques III.* N. N. Li [ed.], AIChE Symposium Series vol. 82 #250 (1986).

Price, P. E., Jr., and Danner, R. P. "Extension and Evaluation of the Minka and Myers Theory of Liquid Adsorption." *Industrial and Engineering Chemistry Research* 27: 506–12 (1988).

Roberts, P. V., Hopkins, G. D., Munz, C., and Riojas, A. H. "Evaluating Two Resistance Models for Air Stripping of Volatile Organic Contaminants in a Countercurrent, Packed Column." *Environmental Science and Technology* 19(2): 164–173 (1985).

Schweitzer, P. A. *Handbook of Separation Techniques for Chemical Engineers.* McGraw Hill, New York (1979).

Semmens, M. J., Qin, R., and Zander, A. "Using a Microporous Hollow Fiber Membrane to Separate VOC's from Water." *J. AWWA* 81(4): 162–167 (1989).

Sheindorf, C., Rebhun, M., and Sheintuch, M. "A Freundlich Type Multicomponent Isotherm." *Journal of Colloid and Interface Science* 79: 136 (1981).

Siler, J. L., and Bhattacharyya, D. "Low Pressure Reverse Osmosis Membranes: Concentration and Treatment of Hazardous Wastes." *Hazardous Waste and Hazardous Materials* 2(1): 45–65 (1985).

Staudinger, J., Knocke, W. R., and Randall, C. W. "Evaluating the Onda Mass Transfer Correlation for the Design of Packed Column Air Stripping." *J. AWWA* 82(1): 73–79 (1990).

Summers, R. S., and Roberts, P. V. "Rate of Humic Substance Uptake During Activated Carbon Adsorption." *Journal of Environmental Engineering* 113(6): 1333–49 (1987).

Tare, V., and Venkobachar, C. "New Conceptual Formulation for Predicting Filter Performance." *Environmental Science and Technology* 19(6): 497–99 (1985).

Taylor, J. S., Thompson, D. M., and Carswell, J. K. "Applying Membrane Processes to Groundwater Sources for Trihalomethane Precursor Control." *Journal of the American Water Works Association.* August, 72–82 (1987).

Treybal, R. E. *Liquid Extraction,* 2nd ed. McGraw-Hill, New York (1963).

U.S. EPA. Treatability Manual: Volume III. Technologies for Control/Removal of Pollutants. U.S. EPA 600-8-80-042c (1980).

VESILIND, P. A. *Treatment and Disposal of Wastewater Sludges.* Ann Arbor Science, Ann Arbor MI (1979).

YU, J., AND NERETNIEKS. I. "Single Component and Multicomponent Adsorption Equilibria on Activated Carbon of Methylcyclohexane, Toluene and Isobutyl Methyl Ketone." *Industrial and Engineering Chemistry Research* 29: 220–31 (1990).

ZANDER, A. K., QIN, R., AND SEMMENS, M. J. "Membane/Oil Stripping of VOC's from Water in Hollow Fiber Contactor." *Journal of Environmental Engineering* 115(4): 768–783 (1989).

5

Chemical Waste Treatment Processes

PROCESS CLASSIFICATION

Chemical treatment processes are those that are accomplished through means of some kind of chemical (as opposed to physical) reaction. The chemical reactions utilized may involve single or multiple phases. For the most part, the chemical processes discussed in this chapter are most amenable to treating soluble-phase contaminants, that is, for the treatment of wastewaters. However, mixtures of solids and liquids (sludges) containing hazardous constituents may also be amenable to some chemical treatment processes. Table 5-1 summarizes the common chemical-treatment processes that can be used to treat hazardous wastes.

NEUTRALIZATION

Introduction

Neutralization is a chemical treatment process that is used to treat acidic or basic wastes, as well as RCRA corrosive wastes. A waste that exhibits the characteristics of corrosivity, as defined in Title 40 of the Code of Federal Regulations (40 C.F.R.), section 261.22, is a D002 characteristic hazardous waste. Under 40 C.F.R., section 261.22, a solid waste is a characteristic corrosive hazardous waste if it is an aqueous waste with a pH of less than or equal to 2, or greater than or equal to 12.5, or if it is a liquid that corrodes steel at a rate greater than a designated corrosion rate. Some listed hazardous wastes, such as spent pickle liquor generated by steel finishing operations (K062), are also corrosive wastes, and require neutralization.

Neutralization of acidic wastes is achieved by reacting the solution with a base,

TABLE 5-1 COMMON CHEMICAL TREATMENT PROCESSES

Treatment process classification	Applications
Neutralization	Neutralization of acidic or basic liquids to minimize corrosivity.
Precipitation	Removal of dissolved heavy metals or other dissolved hazardous inorganic contaminants.
Oxidation/reduction	Oxidative destruction of cyanide or organic contaminants.
	Change of valency of ionic constituents to render them less toxic or amenable to other treatment processes.
	Reduction of dissolved heavy metals to metallic form, and subsequent physical separation from solution.
Ion exchange	Removal of dissolved ionic contaminants.
Stabilization	Stabilization of metal containing sludges by precipitation and encapsulation with cement.

thereby raising the pH to an acceptable level. Conversely, the neutralization of a basic or alkaline waste is achieved by reacting the solution with an adequate quantity of an acid to bring the solution pH to within the acceptable range.

Solution pH

Neutralization is a treatment process designed to change the pH from a value outside the acceptable range to a value that is within the acceptable range. The pH of a solution is defined as the negative common logarithm of the hydrogen ion activity. In dilute solutions, the hydrogen ion activity is closely approximated as the hydrogen ion concentration, and for the remainder of this discussion, it will be assumed the hydrogen ion activity and concentration are synonymous. At equilibrium, the hydroxide ion (OH^{-1}) concentration in aqueous solutions is directly related to the hydrogen ion concentration through the following reaction:

$$H_2O \rightleftharpoons H^+ + OH^- \tag{5.1}$$

Because the concentration of water in nearly all aqueous solutions is practically constant, the equilibrium expression for the dissociation of water into the hydrogen and hydroxide ions can be written as follows:

$$K_W = \{H^+\}\{OH^-\} \tag{5.2}$$

At 25°C, K_W is approximately equal to 1×10^{-14} (mole²/liter²). Knowledge of the pH of a solution at equilibrium therefore provides both the solution hydrogen ion concentration and the hydroxide ion concentration. For solutions where the hydroxide and hydrogen ion concentrations are identical (1×10^{-7} moles/liter at 25°C), the solution is said to be neutral, and the pH is, by definition, calculated to be 7. For all practical purposes, the pH of an aqueous solution can range from near 0 to 14. Solutions with a pH less than the neutral point (pH = 7, at 25°C) are said to be acidic, and have an excess of H^+ ions with respect to OH^- ions, while solutions with a pH value greater than the neutral point are said to be basic, having an excess of OH^- ions as compared to H^+ ions.

Acids and Bases

Various theories have been proposed to model acid and base behavior. The Brönsted–Lowery model defines an acid as a substance which can donate a proton to any other substance, while a base is a substance which can accept a proton. An example of an acid is HCl (hydrochloric acid) which in aqueous solution freely dissociates, donating a proton (H^+) to the solution, as follows:

$$HCl \rightleftharpoons H^+ + Cl^- \tag{5.3}$$

Although the hydrogen ion (proton) is often described to exist in solution as the H^+ ion, free protons cannot exist as a bare ion in solution, and associate with water molecules through hydrogen bonding to form the hydronium ion (H_3O^+) or even higher-order species (for example, $H_7O_3^+$). However, for purposes of discussion, protonated water molecules will be described as H_3O^+. In light of the Brönsted–Lowery model of acids and bases, the dissociation of hydrochloric acid is more appropriately described as:

$$HCl + H_2O \rightleftharpoons H_3O^+ + Cl^- \tag{5.4}$$

where hydrochloric acid donates a proton to the water molecule. Because free protons do not exist in aqueous solutions, the Brönsted–Lowery model requires that all acid-base reactions must occur in pairs, with one specie serving as the acid, and one specie serving as the base. In the dissociation of hydrochloric acid, HCl is the acid, donating a proton to a water molecule, which serves as the base by accepting the proton. The net reaction of an acid with water is to increase the concentration of H_3O^+ (and decrease the concentration of OH^-, thus decreasing the pH).

An example of a reaction of a base with water is the formation of the ammonium ion by reaction of ammonia with water:

$$NH_3 + H_2O \rightleftharpoons NH_4^+ + OH^- \tag{5.5}$$

In the reaction of ammonia with water, water serves as the acid by donating a proton to ammonia, the base. The net effect of the reaction of a base with water is to increase the concentration of hydroxide ions in solution. Because at equilibrium the concentration of hydroxide and hydrogen (hydronium) ions are inversely related through the dissociation constant of water, the net effect of the reaction of a base with water is to decrease the concentration of hydrogen ions, thus increasing the pH.

Acids and bases are categorized as strong or weak depending upon the extent with which they will donate or accept protons. The degree to which a substance dissociates to donate protons, or associates to accept protons, is quantified by an equilibrium constant for the reaction. Consider the general dissociation reaction of an acid:

$$HB \rightleftharpoons H^+ + B^- \tag{5.6}$$

Note that for simplicity, the dissociation reaction has been written to produce the H^+ ion, not the H_3O^+ ion. This simplifying convention will be used in the remaining discussion unless otherwise noted. The equilibrium constant, which is known as the acid dissociation constant, can be written for the above reaction as:

$$K_{HB} = \frac{[H^+][B^-]}{[HB]} \quad \text{and} \quad pH = pK_{HB} + \log \frac{[B^-]}{[HB]} \tag{5.7}$$

where pK_{HB} is defined as the negative logarithm of K_{HB}.

Acids which nearly completely dissociate in water are termed strong acids, and are characterized by pK_{HB}'s which are less than zero. Examples of strong acids are hydrochloric acid, perchloric acid, nitric acid, and sulfuric acid. If 10^{-2} moles of a strong acid are added to 1 liter of pure water, the equilibrium H^+ concentration will be very nearly 10^{-2} moles/liter, the pH will be approximately 2, and virtually none of the added acid will remain associated in the solution. Weak acids do not completely dissociate in water, and are characterized by pK_{HB}'s that are greater than zero. Examples of weak acids are phosphoric acid and hypochlorous acid.

Example 5-1

a) 10^{-3} moles of hydrochloric acid (HCl) are added to one liter of pure water at 25°C. Calculate the resultant equilibrium pH.

b) 10^{-3} moles of hypochlorous acid (HOCl) are added to one liter of pure water at 25°C. Calculate the resultant equilibrium pH.

Answer

a) Hydrochloric acid: Because virtually all of the hydrochloric acid will dissociate, forming 10^{-3} moles/liter of H^+ ions, the equilibrium pH will be 3.

b) Hypochlorous acid: The pK_{HB} for hypochlorous acid (HOCl) is 7.5 at 25°C. The following set of equations can be derived and solved to determine the equilibrium concentrations of H^+, OH^-, HOCl, and OCl^-. (For further details on derivation and solution of aqueous equilibria systems see Stumm and Morgan, 1981; Snoeyink and Jenkins, 1980; or Butler, 1964.)

1. $\dfrac{[H^+][OCl^-]}{[HOCl]} = 10^{-7.5} \dfrac{\text{moles}}{\text{liter}}$ (equilibrium expression)
2. $[HOCl] + [OCl^-] = 10^{-3} \dfrac{\text{moles}}{\text{liter}}$ (mass balance on added acid)
3. $[H^+] = [OH^-] + [OCl^-]$ (electroneutrality condition)
4. $[H^+][OH^-] = 10^{-14} \dfrac{\text{moles}^2}{\text{liter}^2}$ (K_W at 25°C)

Solving this set of equations gives:

$$[H^+] = 5.61 \times 10^{-6} \text{ moles/liter}$$
$$[OH^-] = 1.78 \times 10^{-9} \text{ moles/liter}$$
$$[HOCl] = 9.94 \times 10^{-4} \text{ moles/liter}$$
$$[OCl^-] = 5.61 \times 10^{-6} \text{ moles/liter}$$
$$pH = 5.25$$

Note that although the hypochlorous acid did reduce the pH from a neutral value of 7 to 5.25, only 0.6% of the hypochlorous acid originally added dissociated.

Strong bases dissociate stoichiometrically in water to form hydroxide ions. An example of a strong base is sodium hydroxide. Sodium hydroxide completely dissociates in water, leaving an added hydroxide ion available to serve as a proton acceptor. In contrast, weak bases do not stoichiometrically produce hydroxide. An example of a weak base is the acetate ion (Ac^-). Consider the addition of sodium acetate to pure water. The sodium completely dissociates from the acetate ion in solution, and the acid-base reaction of interest becomes:

$$Ac^- + H_2O \rightleftharpoons HAc + OH^- \tag{5.8}$$

If acetate were a strong base, the above reaction would proceed to completion, leaving virtually no free Ac^- in solution at equilibrium. However, because the acetate ion is a weak base, the equilibrium solution will contain both Ac^- and HAc, and the equilibrium distribution can be described through the basicity constant of the reaction (K_B) defined as:

$$K_B = K_{Ac} = \frac{[HAc][OH^-]}{[Ac^-]} \tag{5.9}$$

Note that the acid dissociation constant and the basicity constant are related through the dissociation constant for water:

$$K_W = K_{HB}K_B = 1 \times 10^{-14} \tag{5.10}$$

For example, the acidity constant at 25°C for acetic acid (HAc) is $10^{-4.7}$, and the basicity constant at 25°C for the acetate ion (Ac^-) is $10^{-9.3}$ (Stumm and Morgan, 1981), with the product of the two constants equal to K_W (10^{-14}).

Strong bases are characterized by relatively small pK_B's (large pK_{HB}'s) while weak bases are characterized by relatively large pK_B's. Table 5-2 provides a list of acidity and basicity constants at 25°C for a number of common acids and bases (Snoeyink and Jenkins, 1980).

Another approach to defining acids and bases was presented by Lewis, who defined an acid as a compound that can accept and share a lone pair of electrons donated by a Lewis base. Because the hydrogen ion accepts lone electron pairs, a Brönsted–Lowery acid or base is also a Lewis acid or base. However, there are Lewis acids or bases that are not Brönsted–Lowery acids or bases. For example, heavy metal ions accept electron pairs through hydrolysis reactions to form metal-hydroxy complexes, and are therefore Lewis acids. However, the heavy metal ion undergoing hydrolysis would not be a Brönsted–Lowery acid because it does not donate a proton in the hydrolysis reaction. Nevertheless, addition of a metal ion that undergoes hydrolysis to a solution will tend to decrease the pH of the solution like any acid. The complexation of the metal ion with hydroxide ions will reduce the equilibrium hydroxide ion concentration, thereby decreasing the solution pH. Other examples of Lewis acids include $SOCl_2$, $AlCl_3$, SO_2, and BF_3.

Acidity, Basicity, and Buffering

The measurement of the pH of a solution provides the concentration of hydrogen and hydroxide ions in the solution. However, except in the simple case where a

TABLE 5-2 ACIDITY AND BASICITY CONSTANTS FOR SUBSTANCES IN AQUEOUS SOLUTION AT 25°C

Acid		$-\log K_{HB} = pK^{HB}$	Conjugate base		$-\log K_B = pK_B$
$HClO_4$	Perchloric acid	−7	ClO_4^-	Perchlorate ion	21
HCl	Hydrochloric acid	≈−3	Cl^-	Chloride ion	17
H_2SO_4	Sulfuric acid	≈−3	HSO_4^-	Bisulfate ion	17
HNO_3	Nitric acid	0	NO_3^-	Nitrate ion	14
H_3O^+	Hydronium ion	0	H_2O	Water	14
HIO_3	Iodic acid	0.8	IO_3^-	Iodate ion	13.2
HSO_4^-	Bisulfate ion	2	SO_4^{2-}	Sulfate ion	12
H_3PO_4	Phosphoric acid	2.1	$H_2PO_4^-$	Dihydrogen phosphate ion	11.9
$Fe(H_2O)_6^{3+}$	Ferric ion	2.2	$Fe(H_2O)_5OH^{2+}$	Hydroxo iron (III) complex	11.8
HF	Hydrofluoric acid	3.2	F^-	Fluoride ion	10.8
HNO_2	Nitrous acid	4.5	NO_2^-	Nitrate ion	9.5
CH_3COOH	Acetic acid	4.7	CH_3COO^-	Acetate ion	9.3
$Al(H_2O)_6^{3+}$	Aluminum ion	4.9	$Al(H_2O)_5OH^{2+}$	Hydroxo aluminum (III) complex	9.1
$H_2CO_3^*$	Carbon dioxide and carbonic acid	6.3	HCO_3^-	Bicarbonate ion	7.7
H_2S	Hydrogen sulfide	7.1	HS^-	Bisulfide ion	6.9
$H_2PO_4^-$	Dihydrogen phosphate	7.2	HPO_4^{2-}	Monohydrogen phosphate ion	6.8
$HOCl$	Hypochlorous acid	7.5	OCl^-	Hypochlorite ion	6.4
HCN	Hydrocyanic acid	9.3	CN^-	Cyanide ion	4.7
H_3BO_3	Boric acid	9.3	$B(OH)_4^-$	Borate ion	4.7
NH_4^+	Ammonium ion	9.3	NH_3	Ammonia	4.7
H_4SiO_4	Orthosilicic acid	9.5	$H_3SiO_4^-$	Trihydrogen silicate ion	4.5
C_6H_5OH	Phenol	9.9	$C_6H_5O^-$	Phenolate ion	4.1
HCO_3^-	Bicarbonate ion	10.3	CO_3^{2-}	Carbonate ion	3.7
HPO_4^{2-}	Monohydrogen phosphate	12.3	PO_4^{3-}	Phosphate ion	1.7
$H_3SiO_4^-$	Trihydrogen silicate	12.6	$H_2SiO_4^{2-}$	Dihydrogen silicate ion	1.4
HS^-	Bisulfide ion	14	S^{2-}	Sulfide ion	0
H_2O	Water	14	OH^-	Hydroxide ion	0
NH_3	Ammonia	≈23	NH_2^-	Amide ion	−9
OH^-	Hydroxide ion	≈24	O^{2-}	Oxide ion	−10

From V. Snoeyink and D. Jenkins, *Water Chemistry,* Copyright 1980 by John Wiley & Sons, Inc. Reprinted by permission of John Wiley & Sons, Inc.

solution contains only strong acids or bases, measurement of the pH of the solution does not indicate the quantity of acid or base in the solution. If a solution contains a weak acid, which can serve as a continued source of hydrogen ions, or a weak base, which can provide a continued sink for hydrogen ions, the measurement of pH is not sufficient to determine the quantity of acid or base present. The true measurement of the quantity of acidic materials present in a solution at a given pH is the acidity, while the true measurement of the quantity of basic materials present in solution at a given pH is the alkalinity. To design neutralization processes, it is therefore necessary to determine the acidity or alkalinity of the waste to be treated.

Acidity, or base-neutralizing capacity, is generally defined as the quantitative capacity of a solution to neutralize a strong base to a designated reference point. Often, the reference point is taken to be a pH of 8.3, because this is the pH of color change of the common indicator phenolphthalein. Operationally, the reference point can simply be the desired effluent pH. Thus, if an acid waste stream is to be neutralized to a pH of 7, the reference point can be taken as pH = 7. Potential sources of acidity include strong mineral acids (nitric, sulfuric, etc.), weak acids (acetic, hypochlorous, etc.), and dissolved carbon dioxide (carbonic acid). In some cases, dissolved metals that complex with the hydroxide ion will serve as acids. Acidity is generally measured in units of (proton) equivalents/liter or milliequivalents/liter (meq/l).

Basicity, also referred to as acid-neutralizing capacity, or alkalinity, is defined as the quantitative capacity of a solution to neutralize a strong acid to a designated reference point. The alkalinity reference point is often taken to be a pH of 4.3, which is the pH of color change of the indicator methyl orange. Again, the reference point is often operationally defined to be the desired effluent pH of the neutralized waste stream. Potential sources of alkalinity include dissolved strong and weak bases, as well as dissolved carbonate and bicarbonate present in solution from the equilibration of the solution with atmospheric carbon dioxide. Because in natural waters and domestic wastewaters much of a solution's alkalinity is due to carbonate species, alkalinity is often expressed in units of mg/l as $CaCO_3$, which is calculated by multiplying the alkalinity in meq/l by 50 mg/meq of $CaCO_3$.

Acidity and basicity are generally determined by titrations of the waste using standard neutralizing agents until the pH of the solution is neutralized to a predetermined pH value. Such titrations are necessary to design the neutralization treatment process. The titration results are a plot of solution pH as a function of the volume (or equivalents) of neutralizing agent added. Figures 5-1 and 5-2 provide generic titration curves for the neutralization of a strong acid with a strong base and a weak base, respectively. Titrations of a strong base with a strong acid or a weak acid can be similarly developed. Note that because the acid being neutralized is a strong acid, the pH changes rapidly near the equivalence point due to the lack of buffering capacity in the solution. This results in a neutralization process that is very difficult to control because very small dosages of neutralizing agent result in dramatic changes in pH. Comparison of the two figures reveals that using a weak base as the neutralizing agent results in a less drastic rate of pH change near the equivalence point. Titrations of solutions containing weak acids or bases result in titration curves that exhibit much less drastic rates of change of pH with added titrant than do titrations of strong acids or bases. Because of this, neutralization of solutions containing weak acids or bases is much easier to control than neutralization processes of strong acids or bases. Titrations of basic solutions are generally done using a solution of sulfuric acid as the titrant, while titrations of acidic solutions are generally done using a solution of sodium hydroxide as the titrant.

Example 5-2

a) 500 ml of an acidic solution is titrated with a strong base to increase the pH from an initial value of 3.9 to a final pH of 8.3. If 5 ml of a 4 N solution of NaOH is the titrant dosage, calculate the acidity in units of meq/l.

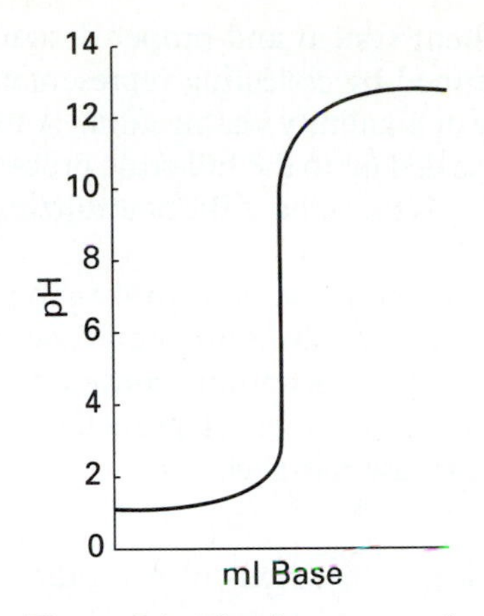

Figure 5-1 Titration curve of strong acid titrated with strong base. (Water Pollution Control Federation; *Pretreatment of Industrial Wastes,* 1981, p. 77. Reprinted by Permission of the Water Environment Federation.)

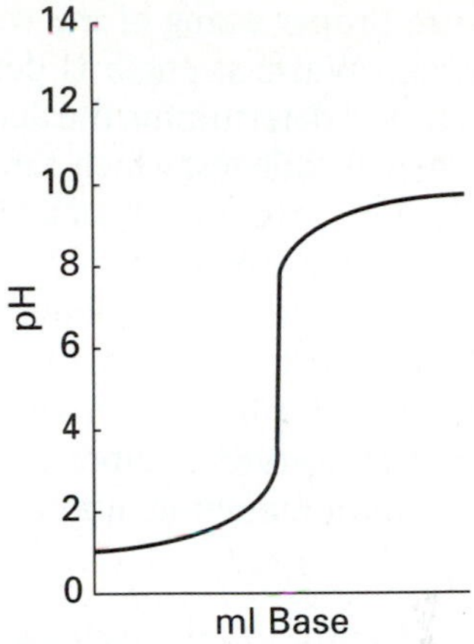

Figure 5-2 Titration curve of strong acid titrated with weak base. (Water Pollution Control Federation; *Pretreatment of Industrial Wastes,* 1981, p. 78. Reprinted by Permission of the Water Environment Federation.)

Answer

$$\text{Acidity} = \left(\frac{5 \text{ ml titrant}}{500 \text{ ml solution}}\right)\left(\frac{4 \text{ eq NaOH}}{1000 \text{ ml titrant}}\right)\left(\frac{1000 \text{ ml solution}}{1 \text{ liter solution}}\right)\left(\frac{1000 \text{ meq}}{1 \text{ eq}}\right)$$

$$= \frac{40 \text{ meq}}{\text{liter}}$$

b) 200 ml of a basic solution is titrated from a pH of 10.9 to a pH of 4.4 by the addition of 4 ml of a 1 N solution of HCl. Determine the alkalinity in units of meq/l and mg/l $CaCO_3$.

Answer

$$\text{Alkalinity} = \left(\frac{4 \text{ ml titrant}}{200 \text{ ml solution}}\right)\left(\frac{1 \text{ eq HCl}}{1000 \text{ ml titrant}}\right)\left(\frac{1000 \text{ ml solution}}{1 \text{ liter solution}}\right)\left(\frac{1000 \text{ meq}}{1 \text{ eq}}\right)$$

$$= \frac{20 \text{ meq}}{\text{liter}} = \left(\frac{20 \text{ meq}}{\text{liter}}\right)\left(\frac{50 \text{ mg CaCO}_3}{\text{meq CaCO}_3}\right) = \frac{1000 \text{ mg CaCO}_3}{\text{liter}}$$

Neutralization Processes

Neutralization can be done on a batch basis, or as a continuous-treatment process. When the corrosive waste is produced intermittently, or if the rate of waste treatment is low, neutralization can be designed as a batch treatment process. Otherwise, the neutralization process must be designed as a continuous-treatment process.

Waste characterization is an important first step in the design of any neutralization process. Both flow rate and waste strength variation must be considered to insure proper sizing of the treatment system and proper design of the pH control system. Waste strength is determined by collecting representative samples of the waste and determining the acidity or alkalinity via titration. A titration can serve as the bench scale test which can be scaled up to the full-scale process if the titrant used to determine the acidity or alkalinity is the same as the neutralizing agent to be used in the full-scale process.

Pretreatment is sometimes needed prior to neutralization. Pretreatment can include filtration, sedimentation, and equalization. Equalization is generally recommended from both process control and economic considerations (Wilk, Palmer, and Breton, 1988). Other common pretreatment steps include cyanide destruction, chromium reduction, and oil and grease removal.

Acidic waste neutralization. Methods of neutralizing acidic wastes include:

- Adding appropriate amounts of strong or weak base to the waste
- Passing acidic waste through limestone beds
- Mixing acidic waste with lime or dolomite lime slurries
- Mixing the acidic waste with a compatible alkaline waste

Caustic agents typically used to neutralize acidic waste are sodium hydroxide (caustic soda), sodium carbonate (soda ash), ammonia, and various limes. The choice of neutralizing agent requires both the consideration of operational issues and economics. Cost comparisons of commercial caustic agents are provided in Table 5-3 (Patterson, 1985). For wastes with mineral acid acidity greater than 5,000 mg/l, high calcium lime or caustic soda are generally used, while for more dilute acid

TABLE 5-3 COST COMPARISON OF COMMERCIAL CAUSTIC AGENTS

Agent	Chemical formula	Basicity factor[a]	Relative treatment cost[b]
Sodium hydroxide	$NaOH$ (50%)	0.687	6.24
Sodium carbonate	Na_2CO_3	0.507	10.98
High-calcium hydrate	$Ca(OH)_2$	0.710	1.86
Dolomite hydrate	$Ca(OH)_2 \cdot MgO$	0.912	1.18
High-calcium quicklime	CaO	0.941	1.00
Dolomite quicklime	$CaO \cdot MgO$	1.110	0.76
High-calcium limestone	$CaCO_3$	0.489	1.40
Dolomite limestone	$CaCO_3 \cdot MgCO_3$	0.564	1.05

[a] A measure of the alkali available in the commercial product for neutralization based upon grams equivalent of CaO per gram reagent.

[b] Based upon one unit of cost for CaO and the treatment effectiveness (i.e., "basicity factor") of each reagent.

From J. W. Patterson, *Industrial Wastewater Treatment Technology,* 2nd ed., p. 365, 1985. Reprinted by permission of Butterworth Heinemann.

wastes, limestone treatment may also be economically feasible (Camp, Dresser, and McKee, 1984).

Limestone ($CaCO_3$) neutralization is commonly used because of limestone's availability and relatively low unit chemical costs. However, this is offset by the low reactivity of lime which results in long treatment residence time requirements (45 minutes or greater), and the large volumes of sludge typically produced when neutralizing with lime. Also, pH values above neutrality cannot be achieved using limestone. Limestone neutralization is generally done by passing the waste through a bed of limestone. Either upflow or downflow systems can be used, but the upflow systems provide better flow distribution. Figure 5-3 provides a flow schematic of an upflow limestone bed neutralization process (WPCF, 1981). Design criteria include a limestone bed depth of 3 ft to 8 ft, and hydraulic loading rates from 1.3 gal/min · ft^2 to 30 gal/min · ft^2 (WPCF, 1981). When treating concentrated acidic wastes, the limestone particles can become coated with precipitate, rendering the particles inactive and adding to sludge production. Because of this, it is recommended that stone diameters of less than 0.074 mm be used in limestone bed neutralization. Furthermore, limestone is not recommended for acidic wastes with mineral acid strengths greater than 5,000 mg/l (Wilk, Palmer, and Breton, 1988). The presence of sulfate in acidic wastes results in a voluminous sludge (gypsum) when limestone is used as the neutralizing agent. Sludge production in limestone beds can cause serious head loss problems. An additional difficulty in limestone bed treatment is that carbon dioxide gas produced in the neutralization can gas-bind the beds.

Acidic wastes can also be neutralized by the addition of lime slurries to the waste. Lime slurries are produced by reacting lime (CaO) with water in a process known as slaking. The slaked lime ($Ca(OH)_2$) slurry is typically produced at a solids

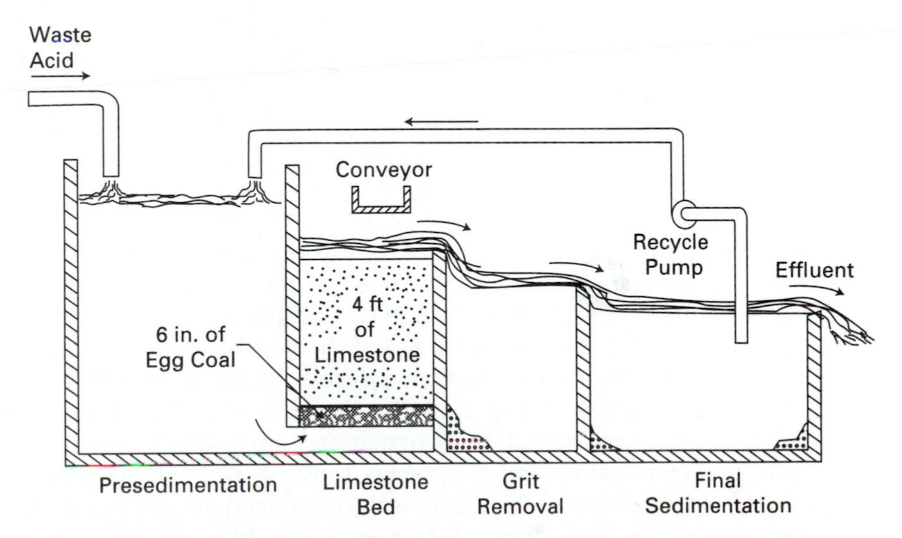

Figure 5-3 Simplified flow diagram of limestone neutralization system. (Water Pollution Control Federation; *Pretreatment of Industrial Wastes,* 1981, p. 86. Reprinted by Permission of the Water Environment Federation.

concentration of 10% to 35%. Acid neutralization using lime slurries requires 15 to 30 minute retention times (Wilk, Palmer, and Breton, 1988). A major disadvantage of using any form of lime is that calcium forms insoluble salts, especially when the treated acidic waste contains sulfate. These insoluble salts can coat pH electrodes, valves, pipes, and pumps causing serious maintenance problems. Also, handling of lime is a nuisance due to the formation of airborne lime dust. An advantage of using lime instead of limestone is that lime is more versatile in terms of neutralization endpoints, with the achievable effluent pH ranging from 6.0 to 12.4.

Although liquid sodium hydroxide is more expensive than many of the other commonly used basic neutralizing agents, it is often used because of its ease of storage and delivery (low equipment capital costs), rapid reaction rate, uniformity of composition, and relatively low volumes of sludge produced. Because of its high cost, sodium hydroxide is often not economical for use in high volume applications. Acid neutralization using sodium hydroxide can be achieved with retention times from 3 to 15 minutes (Wilk, Palmer, and Breton, 1988). This corresponds to smaller feed systems and tankage requirements as compared to lime neutralization. Also, an advantage of sodium hydroxide is that (unlike neutralization with lime) the effluent generally does not contain any sludges. A potential disadvantage of using sodium hydroxide is that because it is a strong base, it does not provide any buffering capacity to a solution. Therefore, using sodium hydroxide to neutralize a strong acid solution would be very difficult to control. Acidic wastes can be overneutralized ($pH > 13$) with caustic soda if the system controls malfunction. Caustic soda is available in both solid and liquid form, although the liquid is generally used because of its ease of handling. Concentrated liquid sodium hydroxide is purchased as a 50% solution which freezes at temperatures below 50° F. To avoid freezing, the storage area and supply lines must be insulated or heated when temperatures are expected to drop below the freezing point. Alternatively, the concentrated caustic can be diluted and stored for use as a 20% solution. Another disadvantage of caustic soda is that it can produce fumes which can burn unprotected skin.

Sodium carbonate is safer than caustic soda, but is less reactive and more expensive to use than caustic soda. Sodium carbonate is generally fed as a slurry because of its low solubility in water. An advantage of using sodium carbonate is that the carbonate provides some buffering capacity to the waste, thereby easing process control. However, the evolution of CO_2 gas can cause foaming problems. Ammonia neutralization has some advantages including ease of handling, imparted buffering capacity, and high reactivity and basicity. However, the resultant neutralized waste may contain objectionable levels of dissolved nitrogen compounds.

Mutual neutralization of acidic and alkaline wastes can be the most economical method of neutralization when compatible wastes are readily available (Camp, Dresser, and McKee, 1984). However, because of flow and waste strength variability, mutual neutralization generally requires large equalization capacities. Also, supplemental neutralizing agents must be available for situations when the acidity and alkalinity of the wastes to be combined are not equal. Mixing alkaline wastes together with metal containing acidic wastes can produce heavy metal hydroxide sludges. Wastes containing cyanide are not amenable to mutual neutralization processes because of the potential for evolution of deadly hydrogen cyanide.

Alkaline waste neutralization. Methods of neutralizing alkaline wastes include:

- Adding appropriate amounts of strong or weak acid to the waste
- Adding compressed carbon dioxide gas to the waste
- Blowing flue gas through the waste
- Mixing the alkaline waste with a compatible acidic waste (already discussed)

The most common method of neutralizing alkaline wastes is by the addition of mineral acids. The two most common acids used to neutralize alkaline wastes are sulfuric and hydrochloric acid. Sulfuric acid is more commonly used because of its lower cost. Neutralization residence times of 15 to 30 minutes are recommended when using sulfuric acid (Wilk, Palmer, and Breton, 1988). A disadvantage of using sulfuric acid is that it will form sludges when reacted with calcium-containing alkaline wastes. Hydrochloric acid is also a commonly used neutralizing agent. Although more expensive than sulfuric acid, hydrochloric acid neutralization will not result in sludge production when neutralizing calcium containing alkaline wastes. Neutralization reaction residence times of 5 to 20 minutes are recommended when using hydrochloric acid as the neutralizing agent (Wilk, Palmer, and Breton, 1988). A disadvantage of using hydrochloric acid is that it can form a highly corrosive acid mist upon reaction. Nitric acid may also be used, but its use is generally prohibited by high costs. The relative costs of mineral acids commonly used to neutralize alkaline wastes are provided in Table 5-4 (Patterson, 1985).

Carbon dioxide can also be used to neutralize alkaline wastes. Carbon dioxide reacts with water to form the weak acid H_2CO_3 (carbonic acid). Reactor retention times of 1 to 1.5 minutes are recommended for carbon dioxide neutralization (Wilk, Palmer, and Breton, 1988). The principal disadvantage of using carbon dioxide is its high cost (approximately 3 to 4 times more expensive than sulfuric acid; Wilk, Palmer, and Breton, 1988). An additional disadvantage of neutralization using carbon dioxide is that the added carbonate can form precipitates with dissolved calcium. However, carbon dioxide neutralization is easy to control because the added carbonate buffers the neutralization process. Carbon dioxide present in flue gas (typically 14%) can be used to neutralize alkaline wastes. The process is economical when an adequate supply of flue gas is available (Evans and Wilson, 1972).

TABLE 5-4 COMPARISON OF COSTS (1974–1975) OF NEUTRALIZING CHEMICALS FOR ALKALINE WASTES

Chemical	Relative cost per ton	lb required/100 lb waste alkali		
		$CaCO_3$	$Ca(OH)_2$	NaOH
Sulfuric acid (50° Be)[1]	1.0	126	170	158
Hydrochloric acid (20° Be)	2.6	232	314	250
Nitric acid (39.5° Be)	3.5	210	284	263

[1] °Be (degrees Baumé) is a measurement of specific gravity. For bulk acids, this measurement provides a convenient measurement of concentration (Perry et al., 1984).

From J. W. Patterson, *Industrial Wastewater Treatment Technology,* 2nd ed., 1985, p. 363. Reprinted by permission of Butterworth Heinemann.

Neutralization Process Design and Control

Neutralization process control systems generally consist of a reagent feed system, a pH monitoring system, and a controller that regulates the reagent feed based upon the measured pH and the desired effluent pH. The control of the neutralization process can be difficult because pH is a logarithmic function of the hydrogen ion concentration. Near the equivalence point, rapid changes in pH, and overneutralization can occur if the control system is not adequately sensitive. This is especially true if the waste has little or no buffering capacity.

Some examples of neutralization process control systems for continuous-neutralization processes include simple on/off reagent feed systems, multi-mode reagent feed systems, and multistage neutralization systems (WPCF, 1981). With the simple on/off reagent feed systems, the reagent is fed to the waste at a constant rate until the reactor pH reaches the desired set point. The reagent feed is disengaged until the pH reaches a second set point which re-engages the reagent feed. This form of control is known as feedback control. For instance, in an acid-neutralization process, sodium hydroxide may be added until the pH reaches 9.0, whereupon the reagent feed is turned off until the pH drops down below 6.0. Figure 5-4 provides a schematic of a simple on/off pH control system. Successful process control using the simple on/off control scheme generally requires reactor retention times greater than 10 minutes (WPCF, 1981). Improvement in the control can be achieved by using two separate reagent feed valves. A large feed valve may be used for the gross reagent feed, and a smaller valve for reagent feed near the endpoint of neutralization.

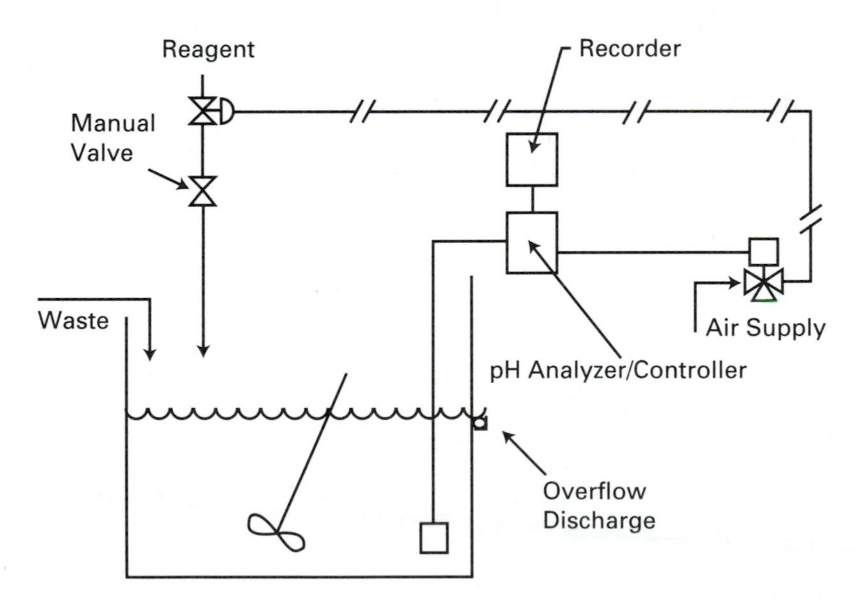

Figure 5-4 Simple on-off control system. (Water Pollution Control Federation; *Pretreatment of Industrial Wastes,* 1981, p. 80. Reprinted by Permission of the Water Environment Federation.)

More precise process control can be obtained using a multimode neutralization process control system. With the multimode control system, the rate of reagent feed is proportional to the magnitude of the difference between the measured pH and the desired effluent pH. Precise neutralization-process control can also be obtained using a two stage neutralization process (Figure 5-5). The first stage of the process performs the gross pH adjustment, while the fine pH adjustment near the endpoint of the reaction is performed in the second stage. Process control can also be performed using a combination of feed-forward and feedback control. In the feed-forward control loop, the rate of reagent feed is controlled by such parameters as influent pH and flowrate. The final pH adjustment is then made using a feedback control loop based upon the pH measured within the neutralization tank.

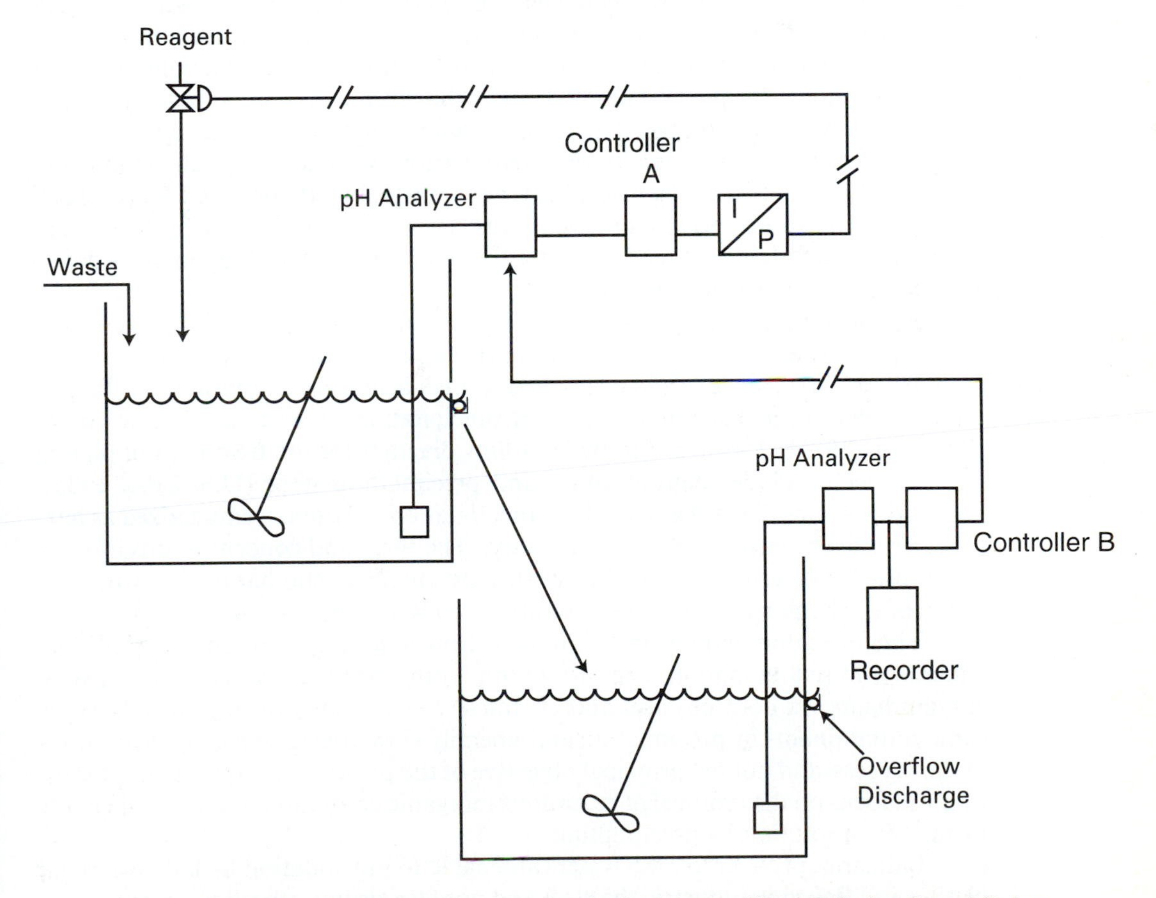

Figure 5-5 Cascade control system-two feedback loops. (Water Pollution Control Federation; *Pretreatment of Industrial Wastes,* 1981, p. 83. Reprinted by Permission of the Water Environment Federation.)

PRECIPITATION

Introduction

Chemical precipitation is the process whereby soluble phase species are removed from solution by adding a precipitant, which results in the formation of an insoluble compound containing the contaminant. The precipitate can then be separated from the wastewater using some physical separation process, such as sedimentation or filtration. Frequently, coagulants or flocculants are added with the precipitant to enhance the separation of the precipitate from the remaining soluble phase. A generalized schematic of a typical precipitation process is provided in Figure 5-6. The unit operations of sedimentation and sludge thickening were already covered in Chapter 4 of this text.

As indicated in Figure 5-6, a precipitation process results in the formation of at least two process streams; the treated effluent, and a sludge containing most of the contaminants originally present in the solution. Properly designed and implemented, the volume of the sludge will be significantly less than the volume of the original solution. The precipitation process is therefore a volume-reduction process, and not a destruction process. The resultant sludge may still contain up to 99% by weight of water and is generally dewatered before ultimate disposal. Often, the dewatered sludges are characteristic hazardous wastes based upon TCLP criteria or are listed hazardous wastes by the industrial process from which they were generated. Such sludges must be further treated prior to landfilling in accordance with current land disposal restrictions.

Applications of precipitation generally focus on the removal of dissolved inorganic ions. Typical examples where precipitation is used in hazardous waste treatment include: hydroxide or sulfide precipitation of heavy metals in solution, precipitation of barium as barium sulfate, and precipitation of silver as silver chloride (Patterson, 1985). Figure 5-7 provides a flow diagram for the treatment of plating wastes, which includes neutralization and precipitation steps (U.S. EPA, 1985). Hazardous wastes that are amenable to precipitation are often characterized as hazardous wastes by virtue of their leachability. The threshold concentrations of inorganic ions for determining whether wastes are characteristic hazardous wastes by virtue of leachability are provided in Table 5-5 (40 C.F.R. 261.24).

The formation and settling of inorganic precipitates (commonly termed flocculation and sedimentation) can also entrap both dissolved and colloidal organic contaminants via both physical and chemical mechanisms. Such removals of organic contaminants by precipitation are generally viewed as beneficial side-reactions of the process, and not the principal objective of the process. Therefore, this discussion will focus on the removal of hazardous inorganic contaminants, and principally metals, from solution by precipitation.

Industries producing wastewaters amenable to precipitation include the metal plating and finishing industry, the steel and nonferrous industry, the inorganic pigments industry, the mining industry, and the electronics industry. Landfill leachate and contaminated groundwaters can also contain hazardous species which are amenable to chemical precipitation.

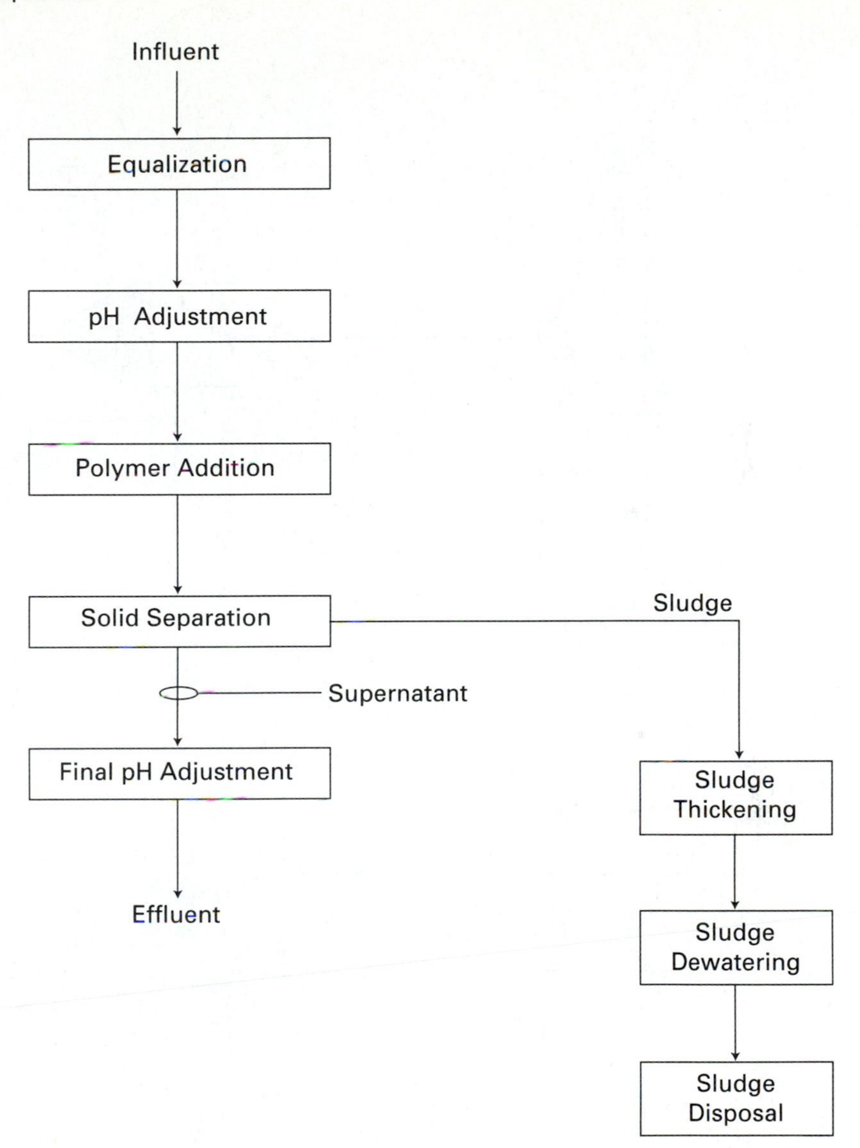

Figure 5-6 Generalized schematic of precipitation process. (Water Pollution Control Federation; *Pretreatment of Industrial Wastes,* 1981, p. 78. Reprinted by Permission of the Water Environment Federation.)

Heterogeneous equilibria. The precipitation process is most simply described by considering the equilibrium between a precipitate and its constituent components in the dissolved phase. Consider the precipitation reaction of the hypothetical cation A^{+n} with the anion B^{-m} to form the precipitate A_mB_n where n and m are the valencies of the cation and anion, respectively.

$$mA^{+n}_{(aq)} + nB^{-m}_{(aq)} \rightleftharpoons A_mB_{n(s)} \tag{5.11}$$

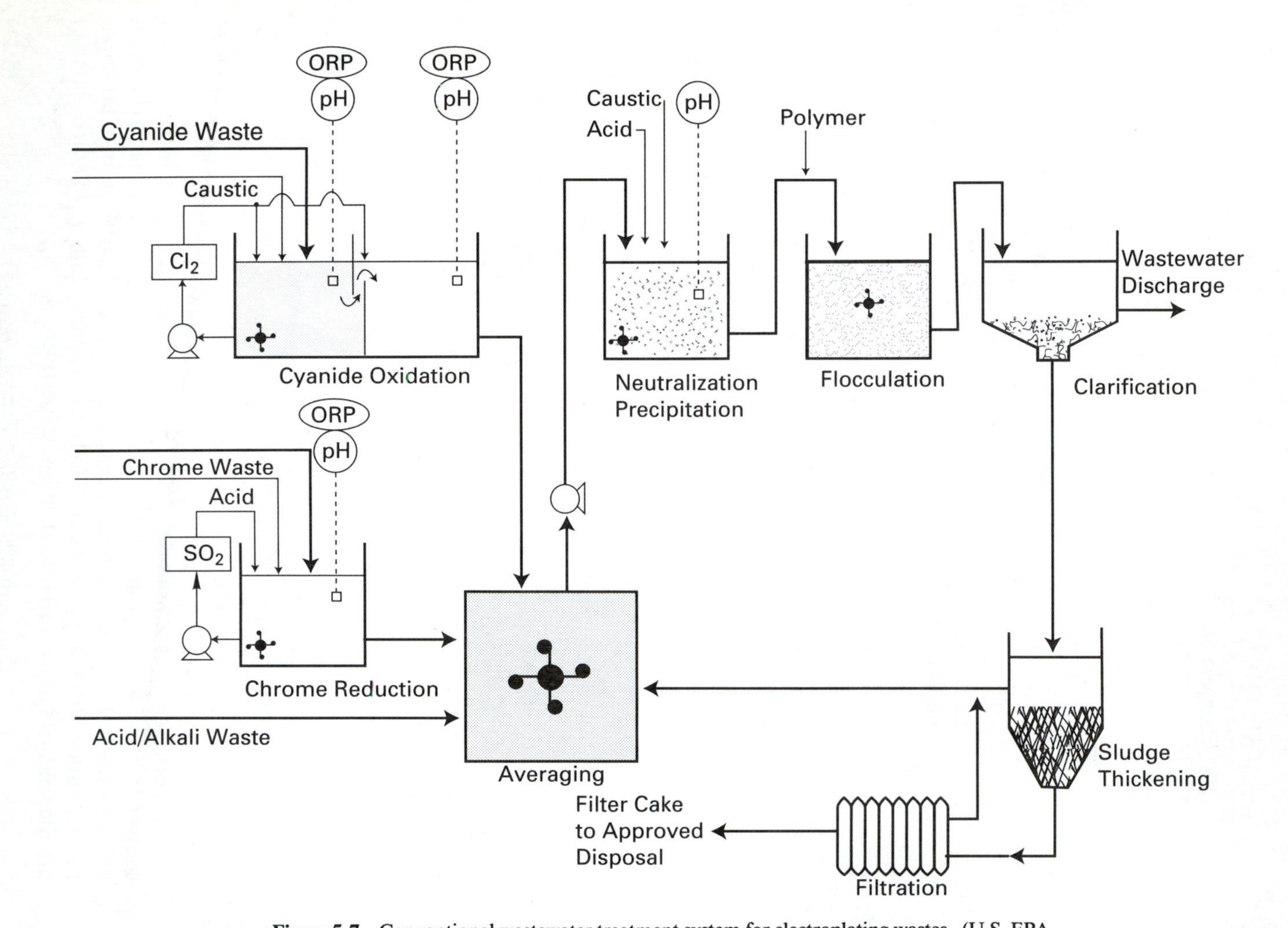

Figure 5-7 Conventional wastewater treatment system for electroplating wastes. (U.S. EPA, *Environmental Regulations and Technology: The Electroplating Industry,* EPA/625/10-85/001. 1985. p. 15.)

TABLE 5-5 TCLP THRESHOLD CONCENTRATIONS FOR INORGANIC IONS

Constituent	Threshold concentration (mg/l)
Arsenic	5.0
Barium	100.0
Cadmium	1.0
Chromium	5.0
Lead	5.0
Mercury	0.2
Selenium	1.0
Silver	5.0

An equilibrium constant can be defined for the above reaction as:

$$K_{eq} = \frac{\{A^{+n}_{(aq)}\}^m \{B^{-m}_{(aq)}\}^n}{\{A_m B_{n(s)}\}} \tag{5.12}$$

where the brackets { } represent activity. For a pure solid precipitate, the activity is generally defined to be unity. Under this condition, the equilibrium expression can be rewritten in terms of the solubility product K_{s0} (product of K_{eq} and the precipitate activity) as:

$$K_{s0} = \{A^{+n}_{(aq)}\}^m \{B^{-m}_{(aq)}\}^n \tag{5.13}$$

where the right hand side of the above equation is termed the ion activity product *(IAP)*. Thus, for the equilibrium between an ideal pure precipitate and the aqueous phase, the *IAP* is a constant (K_{s0}) for a given constant temperature and pressure. If the *IAP* is less than K_{s0}, the system is undersaturated, and a precipitate will not be present. If the *IAP* is greater than K_{s0}, the system is said to be supersaturated. Supersaturation is an indication that the precipitation reaction has not yet reached equilibrium.

Activities for dissolved species can be defined as the product of its molar concentration and an activity coefficient;

$$\{A^{+n}_{(aq)}\} = \gamma_A [A^{+n}_{(aq)}] \tag{5.14}$$

where γ_A is the solution phase activity coefficient for A^{+n}. The activity coefficient accounts for solution phase nonidealities. For dilute solutions, the activity coefficient can be assumed to be one. There are various different expressions for estimating solution phase activity coefficients of electrolytes. Some of the more commonly used activity coefficient expressions are the Debye–Hückel expressions and the Davies equation (Stumm and Morgan, 1981).

It is important to note that the solubility product expression is only valid if all the conditions assumed in its development apply. The expression does not apply when:

- The precipitate has not yet formed because $IAP < K_{s0}$.

- The system has not yet reached equilibrium. Although many precipitation reactions occur practically instantaneously, some take a significant time to reach equilibrium. In such cases, the kinetics of the precipitation process must be known to model the reaction. These kinetic limitations can be in terms of the formation of the precipitate itself, or can be due to changes in the structure of the precipitate from one solid form to another. The different forms of the precipitate are likely to be characterized by different K_{s0}'s.
- The activity of the solid phase is not constant. In cases where the solid phase is a mixture of various precipitates, the assumption of unit solid phase activity is not valid (Vaslow and Boyd, 1952).

The following example illustrates an application of the solubility product expression.

Example 5-3

Cadmium (Cd^{+2} (aq)) can react with the hydroxide ion to form a precipitate of cadmium hydroxide. A sample of wastewater is collected from a 55 gallon drum and analyzed for pH, and heavy metals. The analysis indicates that cadmium is the only metal present at significant concentrations, and that the Cd^{+2} concentration, and the pH are 5 mg/l, and 7.9, respectively. Ignoring nonidealities (i.e., assume the solid phase activity and the solution phase activity coefficients are equal to one), determine whether the bottom of the drum is likely to contain a precipitate of cadmium hydroxide ($Cd(OH)_2$ (s)). Use a pK_{s0} for cadmium hydroxide of 14.35 (pK_{s0} from Stumm and Morgan (1981)).

Answer

A precipitate of cadmium hydroxide will only exist if its solubility is exceeded. Thus, all that is required is to compare the solubility product to the *IAP* for cadmium hydroxide in the drum.

$$[Cd^{+2}] = 5 \text{ mg/l} = 4.45 \times 10^{-5} \text{ moles/liter}$$

$$[OH^{-1}] = K_W/[H^{+1}] = 10^{-14}/10^{-7.9} = 10^{-6.1} = 7.94 \times 10^{-7} \text{ moles/liter}$$

$$[Cd^{+2}] \times [OH^{-1}]^2 = 2.81 \times 10^{-17} < 10^{-14.35} \ (K_{s0} \text{ of } (Cd(OH)_2)$$

Therefore, based on the given data and assumptions, since the *IAP* is less than the solubility product for cadmium hydroxide, the drum should not contain a precipitate of cadmium hydroxide. This does not preclude the possibility that some other metal might have precipitated at this pH. For multicomponent systems, the same sort of analysis would have to be done for each cationic component that forms a hydroxide precipitate. Also, as in all such calculations, it is important to remember the assumptions made. In this case, the key assumption made was that the activity coefficient for the free cadmium ion could be assumed to be one (note that since pH is formally defined in terms of the activity of the hydrogen ion, the above calculation of the hydroxide ion concentration from the pH and K_W actually gave the hydroxide ion activity, and no error was introduced). Activity coefficient corrections will be considered later.

Another situation where the solubility product expression can be used is to calculate the necessary equilibrium concentration of precipitant to achieve a desired equilibrium aqueous phase concentration of a given contaminant. The following example again considers the precipitation of soluble phase cadmium with the hydroxide ion as the precipitant.

Example 5-4

Consider a cadmium containing wastewater which is to be precipitated by the addition of hydroxide, to a free cadmium concentration of 1 mg/l (8.896×10^{-6} moles/liter). Assume an activity coefficient of one for the cadmium ion. Determine the required equilibrium pH to achieve the desired final free cadmium ion concentration. Again, pK_{s0} for cadmium hydroxide can be taken as 14.35.

Answer

From the definition of the solubility product:

$$K_{s0} = 10^{-14.35} = \{Cd^{+2}\}\{OH^{-1}\}^2 \text{ (moles/liter)}^3$$

Because the activity coefficient of cadmium has been assumed to be unity, the activity of the free cadmium can be equated to the concentration. Thus, the required activity of hydroxide ion that must be achieved to reach the target concentration can be calculated as:

$$\{OH^{-1}\} = \sqrt{\frac{10^{-14.35}}{8.896 \times 10^{-6}}} = 2.241 \times 10^{-5} \text{ moles/liter}$$

Because

$$\{OH^{-1}\}\{H^{+1}\} = K_W = 1 \times 10^{-14} \text{ (moles/liter)}^2$$

$$\{H^{+1}\} = 4.463 \times 10^{-10} \text{ (moles/liter)}$$

$$pH = 9.35$$

By raising the pH to 9.35, assuming unit activity coefficient for cadmium, one can reduce the free cadmium concentration to 1 mg/l.

The next logical step would be to determine the precipitant dose needed to bring the pH to 9.35. For the precipitation of cadmium, this question is not as simple as it appears for reasons that will be discussed later.

Nucleation

Once a solution becomes supersaturated, it is possible that the formation of precipitate will occur. However, precipitation generally does not begin until the degree of supersaturation is great enough for the free energy of the precipitate formation reaction (nucleation) to exceed the free energy required to create the surface of the precipitate (Stumm and Morgan, 1981). Formation of a precipitate in a solution with no initial solids present is termed homogenous nucleation. Because the energy required to form a crystalline surface can be significant, a relatively high degree of supersaturation may be required to initiate the precipitation reaction. Once precipitate formation occurs, the existing precipitated solids serve to catalyze further precipitate formation, and the thermodynamics and the kinetics of the precipitation formation become more favorable. The process of precipitate formation where solids are present to catalyze the reaction is known as heterogeneous nucleation.

Generally, heterogeneous nucleation is characterized by faster kinetics than homogeneous nucleation.

Often, foreign solids are present in solutions which can serve as nucleation sites. Alternatively, solids can be added to a solution to induce nucleation. For example, a continuous precipitation process may include the recycle of a fraction of the separated precipitate to the precipitation reaction chamber. The addition of the already formed precipitate (solids recycle) induces heterogeneous nucleation, improving the kinetics of the reaction and allowing a lower degree of supersaturation to be used in the process.

Activity Coefficient Calculations

As noted before, chemical equilibrium is more rigorously defined in terms of the activities of the components of the reactions, and not their molar concentrations. The activity of a chemical specie can be written as the product of an activity coefficient and concentration. For solutions of electrolytes (in this case the cadmium ion will be used for illustration purposes), this is typically represented by the following expression:

$$\{Cd^{+2}\} = \gamma_{Cd}[Cd^{+2}] \tag{5.15}$$

where { } indicates activity, γ is the activity coefficient, and [ ] indicates molar concentration. An activity coefficient must be defined in terms of some reference state, where the value of the activity coefficient goes to unity. For electrolytes, the reference state is typically defined such that the activity coefficient approaches unity as the concentration of the species of interest approaches infinite dilution.

Activity coefficients account for the fact that chemical reactions frequently occur nonideally, with the excess Gibbs free energy term being accounted for by the activity coefficient. For electrolytes, these nonidealities are thought to be principally caused by electrostatic interactions between the various ions in solution. A full discussion of the various theories of activity coefficients are outside the scope of this book. The interested reader should consult standard chemical equilibria and water chemistry textbooks (Butler, 1964; Stumm and Morgan, 1981; Snoeyink and Jenkins, 1980).

There are a number of commonly used expressions for the activity coefficients of electrolytes including the Debye–Hückel, the extended Debye–Hückel, the Davies equation, and the Pitzer equations (AICHE, 1986; Stumm and Morgan, 1981, Pitzer, 1973; Pitzer and Mayorga, 1973). For all these equations, the activity coefficients are principally functions of the solution phase ionic strength, and the charge of the ion. The ionic strength, I, of a solution is defined as:

$$I = \frac{1}{2}\sum_{i}^{n}(C_i z_i^2) \tag{5.16}$$

where C_i is the molar concentration of species i, n is the number of aqueous ionic species, and z_i is the valency of the dissolved species i.

The following example illustrates the use of the Debye–Hückel equation for electrolyte activity coefficients. Note that this equation is one of the simpler activity coefficient expressions, and is strictly only applicable for ionic strengths less than 0.005 (Stumm and Morgan, 1981).

Example 5-5

Consider again the conditions presented in Example 5-4, where it is desired to precipitate cadmium in solution as the insoluble cadmium hydroxide by raising the pH to a sufficient degree to achieve a free cadmium concentration of 1 mg/l. Assume that the following species contribute to the solution phase ionic strength:

$$[Na^{+1}] = 1 \times 10^{-3} \text{ (moles/liter)}$$
$$[NO_3^{-1}] = 2 \times 10^{-4} \text{ (moles/liter)}$$
$$[SO_4^{-2}] = 4 \times 10^{-4} \text{ (moles/liter)}$$

Answer

a) Calculate the solution phase ionic strength.

$$I = \frac{1}{2}([Na^{+1}] + [NO_3^{-1}] + 4[SO_4^{-2}])$$

$$= \frac{1}{2}(1 \times 10^{-3} + 2 \times 10^{-4} + 4(4 \times 10^{-4}))$$

$$= 1.4 \times 10^{-3}$$

b) Use the Debye–Hückel equation to estimate the activity coefficient of free cadmium.

$$\log \gamma = -Az^2\sqrt{I}$$

where $A \approx 0.5$ for water at 25°C

For the cadmium ion, $z = 2$. Therefore, the activity coefficient for the free cadmium ion is calculated as:

$$\log \gamma_{Cd} = -0.5(2)^2\sqrt{0.0014} = -0.0748$$

which gives a value of 0.842 for the activity coefficient for free cadmium. Note that the deviation of the activity coefficient from one is a relative measure of the effect of nonidealities.

c) Determine the required pH to reduce the concentration of free cadmium to 1 mg/l (8.896×10^{-6} moles/liter). Again, assume pK_{s0} for cadmium hydroxide of 14.35.

$$\{Cd^{+2}\} = \gamma_{Cd}[Cd^{+2}] = 0.842[Cd^{+2}]$$

$$K_{s0} = 10^{-14.35} = \{Cd^{+2}\}\{OH^{-1}\}^2$$

$$4.467 \times 10^{-15} = 0.842(8.896 \times 10^{-6})\{OH^{-1}\}^2$$

which gives a required activity of hydroxide ion of 2.442×10^{-5}, which is equivalent to a pH of 9.39. Note that this pH is 0.04 pH units higher than that calculated in Example 5-4, where it was assumed that the activity coefficient for the free cadmium ion was unity. In this case, neglecting the cadmium activity coefficient did not introduce much error in the required pH calculation. However, in more concentrated solutions than the one considered in the example, neglecting solution phase activity coefficients could introduce a significant error in such calculations.

Precipitant Dosage

Another calculation which arises in the design of precipitation processes concerns the required amount of precipitant to achieve a desired degree of removal. The previous examples have only determined the required equilibrium concentration of precipitant that must be achieved in the aqueous phase to effect the desired degree of removal. This calculation is often complicated by the fact that not all of the added precipitant will be available to react with the contaminant, and/or the contaminant may be bound to other ions in solution, and not directly available for reaction with the precipitant. These complications will be discussed later. The following example illustrates the calculation of precipitant dosage for a simple system where such complicating factors can generally be ignored.

Example 5-6

The silver ion can be effectively precipitated from solution as silver chloride (AgCl(s)). The K_{s0} of silver chloride is 1.56×10^{-10}. Given a solution with an aqueous silver concentration of 100 mg/l, determine the mass of sodium chloride that must be added to reduce the aqueous silver concentration to 1 mg/l, if the total solution volume is 55 gallons (208.2 liters).

Answer

$$100 \text{ mg/l } Ag^{+1} = 9.270 \times 10^{-4} \text{ (moles/liter)}$$

$$1 \text{ mg/l } Ag^{+1} = 9.270 \times 10^{-6} \text{ (moles/liter)}$$

At the desired equilibrium composition, the Ag^{+1} concentration will be equal to 9.270×10^{-6} moles/liter. The free soluble chloride concentration will be set by the solubility product expression.

$$[Cl^{-1}]_{aq} = \frac{K_{s0}}{[Ag^{+1}]_{aq}} = \frac{1.56 \times 10^{-10}}{9.27 \times 10^{-6}} = 1.683 \times 10^{-5} \left(\frac{\text{moles}}{\text{liter}}\right)$$

This gives the amount of chloride ion that will be present in the aqueous phase after the precipitation reaction has occurred. Because 99 mg/l of silver ion must be removed from the solution, the amount of chloride ion tied up as silver chloride precipitate will be the molar equivalent of the removed silver ion, since the one mole of silver reacts with one mole of chloride ion to form the silver chloride precipitate. Therefore, the number of moles of chloride ion tied up in the precipitate can be calculated by:

$$\left(\frac{99 \text{ mg } Ag^{+1}}{\text{liter}}\right)\left(\frac{1 \text{ g}}{1000 \text{ mg}}\right)\left(\frac{1 \text{ mole } Ag^{+1}}{107.87 \text{ g}}\right)\left(\frac{1 \text{ mole } Cl^{-1}}{1 \text{ mole } Ag^{+1}}\right) = 9.178 \times 10^{-4} \left(\frac{\text{moles } Cl^{-}}{\text{liter}}\right)$$

Therefore, 9.178×10^{-4} moles of chloride ion will be tied up in the precipitate for each liter of solution treated. The total number of moles of chloride ion that had to be added to bring the free silver ion concentration to 1 mg/liter is the sum of the chloride tied up in the precipitate and the free chloride ion in solution at equilibrium. With the total volume of solution of 208.2 liters, and noting that the chloride ion is added to the solution as sodium chloride (NaCl), the total mass of sodium chloride required to bring the free silver concentration to 1 mg/l can be calculated as:

$$\text{Total chloride} = 9.178 \times 10^{-4} + 1.683 \times 10^{-5}$$

$$= 9.346 \times 10^{-4}\left(\frac{\text{moles}}{\text{liter}}\right)$$

$$208.2 \text{ liters}\left(\frac{9.346 \times 10^{-4} \text{ moles Cl}^{-1}}{\text{liter}}\right)\left(\frac{1 \text{ mole NaCl}}{1 \text{ mole Cl}^{-1}}\right)\cdot\left(\frac{58.5 \text{ g}}{\text{mole NaCl}}\right)$$

$$= 11.38 \text{ grams NaCl}$$

The total dry mass of AgCl precipitate can be calculated from the fact that 9.178×10^{-4} moles/liter of silver ion and chloride ion will combine as the precipitate.

$$208.2 \text{ liters}\left(\frac{9.346 \times 10^{-4} \text{ moles AgCl}}{\text{liter}}\right)\left(\frac{143.37 \text{ grams AgCl}}{\text{mole AgCl}}\right)$$

$$= 27.9 \text{ grams AgCl precipitate}$$

As previously mentioned, precipitant dosage calculations are often complicated by reactions ancillary to the precipitation reaction. Hydroxide precipitation reactions are inherently pH dependent and therefore, acid-base reactions (buffering capacity) must be considered in designing hydroxide precipitation reactions. For example, aqueous systems containing dissolved carbonate (which can exist as $CO_{3\ aq}^{-2}$, $HCO_{3\ aq}^{-1}$, and H_2CO_{3aq}) are buffered against rapid pH changes, and the hydroxide dosage calculation must consider the reactions of the carbonate species with hydroxide. Also, the sulfide ion (S^{-2}), which can be used to precipitate heavy metals, will react with the aqueous hydrogen ion to form HS_{aq}^{-1} and H_2S_{aq}. In such systems, *a priori* calculations of precipitant dosage are difficult and require a thorough knowledge of the chemical makeup of the solution of interest.

Complexation reactions can also complicate precipitant dosage calculations. For example, the heavy metal cadmium can exist in at least the following soluble phase forms; free cadmium (Cd^{+2}), and the hydrolysis products $CdOH^{+1}$, $Cd(OH_2)^0$, $Cd(OH)_3^{-1}$, and $Cd(OH)_4^{-2}$ (Stumm and Morgan, 1981). These hydrolysis products of cadmium are referred to as hydroxy complexes, with the hydroxide ion referred to as the ligand. The formation of the complexes tends to increase the total solubility of the metal precipitate beyond what it would be if the only soluble metal species were the free ions, and therefore, the required precipitant (hydroxide) dosage to achieve a desired equilibrium pH is greater than would be calculated if complexation were ignored. Figure 5-8 illustrates the effect of pH on the theoretical concentrations of the various cadmium species in equilibrium with cadmium hydroxide (solid). Calculations of precipitant dosage and final desired pH must consider the total soluble phase metal concentration (sum of free and complexed metal concentrations), not simply the concentration of free-metal ion. The following example illustrates the effect of complexation on precipitation reactions.

Example 5-7

Consider again the system described in Example 5-4. Assume that a precipitate of cadmium hydroxide is in equilibrium with a solution at a pH of 9.35. Assume further

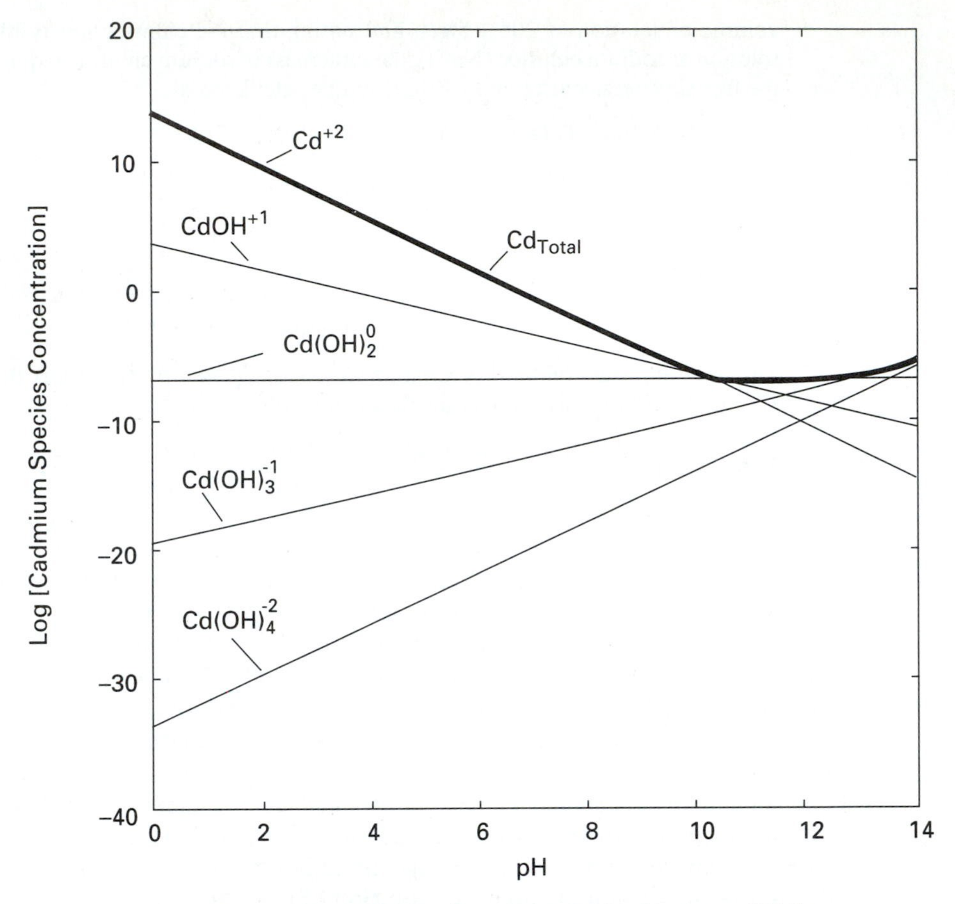

Figure 5-8 Log solubility plot for cadmium-hydroxide system (25°C).

that activity coefficient corrections can be ignored. Given the following, calculate the concentration of all soluble phase cadmium species, and the total soluble-phase cadmium concentration, in equilibrium, with a precipitate of cadmium hydroxide at a pH of 9.35.

Given:

$$Cd^{+2} + H_2O \rightleftharpoons CdOH^+ + H^+ \qquad K_{eq} = 10^{-10.1}$$

$$Cd^{+2} + 2H_2O \rightleftharpoons Cd(OH)_2^0 + 2H^+ \qquad K_{eq} = 10^{-20.4}$$

$$Cd^{+2} + 3H_2O \rightleftharpoons Cd(OH)_3^- + 3H^+ \qquad K_{eq} = 10^{-33.3}$$

$$Cd^{+2} + 4H_2O \rightleftharpoons Cd(OH)_4^{-2} + 4H^+ \qquad K_{eq} = 10^{-47.4}$$

$$Cd(OH)_2(s) + 2H^+ \rightleftharpoons Cd^{+2} + 2H_2O \qquad K_{eq} = 10^{+13.65}$$

Answer

Neglecting activity coefficient corrections, it was shown in Example 5-4 that the free cadmium concentration in equilibrium with a precipitate of cadmium hydroxide at a pH of 9.35 is 1 mg/l (8.896×10^{-6} moles/liter). With this information, the concentration of the other cadmium species is easily calculated. For example, the concentration of $CdOH^+$ is calculated from the equilibrium expression relating Cd^{+2} and $CdOH^+$ as follows:

$$10^{-10.1} = \frac{[CdOH^+][H^+]}{[Cd^{+2}]}$$

$$[H^+] = 10^{-9.35} \text{ moles/liter}$$

$$[Cd^{+2}] = 8.896 \times 10^{-6} \text{ moles/liter}$$

$$[CdOH^+] = \frac{(10^{-10.1})(8.896 \times 10^{-6})}{(10^{-9.35})} = 1.582 \times 10^{-6} \text{ moles/liter}$$

The concentrations of the other cadmium complexes can be similarly calculated using the appropriate equilibrium expressions. The concentration of each of the soluble-phase cadmium species, and the total soluble-phase cadmium concentration for the conditions described are:

$$[Cd^{+2}] = 8.896 \times 10^{-6} \text{ (moles/liter)}$$
$$[Cd(OH)^{+1}] = 1.582 \times 10^{-6} \text{ (moles/liter)}$$
$$[Cd(OH)_2^0] = 1.775 \times 10^{-7} \text{ (moles/liter)}$$
$$[Cd(OH)_3^{-1}] = 5.003 \times 10^{-11} \text{ (moles/liter)}$$
$$[Cd(OH)_4^{-2}] = 8.896 \times 10^{-16} \text{ (moles/liter)}$$
$$[Cd]_{total} = 1.066 \times 10^{-5} \text{ (moles/liter)}$$

Therefore, at a pH of 9.35, the total soluble-phase cadmium concentration is approximately 20% greater than the free cadmium concentration, illustrating that the formation of hydroxide complexes in heavy metal precipitation cannot be ignored.

Many metal ions form complexes with a variety of ligands, including hydroxide, chloride, sulfate, sulfide, EDTA, and cyanide (Smith and Martell, 1976). In cases where the metal is complexed with the precipitant, the formation of the metal complexes increase the solubility of the metal precipitate beyond what would be the case if the free metal ion were the only soluble specie formed. In cases where the metal is complexed with ions other than the precipitant, for example with EDTA or cyanide, the metal cannot be effectively precipitated until the complex is destroyed or until the equilibrium is shifted by excessive precipitant dosages so that the effect of complexation becomes negligible. Often, a strong oxidant such as chlorine or ozone is used in such cases to destroy the complex prior to precipitation.

A full discussion on acid-base chemistry and complexation is outside the scope of this textbook. The interested reader should consult some of the available water chemistry textbooks (Stumm and Morgan, 1981; Snoeyink and Jenkins, 1980). However, because acid-base and complexation reactions cannot generally be ignored in most real systems where precipitation is applied, and because these reactions make *a priori* calculations of precipitant dosage impractical, the design of precipitation processes is generally not done on a theoretical basis alone. Another complication to designing precipitation processes is that many real wastewaters contain multiple contaminants targeted for removal by precipitation. In multicomponent systems, *a priori* calculations of optimal precipitant dosage are extremely difficult to do.

Because of potential chemical interactions, solubility variances, mixing effects, and competition for precipitant in multicomponent systems, actual required precipitant dosages may be greater than predicted by a factor of four based on theoretical solubilities (Cawley, 1981). Therefore, precipitation processes are generally designed and optimized based on bench scale and pilot scale testing using representative waste samples and potential precipitants (and flocculants). Such tests will be discussed later in this chapter.

Coagulation/Flocculation

Successful applications of chemical precipitation require that the precipitate can be efficiently separated from the remaining solution. Many precipitation reactions result in the formation of colloidal precipitates (particle diameters less than 1 μm) that are difficult to separate from the solution. To aid in the settling of precipitates, a coagulation/flocculation step is often necessary. Colloids generally exhibit an electrostatic charge that can result in a balance of repulsive and attractive forces between the particles. This balance of charges creates what is known as a stable suspension which cannot be settled without destabilizing the suspension. To remove the colloidal precipitates by gravity settling, the particles must be induced to collide and combine, thereby growing in size. Colloidal suspensions can be destabilized by the addition of coagulants, which reduce or neutralize the repulsive forces between particles so that the particles will approach and combine. To increase the number and rate of collisions between particles, the suspension is often gently agitated after the addition of the coagulant. The agitation of the destabilized particles causes particle aggregation and the formation of a rapid-settling floc. The coagulant also can serve as a seed for particle growth. If added as a slurry, a coagulant/flocculant can physically enmesh the suspended particles through its own rapid settling in a process known as sweep flocculation. The more common process of gently agitating the destabilized suspension is known as flocculation, and the process of coagulant addition and gentle agitation is frequently termed coagulation/flocculation.

Coagulants are typically classified as inorganic, synthetic organic polymers, or natural organic polymers (Reynolds, 1982). Most applications of coagulants involve inorganic chemicals or synthetic organic polymers. Inorganic coagulants include aluminum sulfate (alum), iron salts (ferrous sulfate, ferric sulfate, ferric chloride), and lime (calcium oxide or calcium hydroxide). Calcium hydroxide therefore can serve as both a precipitant and a coagulant for heavy metals. Inorganic coagulants generally function by forming a hydroxide precipitate with charge characteristics and settling properties which can remove the floc of interest by both destabilizing the suspension and enmeshing the floc within the precipitated coagulant matrix. Synthetic organic polymers are classified as anionic, cationic, or polyampholitic, depending on their characteristic charge. These organic coagulants or coagulant aids function by charge destabilization of the floc and by association of the colloid with the coagulant aid, which then bridges with other polymer molecules to form a larger, settleable floc.

The sequence of precipitation/coagulation/flocculation is often done in two steps. In continuous-flow systems, the precipitant and coagulant are added to the wastewater in a rapid mix tank, which quickly disperses the chemicals within the wastewater. Typical residence times for rapid mix basins range from 10 seconds to 6 minutes (Cawley, 1981; Eckenfelder, 1966 and 1970). However, in hydroxide precipitation systems where the rate of precipitant dosage is controlled by monitoring the pH of the mix tank, a residence time of 10 minutes to 15 minutes is needed for the precipitant feed system to be controlled properly (WPCF, 1981). The flocculation step, where gentle agitation of the wastewater is performed, follows the rapid mix. The agitation induces particle collisions. Adequate residence time must be provided to allow time for particle growth. Flocculation basins are generally designed to provide a residence time of 15 minutes to 30 minutes (Cawley, 1981; Eckenfelder, 1966 and 1970). Options to provide the gentle agitation for flocculation include baffled basins, mechanical agitation, or pneumatic agitation. The agitation cannot be too vigorous or it will shear apart the flocculating particles. A parameter often used in the design of flocculation processes is the velocity gradient of mixing, or G (Camp, 1955; Weber, 1972). Typical velocity gradients for flocculation range from 5 fps/ft to 100 fps/ft (Reynolds, 1982). The dimensionless product of the velocity gradient and the flocculation basin residence time is often termed the Gt value. Typical Gt values for flocculation in drinking water treatment range from $10^4 - 10^5$ (Weber, 1972).

Coprecipitation

The effectiveness of precipitation of dissolved inorganic and organic species can be enhanced through the process of coprecipitation. Coprecipitation involves the simultaneous precipitation of the contaminant of interest with other dissolved species through the addition of coagulants such as iron compounds (such as ferric chloride, ferric sulfate, ferrous sulfate, ferrous hydroxide) or alum, sometimes together with lime to the wastewater. The precipitation of the nontarget species (such as the added iron or aluminum species) improves the effectiveness of target contaminant removal through a number of mechanisms, including adsorption of the target compound to the coprecipitated solids, and sweep flocculation or enmeshment of the target compound within the coprecipitate. Coprecipitation of such metals as arsenic, cadmium, copper, and mercury can result in treated effluent concentrations significantly below the theoretical hydroxide solubilities of these metals.

DESIGN OF PRECIPITATION PROCESSES

Batch versus Continuous. One of the first decisions to be made in designing a precipitation process it to determine whether the process will be a batch or continuous process. For small flow rates, or when production of the waste is discontinuous, batch treatment is often chosen. Batch wastewater treatment systems employing precipitation can be economically designed for flows up to 50,000 gallons/day (Cawley, 1981). Batch-precipitation processes usually employ two tanks, each designed to handle the entire waste volume expected during the treatment period. The two tanks allow one tank to be in operation while the second tank can collect the newly produced wastes. Batch-precipitation tanks serve as the equalization tank, the precipitation reaction tank, and settling tank.

In batch treatment, the waste is first mixed to insure a homogeneous mixture. The chemicals (precipitant and possibly coagulant) are then added to the solution, which is then stirred for approximately 10 minutes (Cawley, 1981). Mixing is then discontinued and the solids are allowed to flocculate and settle for a few hours under quiescent conditions. The clear supernatant is then decanted, while the sludge is removed for further thickening and possibly dewatering. Some sludge may be retained in the tank to serve as a flocculant seed for the next batch.

In continuous-treatment systems, a typical flow scheme includes a chemical feed system, a rapid mix tank, a flocculation tank, a sedimentation tank, and possibly a filtration system. Frequently, some type of control system is used to control the rate of chemical feed. For instance, in continuous hydroxide-precipitation processes, the rate of precipitant dosage is controlled by monitoring the system pH and comparing the measured value with the desired value.

Waste characterization. An important step in designing a precipitation process is to characterize as fully as possible the waste stream. Samples should be collected and analyzed for contaminants of interest, as well as for other parameters which could affect the precipitation process, such as pH, alkalinity, total dissolved solids, presence and concentration of complexing agents, and presence of other specific ions which could exert a precipitant demand of their own. For example, dissolved iron could be present in solution with such RCRA heavy metals as cadmium or chromium. Although the presence of iron may not violate discharge or disposal regulations, the iron will exert a hydroxide demand in a hydroxide precipitation process. A review of the processes that create the waste or wastewater may be worthwhile to determine what chemical species are likely to be present. Such considerations can reduce the analytical costs related to characterization by reducing the number of analytes that must be tested for.

If the waste stream is produced from a continuous process, but is likely to produce a waste of varying concentrations, both grab and composite samples should be collected to determine the average and variability of the contaminant concentrations. Composite samples may also be warranted in cases where the waste is present in multiple lots, such as drums.

If waste or wastewater is produced continuously, the rate of waste or wastewater generation should be characterized both in terms of an average, minimum, and maximum. Waste streams that are generated at variable rates may require flow equalization prior to treatment. Equalization may also be warranted if the concentration of the waste stream varies significantly. Dampening of the flow and/or waste strength by equalization will improve the efficiency of the precipitation process.

Determine treatment standards. An important step in the design of any treatment process is to determine the treatment criteria. If numerical discharge or disposal limits are applicable, knowledge of the limits is essential in choosing a precipitant, and designing the process to achieve the numerical limits.

Choose precipitant. For some ions, there may be more than one precipitant that can be used to precipitate the ions. For instance, heavy metals can generally be precipitated by both hydroxide or sulfide. As will be presented later, the solubility of metal sulfides is less than the solubility of metal hydroxides. Therefore, stringent treatment standards may be achievable by sulfide precipitation but not by hydroxide precipitation. However, if both hydroxide and sulfide can achieve the treatment standards, considerations such as cost, ease of process control, and safety may make hydroxide the preferred precipitant.

Bench scale testing. After choosing potential precipitant(s), bench scale testing, often called jar testing, is generally done to determine the best precipitant and optimal precipitant dosage. The first step in the jar test is to gather representative samples of the wastewater to be treated. Various precipitants and precipitant dosages are added to the samples. After the precipitate has settled out, the contaminant and precipitant concentration of the supernatant are measured. The choice of precipitant, and the optimal precipitant dosage can thus be determined.

Jar tests can also be used to determine the rate of precipitant formation, and the settling characteristics of the precipitate. If the precipitate settles rapidly leaving a clear supernatant, then coagulant addition may not be necessary. Often, however, a coagulant is needed to aid in the precipitate settling. Jar tests can be used to test various coagulants and coagulant dosage to optimize the settling process. After choosing the coagulant and coagulant dosage, settling tests can be conducted to gather the necessary data to design a sedimentation basin for floc removal. Also, sludge dewatering capability must be tested to aid in the design of the final dewatering step. Testing of sludge dewatering characteristics is important, because a large fraction of the overall cost of the precipitation process is due to sludge dewatering and handling. Sedimentation and sludge dewatering are discussed in more detail in Chapter 4. For further details on jar tests, consult Black et al. (1957), Black and Harris (1969), and Camp (1968 and 1952).

Pretreatment. In some cases, the precipitation reaction can only proceed after the waste is pretreated. For example, metals in solutions containing cyanide generally cannot be precipitated without first destroying the cyanide ion because most metals form very strong complexes with cyanide. Cyanides can be destroyed (oxidized) by alkaline chlorination (U.S. EPA, 1988). Other examples of pretreatment prior to precipitation are the oxidation of the complexing agent EDTA, air stripping of the complexing agent ammonia, and reduction of hexavalent chromium to trivalent chromium (Patterson, 1985).

Common Precipitation Applications

Metals. Metals such as barium, cadmium, chromium (Cr^{+3}), mercury, nickel, lead, selenium, and silver can be removed from aqueous wastes by precipitation (Patterson, 1985). Most of these metals can be precipitated as either hydroxides or sulfides. The exceptions are selenium, which can be reacted with sulfur dioxide to precipitate elemental selenium; barium, which can be precipitated as barium sulfate; silver, which can be precipitated as silver chloride or as silver sulfide, but cannot be precipitated as the hydroxide; and mercury, which can be precipitated as a sulfide, but cannot be precipitated as a hydroxide. Cadmium and lead can also be precipitated as the metal carbonates (Patterson et al., 1977). Hexavalent chromium (Cr^{+6}) must first be reduced to trivalent chromium prior to hydroxide precipitation (U.S. EPA, 1974). Reducing agents commonly used to reduce hexavalent chromium to trivalent chromium are sulfur dioxide, sodium bisulfite, sodium metabisulfite, sodium hyrosulfite, or ferrous sulfate.

Some metals can be removed from solution by reaction with the reducing agent sodium borohydride, which reduces the metal to the elemental form. However, this reaction is actually an oxidation-reduction reaction, and not a precipitation reaction. Sodium borohydride reduction of metals will be discussed further in the oxidation-reduction section of this chapter.

Hydroxide Precipitation. Hydroxide precipitation is commonly used to remove heavy metals from solution (U.S. EPA, 1983). Precipitation is induced by adding to the solution of interest an alkaline reagent which raises the pH of the solution (increases free hydroxide concentration) such that the solubility product of the metal hydroxide is exceeded. A general expression for the precipitation of a metal ion by hydroxide is as follows:

$$M^{+n} + nOH^- \rightleftharpoons M(OH)_{n(s)} \quad (5.17)$$

The precipitated metal hydroxide is then removed from solution by some physical process such as sedimentation and/or filtration. The resultant free-metal concentration is dependent on the metal(s) present, the final pH, and the presence of other materials, such as complexes, which can inhibit precipitation. Figure 5-9 provides the theoretical solubilities of several heavy metal hydroxides as a function of pH (U.S. EPA, 1983). As indicated in Figure 5-9, metal hydroxides are amphoteric, that is, their solubilities increase with both increasing and decreasing pH around a pH of minimum solubility. The pH of minimum solubility varies from metal to metal, and with other factors such as temperature and solution ionic strength. Equilibrium constants for the precipitation and complexation reactions of various metals with hydroxide, and other ligands, can be found in Smith and Martell (1976).

Because metal hydroxides are amphoteric, there are some limitations of hydroxide precipitation for removal of soluble heavy metals. The most obvious limitation is the existence of a minimum solubility. If the metal concentration at the point of minimum solubility is greater than the effluent or treatability standard, hydroxide precipitation may not be able to reduce the metal concentrations to below compliance standards. However, compliance could be met through a combination of precipitation (as hydroxide) and adsorption/entrapment with other coprecipitated hydroxide precipitates such as ferrous or ferric hydroxide (Patterson, 1985). Another limitation to hydroxide precipitation is that in solutions containing multiple metals, a compromise on the final pH must be made, because the pH of minimum solubility varies from metal to metal. A final pH that results in a satisfactory effluent concentration for one metal may not result in a satisfactory effluent concentration for

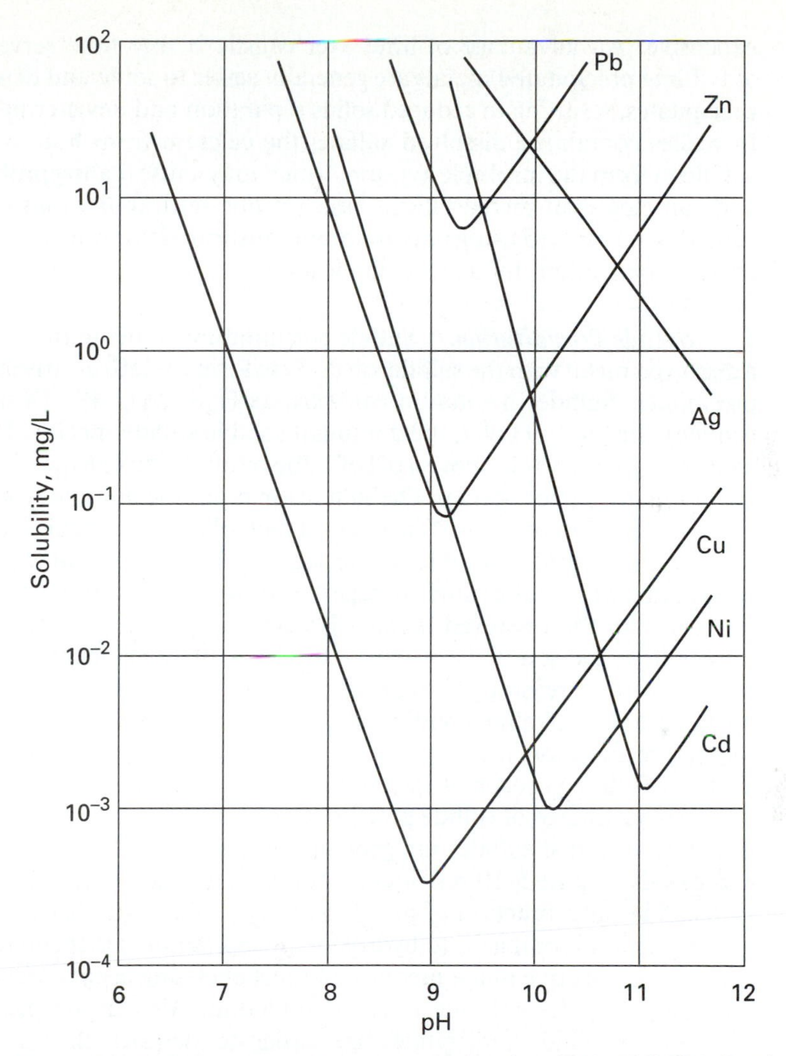

Figure 5-9 Solubilities of metal hydroxides as a function of pH. (U.S. EPA, *Development Document for Effluent Limitations Guidelines and Standards for the Metal Finishing Point Source Category,* EPA 440/1-83-091, June 1983.)

another metal. Finally, sludges produced by hydroxide precipitation are generally more difficult to dewater than metal sulfide or metal carbonate precipitate sludges. However, because hydroxide precipitation is relatively safe, and generally more cost effective than other options, hydroxide precipitation is frequently chosen if theoretically viable.

The two most common sources of hydroxide are sodium hydroxide (caustic) and calcium hydroxide (lime). Caustic is more expensive than lime (on a unit cost basis), but easier to handle and deliver to solution. Lime, in the form of quicklime (CaO), must be slaked (to form $Ca(OH)_2$), slurried, and constantly agitated to avoid caking. Hydrated lime $Ca(OH)_2$) is easier to handle than quicklime, but is more expensive. An advantage of lime over caustic is that lime serves as a coagulant aid. Lime precipitated metals are generally easier to settle and dewater than caustic precipitates, resulting in reduced solids separation and dewatering costs. However, in wastes containing dissolved sulfate, the calcium from lime will react with the sulfate to form the insoluble gypsum, which may cause scaling problems in pipelines and can clog dual media filters. Tests on floc settleability and sludge dewatering should be conducted using both lime and caustic to determine which is the most cost effective precipitant for a given application.

Sulfide Precipitation. Sulfide precipitation of metals involves the reaction of a dissolved metal with the sulfide ion (S^{-2}) to form a relatively insoluble metal sulfide precipitate. Sulfide in solution can exist as H_2S (aq), HS^- (bisulfide), or as S^{-2} (sulfide). Below a pH of 7, the dominant soluble sulfide specie is H_2S (aq) ($pK_{HB} = 7.02$; $T = 25°C$), while above a pH of 7, the bisulfide ion ($pK_{HB} = 13.9$; $T = 25°C$) is the dominant sulfide specie. The sulfide ion is not the dominant specie until the pH is above approximately 14. Since such a high pH is rarely achieved, sulfide is generally not the dominant soluble sulfide species. Therefore, sulfide precipitation must be operated with a total sulfide dosage in excess of the stoichiometric amount needed to react with the dissolved metals. Because of the pH dependency of the sulfide distribution, the solubility of metal sulfides is pH dependent, with solubilities decreasing with increasing pH. Therefore, sulfide-precipitation processes are generally operated under alkaline conditions. Operating at alkaline pH is also necessary in sulfide-precipitation processes to minimize the evolution of H_2S (g), which will evolve if H_2S (aq) becomes the dominant soluble sulfide species.

An advantage of sulfide precipitation over hydroxide precipitation is that the solubility of metal sulfides are generally much lower than the solubility of metal hydroxides. Figure 5-10 provides a comparison of solubilities of metal hydroxides and sulfides as a function of pH (U.S. EPA, 1983). Another advantage of sulfide precipitation as compared to hydroxide precipitation is that sulfide precipitation is relatively insensitive to the presence of most chelating agents because excess sulfide can be added to drive the reaction to completion. Also, in comparison to hydroxide precipitates, sulfide precipitates are easier to dewater than hydroxide sludges. Finally, the sulfide precipitation reaction can remove chromates and dichromates (Cr^{+6} compounds) without the separate reduction to Cr^{+3} that must be done in hydroxide precipitation (U.S. EPA, 1977). However, these advantages of sulfide precipitation are offset by the potential for the evolution of noxious hydrogen sulfide gas, and discharge of excess soluble sulfide. Also, sulfide precipitates tend to be colloidal, and therefore coagulant addition is generally required to separate the sulfide precipitates from solution (Bhattacharyya and Chen, 1986). Furthermore, metal sulfide sludges have the potential to be oxidized after disposal, resulting in the conversion of sulfide to sulfate. Because most metal sulfates are quite soluble, sulfide sludge oxidation can result in the remobilization of the metal from the sludge. The mobilized metal ions could then contaminate the soil and groundwater if leachate from the landfill is not contained.

Conventional sulfide precipitation is performed by adding to the waste a source of sulfide, such as hydrogen sulfide gas (under alkaline conditions), sodium sulfide, or

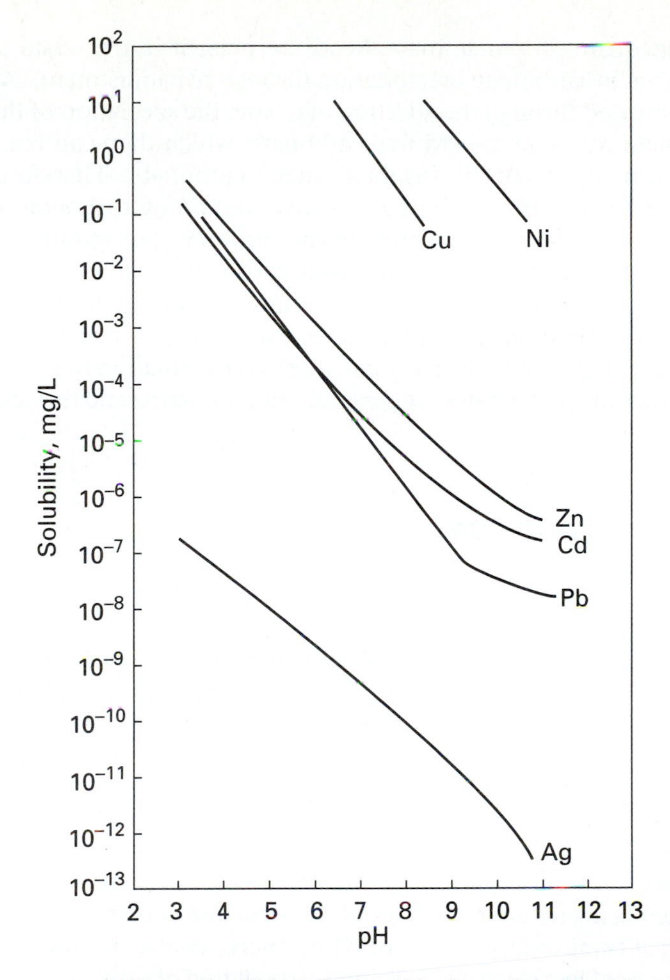

Figure 5-10 Solubilities of metal sulfides as a function of pH. (U.S. EPA, *Development Document for Effluent Limitations Guidelines and Standards for the Metal Finishing Point Source Category,* EPA 440/1-83-091, June 1983.)

sodium bisulfide. As noted before, the pH of the waste must be maintained above a pH of eight to avoid evolving hydrogen sulfide gas. A proprietary sulfide-precipitation process, known as the Sulfex process (U.S. EPA, 1977), has been developed which is less susceptible to hydrogen-sulfide evolution and discharge of excess soluble sulfide. The Sulfex process utilizes precipitated ferrous sulfide (FeS) as the source of the sulfide ion. Because FeS is unstable, it is usually generated on-site just prior to usage by mixing a soluble sulfide (such as NaHS) with ferrous sulfate and lime. The resultant slurry is then added to waste in the precipitation tank. As the dissolved metals in the waste react and form precipitates with the free sulfide in solution, FeS will dissolve to replenish the solution with sulfide ions so that equilibrium between the FeS and the solution is maintained. The presence of the FeS precipitate helps control the concentration of free sulfide in solution, thereby controlling the potential for evolution of hydrogen sulfide or discharge of excess sulfide in the effluent.

Carbonate Precipitation. Removal of some metals from solution can be effectively achieved through carbonate precipitation. In particular, carbonate precipitation can be effective for dissolved cadmium, nickel, and lead (Patterson,

1985). In some cases, adequate carbonate may already be present in the waste to result in significant removal by carbonate precipitation through pH adjustment. As the pH of the waste is increased through the addition of a base, the speciation of the naturally present carbonate will tend toward free carbonate, which then can react with the metal ion to form the relatively insoluble metal carbonate. Therefore, hydroxide precipitation of metals from solutions already containing carbonate is often a coprecipitation process, where both hydroxide and carbonate precipitates are formed. If no carbonate is present in the waste, inorganic carbonate, such as sodium carbonate, can be added to form the metal carbonate precipitate. Although effective carbonate precipitation requires an alkaline pH, the effectiveness of the process is reduced at pH values greater than approximately ten, due to the formation of metal-hydroxy complexes. Carbonate precipitates are generally easier to settle and dewater than hydroxide precipitates.

CHEMICAL OXIDATION AND REDUCTION

Introduction

There are various different types of oxidation and reduction (redox) reactions, including thermal oxidation (incineration), catalytic oxidation, biological oxidation and reduction, and chemical oxidation and reduction. This section focuses on chemical redox reactions used in hazardous waste treatment. Oxidation-reduction reactions, or redox reactions, are reactions which involve the transfer of electrons between the redox coupled species. Oxidation of a substance results in the loss of electrons from that substance, while reduction of a substance results in a gain of electrons in that substance. An oxidant is a substance that causes the oxidation of another specie, while itself being reduced. Conversely, a reductant causes the reduction of a specie, while itself being oxidized. Because free-energy considerations generally preclude the existence of free electrons in solution, oxidation of a dissolved ion or compound is accompanied by the reduction of another dissolved ion or compound in the same solution.

An example of a redox reaction used in hazardous waste treatment is the oxidation of cyanide in aqueous wastes by chlorine. This reaction is termed an *oxidation* reaction for convenience, because the contaminant of interest, CN^{-1}, is *oxidized* by chlorine, which is reduced. Therefore, although the redox reaction of chlorine oxidation of cyanide involves both an oxidation and a reduction reaction, because the contaminant of interest is oxidized, this reaction is typically termed an oxidation reaction. Another example of a redox reaction used in hazardous waste treatment is the reduction of hexavalent chromium (Cr^{+6}) by reaction with sulfur dioxide. In this reaction, Cr^{+6} is reduced to Cr^{+3}, while sulfur dioxide (SO_2) is oxidized to the sulfate ion (SO_4^{-2}). Similarly, this reaction is typically termed a *reduction* reaction because the pollutant of interest, Cr^{+6}, is *reduced.*

Stoichiometry of Redox Reactions. The stoichiometry of redox reactions can be understood by viewing the overall redox reaction as the sum of two individual oxidation and reduction reactions. Consider the reaction where the ferrous ion (Fe^{+2}) is oxidized to the ferric ion (Fe^{+3}) by oxygen:

$$O_2 + 4Fe^{+2} + 4H^{+} = 4Fe^{+3} + 2H_2O \tag{5.18}$$

This reaction can be viewed as the sum of the following half-reactions:

$$4Fe^{+2} = 4Fe^{+3} + 4e^{-} \qquad \text{oxidation}$$

$$O_2 + 4H^{+} + 4e^{-} = 2H_2O \qquad \text{reduction} \tag{5.19}$$

In the oxidation half-reaction, four ferrous ions are oxidized to four ferric ions with the accompanying loss of four electrons, and the change of the oxidation state of iron from +2 to +3. The reduction half-reaction involves the oxygen molecule reacting with four hydrogen ions, and gaining four electrons, to form two water molecules. The oxidation state of the oxygen molecule is changed from zero (0) to −2. For each half-reaction, the electron transfer is balanced by the change in oxidation state of atom or molecule donating or accepting the electrons. The addition of the two half-reactions results in the overall reaction listed above. Also, the electrons transferred in the two half-reactions result in no net free electrons, which is a necessary condition of redox reactions. Oxidation half-reactions are those that result in the loss of electrons, while reduction half-reactions result in the gain of electrons.

Oxidation State. Redox reactions result in changes of the oxidation states of the reactants and products. For mono-atomic species, the oxidation state is simply the electronic charge. However, for covalently bonded compounds, the oxidation state of each element in the compound depends upon the relative electronegativity of the other elements included in the compound. For such compounds, the oxidation state of each atom is calculated as the charge remaining on the atom after each shared pair of electrons is completely assigned to the more electronegative of the two atoms sharing the electrons. When the electronegativity of the two atoms sharing the electrons is the same, the electron pair is split between the two atoms. For molecules, the sum of oxidation states is equal to zero. The sum of the oxidation states for multi-atomic ions is equal to the formal charge of the ion (Stumm and Morgan, 1985).

Balancing Redox Reactions. There are various different approaches to balancing redox reactions. These different approaches can be studied in standard physical chemistry or aquatic chemistry textbooks (Stumm and Morgan, 1981; Snoeyink and Jenkins, 1980; Maron and Lando, 1974).

Equilibrium Calculations. Although electrons do not freely exist in solution, mass action law equilibrium expressions can be written with the hypothetical activity of a free electron included in the expression. Given a reference half-reaction, equilibrium constants can then be calculated for other half-reactions. For redox reactions, the reference half-reaction is the reduction of the hydrogen ion to form hydrogen gas at 25°C:

$$2H^{+} + 2e^{-} = H_{2(g)} \qquad \text{Log}K = 0 \tag{5.20}$$

where the equilibrium constant is expressed as:

$$K = \frac{\{H^+\}^2\{e^-\}^2}{\{H_{2(g)}\}} = 1 \tag{5.21}$$

With equilibrium constants available for half-reactions, equilibrium constants for redox reactions can be calculated. The following example illustrates the calculation of the equilibrium constant for a redox reaction using the half-reaction equilibrium constants.

Example 5-8

Calculate the equilibrium constant for oxidation of ferrous iron to ferric iron by oxygen given the following half-reactions:

$$Fe^{+2} \rightleftharpoons Fe^{+3} + e^- \qquad \text{Log}K = -13.01$$

$$\frac{1}{2}O_{2(g)} + 2H^+ + 2e^- \rightleftharpoons H_2O \qquad \text{Log}K = 41.55$$

Answer

To balance the electron transfer, the first half-reaction can be multiplied by two, and the resulting expression can be added to the second half-reaction, resulting in the following:

$$2Fe^{+2} + \frac{1}{2}O_{2(g)} + 2H^+ \rightleftharpoons Fe^{+3} + H_2O$$

$$\text{Log}K = 41.55 - 26.02 = 15.53$$

At equilibrium, the hypothetical free electron activity of the reactants and products are equal, and the electron activity term drops out of the equilibrium expression. Redox half-reactions, however, allow one to define relative electron activities at equilibrium.

Example 5-9

Calculate the hypothetical free-electron activity for the half-reaction of chlorine gas reduction to chloride ion, given the following:

$$Cl_{2(g)} + 2e^- = 2Cl^- \qquad \text{Log}K = 46 \ (@25^\circ C)$$

Answer

The hypothetical electron activity at equilibrium, or p*e*, is a measure of the relative tendency of a solution to donate or accept electrons. In a solution with a high p*e* (few electrons), the tendency for oxidation (loss of electrons) is high, whereas in a solution with a low p*e* (high electron activity), the tendency for reduction (gain of electrons) is high. Given the equilibrium composition of the oxidized and reduced components of a redox half-reaction, the p*e* value for the system can be calculated.

$$K = \frac{\{Cl^-\}^2}{\{Cl_{2g}\}\{e\}^2}$$

$$-\text{Log}\{e\} = \frac{1}{2}\log K - \log\{Cl^-\} + \frac{1}{2}\log\{Cl_{2(g)}\}$$

The concept of electron activity, and pe, is analogous to the concept of the hydrogen ion activity in aqueous systems, and pH. With pe° defined as $(\text{Log}K)/n$ (where n is the number of electrons transferred), and pe defined as $-\log\{e\}$, the following expression for pe can be derived:

$$pe = pe^0 - \log\{Cl^-\} + \frac{1}{2}\log\{Cl_{2g}\} \tag{5.22}$$

The following example illustrates the calculation of a numerical value for pe for the case where the equilibrium concentrations of the reduced and oxidized species in the half-reaction are known.

Example 5-10

Consider again the oxidation of ferrous iron to ferric iron. Given the following information, calculate the pe of the equilibrium solution. Ignore activity coefficient corrections.

$$[Fe^{+2}] = 10^{-2}\ M$$

$$[Fe^{+3}] = 10^{-4}\ M$$

$$Fe^{+3} + e^- \rightleftharpoons Fe^{+2} \qquad \text{Log}K = 13.01$$

Answer

$$pe = \frac{\log K}{1} + \log[Fe^{+3}] - \log[Fe^{+2}]$$

$$pe = 13.01 - 4 + 2 = 11.01$$

Electrode Potential. The equilibria of redox half-reactions are often expressed in terms of an electrode potential which is related to the reaction equilibrium constant and stoichiometry by the Peters–Nernst equation (commonly known as the Nernst equation). The general expression for the electrode potential for a redox reaction is:

$$E_H = E_H^0 + \frac{RT}{nF}\ln\frac{\Pi_i\{\text{ox}\}^{n_i}}{\Pi_j\{\text{red}\}^{n_j}} \tag{5.23}$$

where {ox} represents the activities of the oxidized components, {red} represents the activities of the reduced components, F is Faraday's constant, n is the number of electrons transferred in the reaction, R is the ideal gas constant, T is absolute temperature in °K, i and j are indices for the oxidized and reduced reaction components, E_H is the redox potential for the reaction (where suffix H indicates that the potential scale is the hydrogen scale), and E_H^0 is the standard redox potential for the reaction (i.e., the potential for the reaction if all substances involved in the reaction were in their standard states at unit activity). The standard redox potential is related to the half-reaction equilibrium constant by the following:

$$E_H^0 = \frac{-\Delta G^0}{nF} = \frac{RT}{nF}\ln K \tag{5.24}$$

Example 5-11

Calculate the redox potential and the standard redox potential for the reduction of the cupric ion to the cuprous ion (at 25°C) given the following:

$$[Cu^{+}] = 10^{-2}\ M$$

$$[Cu^{+2}] = 10^{-3}\ M$$

$$Cu^{+2} + e^{-} \rightleftharpoons Cu^{+} \qquad \log K = 2.7$$

Answer

The standard redox potential is calculated using Equation 5.24 to be:

$$E_H^0 = \frac{0.059}{1} \log K = 0.16 \text{ volt}$$

At the given equilibrium concentrations of the cuprous and cupric ions, the redox potential is calculated using Equation 5.23 to be:

$$E_H = E_H^0 + 0.059 \log \frac{[Cu^{+2}]}{[Cu^{+}]}$$

$$E_H = 0.16 \text{ V} - 0.059 \text{ V} = 0.101 \text{ V}$$

The redox potential is related to pe and pe^0 by the following:

$$pe = \frac{F}{2.3RT} E_H$$

$$pe^0 = \frac{F}{2.3RT} E_H^0 \qquad (5.25)$$

As noted before, the standard reference reaction for redox equilibria is the reduction of the hydrogen ion to form hydrogen gas. The standard electrode potential for this reaction at 25°C is zero. There have been, however, two different conventions used to designate the sign of the potential of other redox reactions with respect to the reference condition. The sign convention used in this text (and the IUPAC convention) is that all redox half-reactions are written as reduction reactions with a sign that corresponds to the sign of $\log K$ of the redox reaction. Some texts use the alternative convention of writing all redox half-reactions as oxidation reactions. The absolute values of the standard electrode potentials are the same in both cases, but the sign is opposite. Therefore, when consulting different texts to obtain standard electrode potentials, one must take care to determine if the potentials are standard reduction potentials, or standard oxidation potentials. Either convention can be used, as long as one uses the chosen convention consistently in all calculations. Standard redox potentials are available from a number of sources, including Sillen and Martell (1964) and Latimer (1952).

Redox potentials can be measured in solution using appropriate redox electrodes. Platinum wire electrodes are most commonly used to make redox potential measurements in solution (Orion Research, 1982). Other materials can also be used to make redox potential measurements, including wax impregnated graphite, vitreous carbon, and titanium oxide (Walton-Day et al., 1990). Redox electrodes measure the potential which arises in solution due to the presence of ions of a substance in two different stages of oxidation. For example, a solution containing both the ferric and ferrous iron ions will exhibit a redox potential related to the standard electrode potential for the oxidation of ferrous iron to ferric iron, and the relative activities of the two iron species. This potential is defined through the Nernst equation (Equation 5.23). Practically, the potential is actually measured through the use of a reference electrode in combination with the redox electrode, which completes the electrochemical circuit in solution. The potential measured by the redox electrode is "referenced" to the known potential of the reference electrode.

Interpretation of measured solution redox potentials is often difficult due to a number of reasons. One difficulty is that many solutions of interest contain multiple redox active substances, all of which contribute to the observed redox potential. Also, the interpretation of redox potentials through the Nernst equation implies that the solution is at equilibrium. This assumption is very often not valid. Therefore, care must be taken in interpreting measured redox potentials. The utility of such measurements is often for more qualitative tasks such as controlling the rate of adding an oxidant to solution in a continuous-treatment process.

Kinetic Considerations. Although some redox reactions occur rapidly enough to allow design through equilibrium calculations, many redox reactions are relatively slow, even in the presence of catalysts. Although the free energy of such reactions may be favorable, without the presence of a catalyst to reduce the activation energy of the reaction, the kinetics of the reaction may be too slow for the reaction to be practically applied. For example, the kinetics of the oxidation of synthetic organic compounds using hydrogen peroxide can be accelerated by the presence of dissolved metal ions (Sedlak and Andren, 1991 a, b). Therefore, design of continuous oxidation-treatment systems often requires consideration and knowledge of the kinetics of the redox process as well as the potential equilibria state. Few generalizations on redox kinetics can be made. However, it is generally observed that increasing the reaction temperature increases the kinetics of redox reactions. Also, redox reaction kinetics are generally increased with increasing reactant concentrations.

Effect of pH. Many redox systems are affected by the solution pH, both in terms of the reaction kinetics and system equilibria. For example, when chlorine is used as an oxidant in aqueous systems, the predominant forms of chlorine are hypochlorous acid (HOCl) and its conjugate base, the hypochlorite ion, OCl^-, and the chloride ion. Because the relative concentrations of hypochlorous acid and the hypochlorite ion are a function of pH, and because the redox equilibria between hypochlorite and the chloride ion are a function of the concentration of the hypochlorite species through the Nernst equation (5.23), the redox equilibria are dependent upon the pH. The relationship between oxidant or reductant speciation, the redox potential, and the pH can be described through what are known as *pe*-pH diagrams (Stumm and Morgan, 1985; Snoeyink and Jenkins, 1980). An example *pe*-pH diagram for the aqueous chlorine system is provided in Figure 5-11 (Snoeyink and Jenkins, 1980).

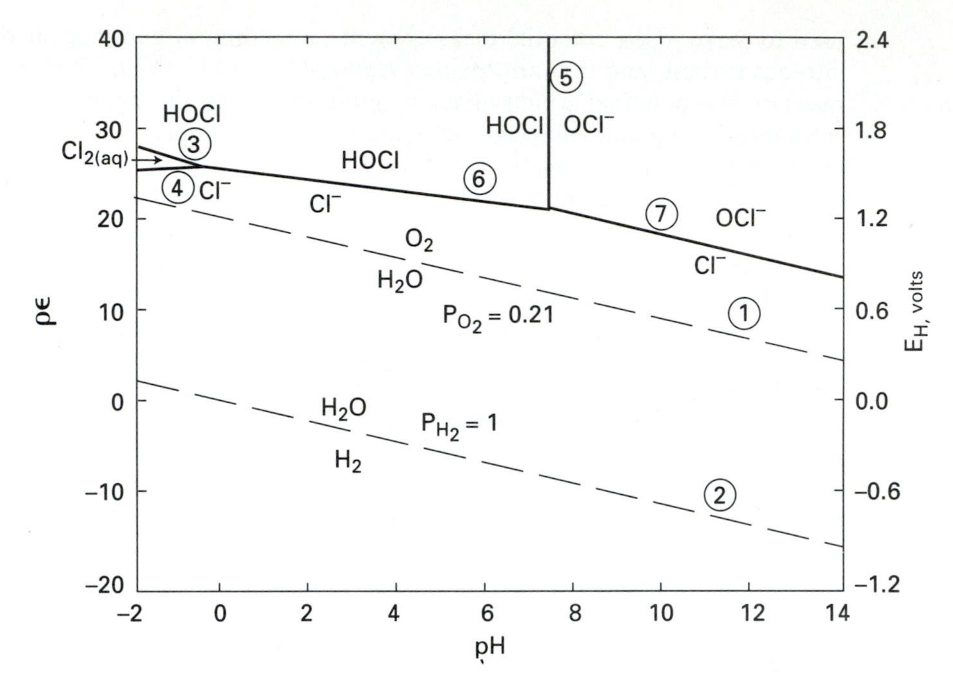

Figure 5-11 The pε-pH diagram for aqueous chlorine; 25°C, $C_{T,Cl} = 1 \times 10^{-4} M$. (Snoeyink and Jenkins, *Water Chemistry*, Copyright 1980, p. 350, reprinted by permission of John Wiley & Sons, Inc.)

Redox Reactions in Hazardous Waste Treatment

Design considerations. ***Reaction Driving Force.*** One of the first issues to consider in designing an oxidation/reduction reaction for hazardous waste treatment is whether there is a sufficient equilibrium driving force to sustain the desired reaction. For example, a given oxidant has the potential to oxidize a given hazardous constituent to a significant degree only if the equilibrium constant for the proposed overall *redox* reaction is substantially greater than unity. However, a large equilibrium constant does not guarantee that the proposed reaction can be successfully used to treat a given waste. The kinetics of the process may limit the successful application of the given reaction. Also, the redox reaction of interest will generally occur simultaneously with other reactions (oxidation/reduction reactions with other waste constituents, precipitation reactions, acid-base reactions) which can interfere with the desired reaction. One must consider whether the *overall* equilibrium status of the system is favorable for achieving the desired reaction.

Stoichiometry. The stoichiometry of the given desired reaction can allow one to make preliminary estimates of the required oxidant or reductant dosage. However, dosages in excess of the theoretical quantity are often required due to kinetic limitations and oxidant or reductant demand from waste components other than the hazardous constituent of interest. Because of the limitations of *a priori* dosage calculations, redox reactions for waste treatment must generally be designed based upon bench-scale or pilot scale data.

Batch Versus Continuous. The decision to operate a redox treatment system in batch or continuous mode depends upon such issues as waste flowrate and waste flow and strength variability. Batch treatment is usually preferred when waste flowrates are low or highly variable in flow and/or composition. However, flow and waste strength equalization can be used to allow continuous treatment of wastes that vary in flowrate or composition. Continuous treatment systems generally require more sophisticated system monitoring and control than batch treatment systems.

Kinetic Considerations. Redox process design often requires consideration of the reaction kinetics. Understanding of the kinetics is required to determine required reactor residence times, reactor volume, oxidant or reductant dosage, and expected extent of treatment. Although the kinetics of some redox processes have been characterized adequately to allow reasonably accurate *a priori* process design, the kinetics must often be determined and included in the process design through bench or pilot scale testing on the waste of interest.

System Control and Monitoring. Monitoring and control of redox processes often involves pH and oxidation-reduction potential *(ORP)* measurement. In continuous systems, *ORP* measurements can be used as a part of a control loop to control the rate of oxidant or reductant feed. Measurement of the pH is often important in terms of optimizing the reaction kinetics, and in terms of safety considerations. For example, cyanide treatment systems must include pH monitoring to protect against low pH conditions that would cause evolution of toxic hydrogen cyanide gas. The pH measurement is sometimes used in the control loop to regulate addition of bases or acids to maintain optimum pH.

Safety Considerations. Redox reactions are often exothermic reactions. Therefore, care must be taken to insure that the heat generated by exothermic redox reactions can be accommodated safely. Chlorine gas, which is often used as an oxidant and which is toxic, must be handled carefully to insure worker safety. Also, as mentioned before, cyanide waste treatment systems must include pH monitoring to protect against evolution of hydrogen cyanide gas.

Applications of Redox Processes in Hazardous Waste Treatment

The following lists some common applications of redox processes that are used in the treatment of hazardous wastes. More detailed discussions on some of the most common applications of redox processes in hazardous waste treatment are provided below.

Contaminants Amenable to Treatment Via Redox Reactions

Contaminants amenable to treatment via oxidation include:

- cyanide wastes
- phenols
- pesticides
- other synthetic and volatile organic compounds
- sulfides

Contaminants amenable to treatment via reduction include:
hexavalent chromium
heavy metals (e.g., cadmium, mercury, lead, hexavalent selenium)

Common Oxidants and Reductants

Chemicals typically used as oxidants in oxidation processes include:
chlorine (chlorine gas, hypochlorite salts)
chlorine dioxide
hydrogen peroxide
potassium permanganate
oxygen
ozone
ozone Advanced Oxidation Processes *(AOP*s)

Some typical reductants used in reduction processes include:
ferrous iron salts
sulfur dioxide
sodium metabisulfate
sodium bisulfate
sodium sulfite
sodium hydrosulfite
sodium borohydride

Cyanide Destruction

A number of cyanide containing wastes have been designated as listed hazardous wastes, including wastewater treatment sludges from electroplating operations (F006), spent cyanide plating bath solutions from electroplating operations (F007), plating bath residues from cyanide plating operations (F008), and spent stripping and cleaning bath solutions from electroplating operations which include the use of cyanide (F009). Numerous other cyanide compounds are P-listed or U-listed RCRA hazardous wastes. Other industrial processes which produce cyanide wastes include mineral extraction and mining, coke furnace operations, and catalytic cracking of petroleum. A typical method of treating cyanide containing wastes is to destroy the cyanide by oxidation processes. Cyanide is often present in wastes along with heavy metals. Because cyanide forms complexes with many heavy metals, the treatment of cyanide wastes can be particularly difficult. Incomplete oxidation of the cyanide can result in discharge of cyanide in the treated aqueous effluent, or association of the cyanide in the sludges that are often generated by precipitation of the heavy metals after the cyanide treatment step. Such sludges cannot be land disposed if the cyanide concentrations exceed the allowable concentrations. For instance, F006 wastes are currently subject to land disposal restrictions if the (total) cyanide concentration in the (non-wastewater) sludge exceeds 590 mg/kg, or if the concentration of cyanide amenable to treatment within the sludge exceeds 30 mg/kg (Federal Register p. 22, 689, June 1, 1990).

Cyanide is most commonly destroyed using chlorine gas with sodium hydroxide (alkaline chlorination), or hypochlorite salts (Patterson, 1985). Other conventional oxidants that can be used to oxidize cyanide include potassium permanganate, hydrogen peroxide (De Renzo, 1981), and ozone (Gurol et al., 1985). Highly concentrated cyanide wastes can be treated by electrolytic decomposition.

Alkaline Chlorination. Treatment of dilute aqueous cyanide wastes (< 1,000 mg/l total cyanide) is typically achieved by alkaline chlorination (Hassan et al., 1991). The reactive oxidant in alkaline chlorination is the hypochlorite ion, which can be added directly to the waste as a hypochlorite salt, or can be produced by adding both chlorine gas and sodium hydroxide to the waste stream. Although the chemical costs for hypochlorite salts are greater than the chemical costs for gaseous chlorine, hypochlorite salts are frequently used because of the greater equipment costs and safety concerns associated with using chlorine gas. Complete destruction of cyanide by alkaline chlorination is a two stage process. The first step converts cyanide to the cyanate ion after first forming the intermediary cyanogen chloride (CNCl) in a two reaction sequence as follows:

$$CN^- + H_2O + OCl^- \rightleftharpoons CNCl + 2OH^- \tag{5.26}$$

$$CNCl + 3OCl^- \rightleftharpoons OCN^- + Cl^- + H_2O \tag{5.27}$$

The conversion of cyanide to cyanate is generally done at a pH of 10 or greater to maximize the rate and extent of the pH dependent conversion of cyanogen chloride to cyanate. Typical reaction times for the oxidation of cyanide to cyanate range from 30 minutes to 2 hours (Crowle, 1971; WPCF, 1981). Reported chlorine dosages necessary for conversion of cyanide to cyanate range from 1.75 lbs chlorine/lb cyanide to 8 lbs chlorine/lb cyanide (Patterson, 1985). The wide range of reported chlorine requirements is likely to vary in the composition of the wastes treated. For example, heavy metals often present in cyanide wastes will exert an oxidant demand above that required for cyanide oxidation. De Renzo (1981) reports that 8 lbs of chlorine and 7.3 lbs of sodium hydroxide are required to convert 1 lb of cyanide to cyanate. The second step of the alkaline chlorination process is the oxidation of cyanate to carbon dioxide and nitrogen gas. The hypochlorite remaining in solution after the conversion of cyanide to cyanate will oxidize cyanate as follows:

$$2OCN^- + 3OCl^- + H_2O \rightleftharpoons 2CO_2 + N_2 + 3Cl^- + 2OH^- \tag{5.28}$$

However, because the kinetics of this reaction are slow at pH values greater than 10, the solution pH is generally decreased to approximately 8.5 to increase the rate of cyanate oxidation. Reported reaction times for the chlorine oxidation of cyanate range from 10 minutes to 1 hour at a pH of approximately 8.5 (WPCF, 1981; Patterson, 1985). Cyanate can also be destroyed by acid hydrolysis at a pH of approximately 2–3 and reaction times on the order of 5 minutes. However, the resultant treated waste must then be neutralized before discharge, increasing the treatment costs and the total dissolved solids concentration of the effluent.

Alkaline chlorination can be conducted either as a batch process or as a continuous process. For flow rates less than 15 gpm, batch treatment is commonly performed. Batch treatment systems are available as "package" units from various vendors. Control of the alkaline chlorination process can be achieved through feedback control loops based upon measured values of the waste pH and the oxidation-reduction potential.

Although the alkaline chlorination process can achieve effluent free cyanide concentrations as low as 1 mg/l (Palmer et al., 1988), the process is frequently complicated by the presence of soluble iron in the waste. Dissolved iron, as well as other heavy metals, form strong complexes with the cyanide ion. Much of the complexed cyanide will not be oxidized by the chlorine, and therefore will be discharged in the treated effluent, or become associated with subsequent precipitation sludges. High levels of cyanide in the sludge may exclude the sludge from land disposal, and can present a safety problem due to the potential release of hydrogen cyanide and cyanogen-chloride gas.

Other Cyanide Oxidation Processes. Oxidants other than chlorine (hypochlorite) can be used to destroy cyanide. Hydrogen peroxide and potassium permanganate have been used to oxidize cyanide (De Renzo, 1981). Other demonstrated or promising methods to oxidize cyanide include wet air oxidation and sulfur-based oxidation (Palmer et al., 1988), supercritical water oxidation (U.S. EPA, 1992), photocatalytic oxidation using UV-irradiated TiO_2 (Weathington, 1988; Bhakta et al., 1992), and ozonation (Gurol et al., 1985), and various so-called advanced oxidation processes which produce hydroxyl radicals to achieve oxidation. Most of these technologies are also currently being applied or investigated for the destruction of synthetic organic compounds, and a general discussion on a number of these technologies will be provided later in this section.

Chromium. Wastes containing chromium can be classified as characteristic hazardous wastes (D007) if the total chromium concentration in the waste or TCLP waste extract exceeds 5.0 mg/l. The treatment standards for D007 wastes are 5.0 mg/l (total chromium) for wastewaters, and 5.0 mg/l (total chromium in waste extract) for non-wastewaters (40 C.F.R., 268.41, 268.43). Although chromium can exist in solution as either hexavalent chromium (Cr^{+6}) or trivalent chromium (Cr^{+3}), the hexavalent form of chromium is most commonly encountered in industrial and hazardous wastes as chromate (CrO_4^{-2}) or dichromate ($Cr_2O_7^{-2}$). Although trivalent chromium can be removed from solution by hydroxide precipitation, hexavalent chromium is not amenable to hydroxide precipitation. Chromium wastes containing hexavalent chromium are typically treated by reduction of the hexavalent chromium to trivalent chromium, followed by hydroxide precipitation of the less soluble trivalent chromium at a pH between 8 and 9.

Reduction of hexavalent chromium can be accomplished using a variety of reductants including sulfur dioxide, sodium bisulfite, sodium metabisulfite, acidic sulfite (H_2SO_3), and ferrous sulfate (Patterson, 1985; De Renzo, 1981; Siegel and Clifford, 1988). Effluent hexavalent chromium concentrations below 1 mg/l can be achieved by chemical reduction under optimal treatment conditions, and required reaction times are generally less than one hour (Patterson, 1985). It is generally recommended to lower the waste stream pH to two to three prior to reduction to optimize the kinetics of the reduction reaction. However, although Siegel and Clifford (1988) confirmed that a relatively low pH was the optimal condition for hexavalent chromium reduction using sodium sulfite, it was found that the extent of the reduction was optimized at a pH of 5 to 8 when ferrous sulfate was used as the reductant.

Oxidation of Organic Compounds

Chemical oxidation of dissolved organic compounds is increasingly being used because the technology has the capability of converting organic compounds to innocuous inorganic compounds (carbon dioxide, water) in contrast to technologies such as air stripping, carbon adsorption, or membrane processes, which only serve to separate the organic compounds from the aqueous phase (air stripping, carbon adsorption), or concentrate the compounds (membrane processes). Various oxidants can be used to oxidize hazardous organic compounds, including hydrogen peroxide, potassium permanganate, oxygen, and ozone. Although chlorine (gaseous chlorine or hypochlorite salts) can also be used as an oxidant, because incomplete chlorine oxidation can result in the formation of chlorinated by-products which may also be toxic and/or regulated, use of chlorine for the oxidation of hazardous wastes is not common. Relative oxidation potentials of various oxidants are provided below in Table 5-6 (Rice, 1981).

Organic compounds can also be oxidized using processes commonly termed advanced oxidation processes (*AOP*s) which utilize the strongly oxidizing hydroxyl free radical as the oxidant. Oxygen is generally not a strong enough oxidant to oxidize most organic compounds. However, wet oxidation and supercritical water oxidation utilize dissolved oxygen at elevated temperatures and pressures to oxidize organic compounds. The supercritical water oxidation process is similar to wet air oxidation, except that the process is operated at a temperature and pressure above the critical point of water (supercritical). Other more novel oxidation processes being developed for oxidation of hazardous organic compounds include photolysis and photocatalytic oxidation.

Wet Oxidation

Wet air oxidation, or wet oxidation, was originally patented by Zimmerman in 1958 (Zimmerman, 1958), and the Zimpro process (Zimpro, Rothschild, Wisconsin) has been used for the conditioning of municipal sludges by wet oxidation since the 1960s. Laughlin et al. (1983) summarized the operation of a successful industrial

TABLE 5-6 STANDARD OXIDATION POTENTIAL OF VARIOUS OXIDANTS (25°C)

Species	Oxidation potential (volts)	Relative oxidation potential
Fluorine (F_2)	3.06	2.25
Hydroxyl radical (OH)	2.80	2.05
Atomic oxygen (O)	2.42	1.78
Ozone (O_3)	2.07	1.52
Hydrogen peroxide (H_2O_2)	1.77	1.30
Perhydroxyl radical (HO_2)	1.70	1.25
Hypochlorous acid (HOCl)	1.49	1.10
Chlorine (Cl_2)	1.36	1.00

waste treatment facility which utilized wet oxidation for destruction of organic wastes containing COD levels of approximately 7%. However, application of wet oxidation for the destruction of hazardous organic wastes is still an emerging technology, with most available information based on bench scale or pilot scale testing (Breton et al., 1987; Spivey et al., 1986; Unterberg et al., 1988). Wet oxidation followed by activated carbon adsorption is an acceptable treatment technology for numerous listed organic hazardous wastes that have technology-based treatment standards under RCRA (Fed. Reg. Vol. 55, No. 21, January 31, 1991). However, carbon adsorption is required after wet oxidation because wet oxidation does not provide complete destruction of the organic compounds in the waste. Wet oxidation has also been used as a pretreatment step for subsequent biological treatment of organic wastes (Surprenant et al., 1988) and has been applied in a few industrial cyanide treatment processes (Palmer et al., 1988).

The wet-oxidation process utilizes dissolved oxygen to oxidize dissolved, and in some cases suspended, organic contaminants in aqueous wastes (Spivey et al., 1986). At 25°C and 1 atm pressure, the aqueous solubility of oxygen is approximately 8 mg/l in equilibrium with air and 40 mg/l in equilibrium with pure oxygen. These dissolved oxygen concentrations are generally insufficient to oxidize most organic compounds. However, at elevated temperature and pressure, dissolved oxygen can oxidize many organic compounds effectively. The elevated temperatures and pressures that are maintained in the wet-oxidation process increases the oxygen solubility and provides adequate energy to drive the oxidation reaction to completion. Because the process requires elevated temperatures and pressures, wet oxidation is well suited to wastes that are already available at elevated temperatures and pressures.

The wet-oxidation process is generally operated by pumping the waste stream into a heated and/or insulated pressure vessel (boiler) where compressed air or oxygen is added. Reaction temperatures of 150°C to 300°C and total pressures of 10 atm to 70 atm are typical. Economic operation of the wet oxidation processes requires that the waste organic concentration is high enough (> 10,000 mg/1 COD) to generate an adequate heat of oxidation to overcome process heat losses and sustain the process temperatures and pressures. However, highly concentrated organic wastes or organic sludges (greater than 30% organics) are generally more economically treated by incineration (Unterberg et al., 1988). Often, the wet oxidation process utilizes a heat exchanger to reclaim heat from the treated effluent by preheating the influent waste stream. Required process residence times vary depending upon the degree of destruction desired and the waste composition, and are generally determined by bench scale or pilot scale testing. Catalysts can be used to increase the rate of reaction, and therefore decrease the required reactor residence time. The applicability of wet oxidation for volatile organic compound oxidation is complicated by the potential need for off-gas treatment.

Supercritical Water Oxidation

Supercritical water oxidation (*SCWO*) is an oxidation process similar to wet oxidation, except that the process is operated at temperatures and pressures above the critical point of water (374°C and 218 atm). The properties of water change dramatically under supercritical conditions. Some of the changes that water undergoes under supercritical conditions are the disappearance of hydrogen bonding, improved heat transfer characteristics, increased oxygen solubility, and reduced density. Because of the lack of hydrogen bonding, supercritical water behaves similarly to a moderately polar solvent. Therefore, organic substances are practically completely miscible in supercritical water, while inorganic salts are nearly insoluble.

A general schematic of the *SCWO* process is provided in Figure 5-12. The *SCWO* process involves contacting the waste (generally aqueous) and oxidant (air, oxygen, or hydrogen peroxide) inside a reactor where conditions of high temperature and pressure are controlled and maintained. Within the *SCWO* reactor, oxidation occurs and the existing and produced inorganic salts precipitate out of solution. The heat produced from the oxidation reaction serves to maintain the reactor temperature. The reactor effluent then is fed into a solid/liquid separator to separate out the precipitated salts. A portion of the effluent is often passed through a heat exchanger to pre-heat the influent waste, while the remainder of the effluent can be passed through an additional heat exchanger, such as a steam generator, for energy recovery. Upon cooling, the effluent will separate into aqueous and gaseous phases. The gas phase will contain unreacted oxidant, carbon dioxide, and nitrogen. Further energy recovery can be achieved by passing the effluent gas phase through a turbine.

Both surface (pressure reactor) and subsurface *SCWO* systems are currently being developed (U.S. EPA, 1992). The surface *SCWO* systems employ high pressure reactors constructed of such materials as titanium, stainless steel, Hastelloy C-276, and Monel 400. Subsurface *SCWO* systems are basically deep injection shafts which utilize hydraulic head to provide the necessary high pressures to reach supercritical conditions. A generalized schematic of a deep shaft *SCWO* system is provided in Figure 5-13. Subsurface *SCWO* reactors as deep as 5,200 ft have been tested. A purported advantage of the deep shaft *SCWO* design is that reactor failure will not expose operators to any danger because the subsurface will absorb the force of any explosion.

SCWO generally outperforms wet oxidation in both destruction efficiency and required reactor residence times. Bench-scale and pilot-scale testing has demonstrated destruction efficiencies of greater than 99.99% for a variety of organic compounds with reactor residence times ranging from one minute to 20 minutes. For example, bench scale tests of *SCWO* using a 210 cm^3 continuous flow reactor at

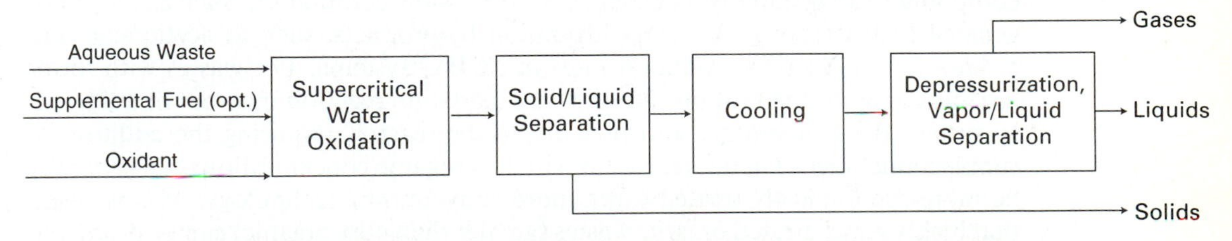

Figure 5-12 Schematic of supercritical water oxidation process. (U.S. EPA, *Supercritical Water Oxidation: Engineering Bulletin,* EPA/540/S-92/006, September 1992, p. 2.)

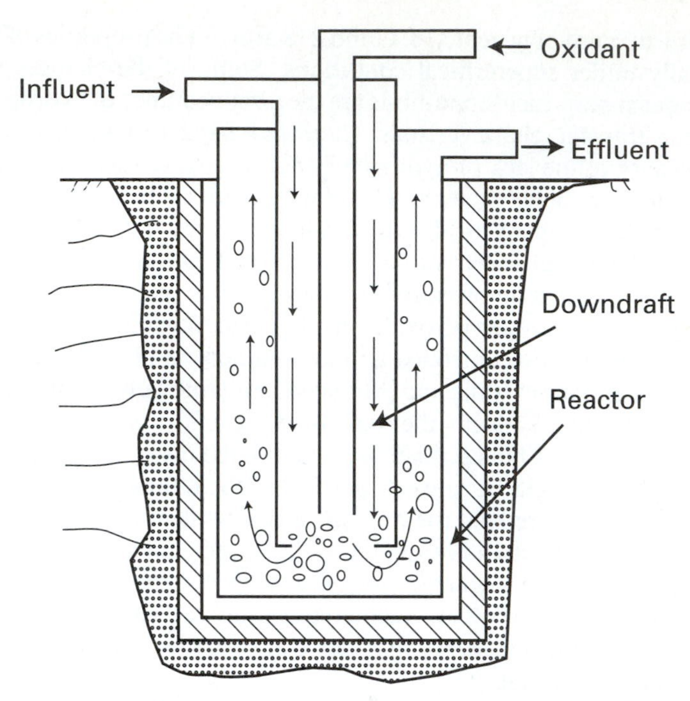

Figure 5-13 Subsurface supercritical water oxidation reactor. (U.S. EPA, *Supercritical Water Oxidation: Engineering Bulletin,* EPA/540/S-92/006, September 1992, p. 2.)

throughput flowrates of 120 cm^3/min indicated destruction efficiencies in excess of 99% for priority organic pollutants present in a dinitrotoluene process wastewater (Li et al., 1993). However, some studies have indicated that *SCWO* may not always provide complete destruction of organics, resulting in organic by-products, including dioxins, which are of environmental concern (Thornton et al., 1991). *SCWO* is currently a developing technology, with no full-scale *SCWO* systems in operation.

SCWO is potentially applicable to liquid wastes, sludges, and slurried solid wastes at waste organic concentrations from 3% to 40%, and is most applicable for wastes of approximately 10% organic (Breton et al., 1987). Although most organic compounds can be rapidly oxidized by *SCWO*, some compounds, such as polyhalogenated hydrocarbons, and some oxidation by-products, such as acetic acid and ammonia, exhibit slow oxidation rates in *SCWO* systems. For wastes with lower organic contents, the heat produced by the oxidation reaction may not be sufficient to sustain the temperature and pressure of the reactor, requiring the addition of supplemental fuel. Therefore, wastes with low organic concentrations can generally be more economically treated using more conventional technology. It is believed that highly concentrated organic wastes (greater than 40% organic) can be destroyed by incineration more economically than by *SCWO* (Breton et al., 1987).

Ozone and Advanced Oxidation Processes

Ozone (O_3) is a strong oxidant that has been used for many years for disinfection and taste and odor control of drinking water, and for tertiary treatment of municipal wastewaters (Rosen, 1973; Singer, 1990). Because of its strong oxidization potential, ozone has been used for the destruction of cyanides in plating wastes (Walker and Zabban, 1953), and destruction of various organic compounds in industrial wastewaters (Rice and Browning, 1981). Oxidation of compounds through the use of ozone can be the result of a direct oxidation reaction of molecular ozone, or through various indirect pathways in which free radicals (principally the hydroxyl radical) produced from ozone decomposition serve as oxidants (Hoigne and Bader, 1976). In fact, because free radical initiators (such as hydroxide ion) are present in nearly all real aqueous systems, oxidation processes which involve ozone nearly always involve oxidation of compounds by both molecular ozone and free radicals produced from ozone. Ozone oxidation, along with oxidation processes that use various combinations of ozone, hydrogen peroxide, and ultraviolet light to stimulate production of the strongly oxidizing hydroxyl radical, are increasingly being considered as alternative technologies for the treatment of hazardous organic-containing wastes, wastewaters, and contaminated groundwaters. Oxidation processes that are engineered to produce hydroxyl radicals to effect the oxidation process are commonly termed advanced oxidation processes *(AOPs).*

Ozone is an unstable gas that cannot be shipped or stored, and therefore must be produced on-site at the point of application. Ozone is generally produced by the use of corona discharge technology, which converts oxygen to ozone. Electrons within the corona discharge split oxygen-oxygen double bonds, which react with other oxygen molecules to form ozone gas. Figure 5-14 provides a schematic of a typical double tube ozone generator. Ozone can be produced in the gas phase at concentrations up to 3% by weight using air as the oxygen source, and 7% by weight using pure

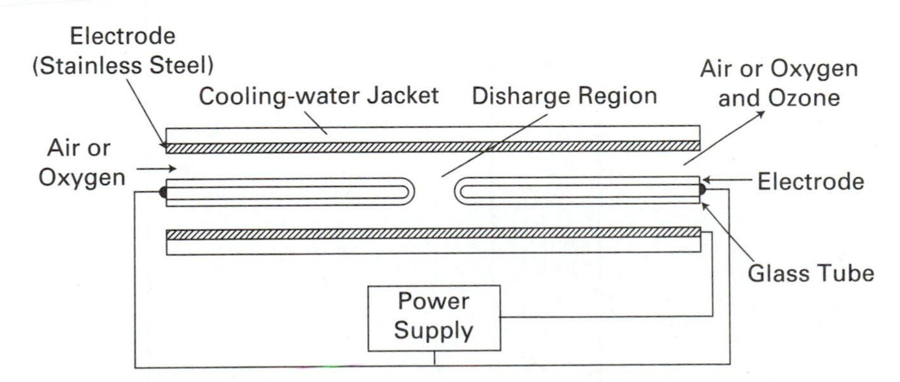

Figure 5-14 Double-tube ozone generator. (W.H. Glaze, "Drinking Water Treatment with Ozone," *Environmental Science and Technology,* Vol. 21, No. 3, 1987, p. 226, reprinted with permission from the American Chemical Society. Copyright 1987, American Chemical Society.)

oxygen as the oxygen source. Large scale ozonators can produce up to 600 Kg ozone/day (Glaze, 1987).

Various commercial designs are available to transfer the gas phase ozone to the liquid phase. These include bubble aeration, mechanical aeration, and spray, packed or tray type towers. Figure 5-15 provides a schematic of one of the more commonly applied methods of transferring ozone from the gas phase to the liquid phase: the counter-current sparged column. The concentration of ozone in a solution equilibrated with a gas phase containing ozone is proportional to the gas phase concentrations, in accordance with Henry's law. The Henry's law constant for ozone at 25°C is 0.082 atm $\times$ m^3 $gmole^{-1}$ (Glaze, 1987). However, ozone will rapidly decompose to oxygen in aqueous solutions containing impurities such as organic compounds or particulates. Even in "pure" water, ozone will decompose at significant rates by reacting with the hydroxide ion, with the kinetics of the decomposition dependent upon ozone concentration and pH (Grasso and Weber, 1989). The relative stability of ozone in natural waters and double-distilled water is provided in Figure 5-16 (Rice and Netzer, 1982). For instance, ozone half-lifes of about 18 minutes in groundwater and less than 10 minutes in lake waters have been reported, while half-lifes in double distilled water were observed to be approximately 85 minutes (U.S. EPA, 1986).

Because of the high reactivity of ozone, equilibrium conditions are generally not achieved in practical applications of ozone oxidation, and the design of ozone-

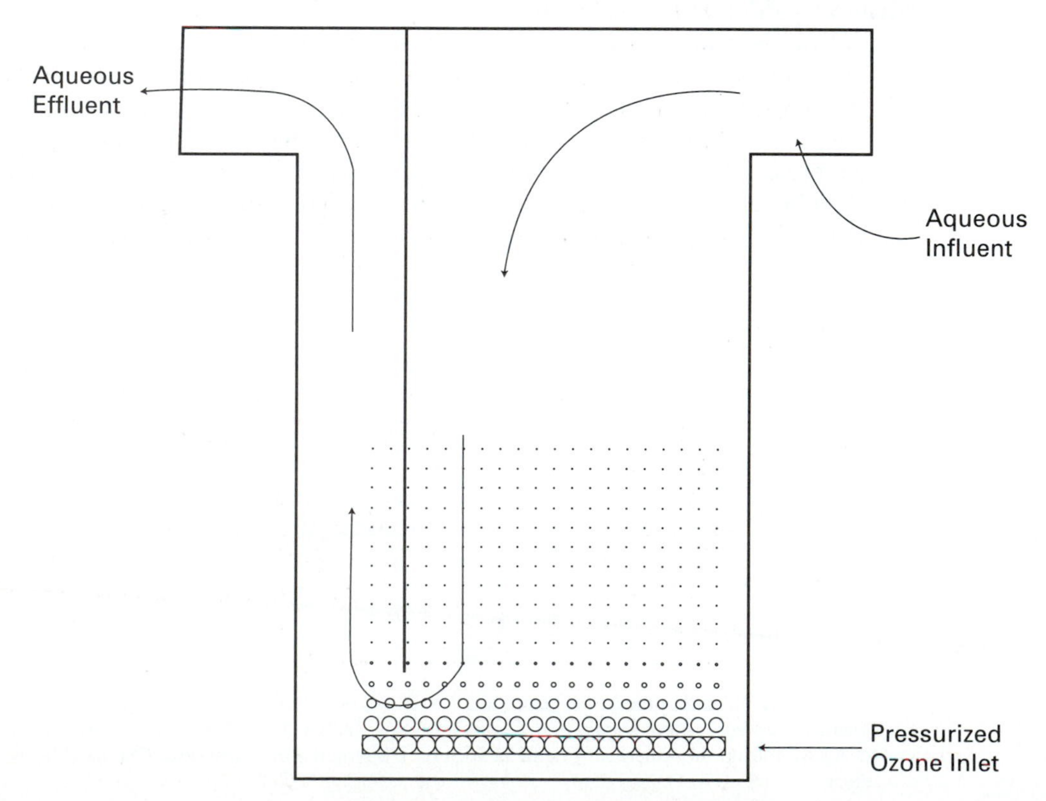

Figure 5-15 Counter-current sparged columns.

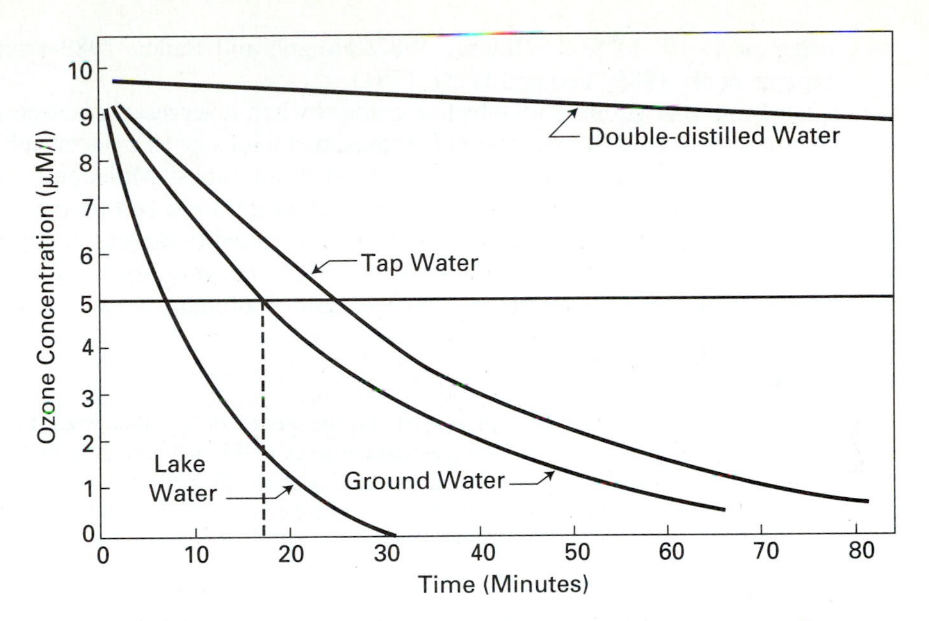

Figure 5-16 Decomposition rates of ozone in various waters (20°C) (R. Rice and A. Netzer, *Handbook of Ozone Technology and Applications,* Vol. 1, 1982, Ann Arbor Science, Reprinted by permission of authors.)

oxidation processes often requires consideration of the reaction kinetics. Most ozone-oxidation processes involve direct contact of ozone/air or ozone/oxygen mixtures with the solution of interest. In some cases, the rate of ozone utilization can exceed the rate of ozone mass transfer from the gas phase to the liquid phase, and therefore, the rate of oxidation becomes mass-transfer limited. For this reason, the rate of mass transfer of ozone from the gas phase to the liquid phase must also often be considered in modeling and optimizing the process.

Molecular Ozone

Molecular ozone is a strong oxidant, stronger than such conventional oxidants as chlorine, permanganate, or hydrogen peroxide, with a standard oxidation potential (25°C) of 2.07 volts, based upon the following half-reaction:

$$O_3 + 2H^+ + 2e^- \rightleftharpoons O_2 + H_2O \qquad E_0 = 2.07 \text{ V} \tag{5.29}$$

It is important to note that although the oxidation potential for ozone is relatively large, this only provides an indication of the equilibrium potential for ozone to oxidize a reduced species. The practicality of ozone as an oxidant is dependent upon the oxidation kinetics of molecular ozone and the potential for formation of the hydroxyl radical, which in some cases is the ultimate oxidant. Kinetic studies of molecular ozone oxidation have generally demonstrated a bimolecular reaction mechanism (i.e., first order kinetics to both ozone and reduced species concentration). Molecular ozone is a very selective oxidant, with second order rate constants varying over 12 orders of magnitude, and typical rate constants on the order of $1-10^3$ $M^{-1}s^{-1}$ (Hoigné, 1982; Hoigné and Bader, 1983-1 and 1983-2; Hoigné et al., 1985; Yao and Haag, 1991).

Molecular ozone is an effective oxidant when it serves as an electron transfer acceptor, such as in the oxidation of reduced metal ions, or as an electrophile, such as in the oxidation of phenol and other activated aromatics. Molecular ozone is also effective as a dipole addition reagent (through addition to carbon-carbon multiple bonds). Molecular ozone has also been demonstrated to be reasonably effective in oxidizing cyanide to cyanate (Gurol et al., 1985). However, molecular ozone demonstrates inadequate kinetics in the oxidation of aliphatic organic halides (Glaze, 1987).

Example 5-12

Hoigné and Bader (1983-1) reported that the second order rate constant for the oxidation of trichloroethylene by molecular ozone is 17-$M^{-1}s^{-1}$. Assume that a solution of TCE at 200 mg/l is being oxidized by sparging ozone through the solution, such that the concentration of ozone in solution is maintained at 5 mg/L. For what period of time must the sparging be conducted to reduce the TCE concentration to 5 mg/l?

Answer

Because the ozone concentration is maintained constant, the reaction is pseudo-first-order to TCE. Therefore:

$$\frac{d[\text{TCE}]}{dt} = K'_{\text{TCE}}\,[\text{TCE}]$$

where

$$K'_{\text{TCE}} = K_{\text{TCE}}[\text{O}_3] = \text{constant}$$

$$\ln\frac{[\text{TCE}]_t}{[\text{TCE}]_0} = -K'_{\text{TCE}}\,t$$

After converting the ozone concentration to molarity units, the required time is calculated to be 2,083 seconds (34.7 minutes). The above calculation assumes that none of the TCE is volatilized from the solution through the sparging. In reality, some volatilization of TCE would occur. Note that to oxidize TCE to the drinking water standard of 0.005 mg/l under these conditions would require approximately 100 minutes.

Ozone Mass Transfer

In the absence of any aqueous phase chemical reactions involving ozone, the kinetics of ozone mass transfer from the gas phase to the liquid phase can be described as follows:

$$\frac{dC}{dt} = K_L a(C^* - C) = K_L a\left(\frac{P}{\text{H}} - C\right) \qquad (5.30)$$

where C is the concentration of ozone in the aqueous phase, C^* is the liquid phase ozone concentration in equilibrium with the concentration of ozone in the gas phase, P is the partial pressure of ozone in the gas phase, H is the Henry's law constant for ozone, and K_La is the overall mass transfer coefficient. The mass transfer coefficient is dependent upon a number of factors, including the method of ozone transfer, and the temperature. Mass transfer coefficients for ozone are often provided by equipment manufacturers. Alternatively, mass transfer coefficients can be determined through equipment testing, or estimated by empirical correlation. Numerous correlations are available for estimating mass transfer coefficients in packed towers (Reid et al., 1977; Ball et al., 1984; Roberts et al., 1985). For bubble aeration, values of K_La typically range from 0.3 min^{-1} to 2.0 min^{-1} (Gurol and Nekouinaini, 1985; Matter-Müller et al., 1981). For gas-sparged turbine reactors, Grasso et al. (1990) presented a model to predict K_La as a function of various system parameters, including gas flow rate and impeller speed. Observed K_La values ranged from 0.15–0.45 min^{-1}.

In natural waters and wastewaters, ozone decomposition reactions and ozone depletion through oxidation of reduced species will deplete aqueous ozone, thus increasing the driving force for mass transfer. In such cases, the ozone mass transfer process can be described as:

$$\frac{dC}{dt} = K_La\left(\frac{P}{H} - C\right) - k_0C[OH^-] - \sum_{i}^{n} k_iS_iC \tag{5.31}$$

where k_0 is the second order rate constant for hydroxide initiated ozone decomposition, and the final term in the equation accounts for all chemical reactions which deplete ozone by reaction with substrates S_i. [k_0 has been estimated to be 70 M^{-1} sec^{-1} (Staehelin and Hoigné, 1985)].

Molecular Ozone Process Design

Design of ozone-oxidation systems based upon the equilibria and kinetics of reactions involving molecular ozone alone is often not valid because of the propensity for ozone to react with generally present free radical initiators that produce such radicals as the hydroxy radical (Glaze, 1986; Staehelin and Hoigné, 1985). Direct oxidation by molecular ozone dominates under conditions that hinder ozone decomposition or that result in the scavenging of free radicals. These conditions are low pH, presence of hydroxyl radical scavengers such as carbonate and bicarbonate, and the lack of other solutes that are initiators or promoters of ozone decomposition and formation of free radicals (Hoigné and Bader, 1976). In most "real" solutions, significant concentrations of ozone decomposition initiators and promoters will be present. Therefore, understanding of oxidation processes utilizing ozone for treatment of dissolved hazardous constituents requires consideration of the products of ozone decomposition, and the effect on the overall reaction kinetics and stoichiometry. This fact has in the past been overlooked in some studies, which have interpreted the the results of ozone oxidation kinetic studies assuming oxidation by molecular ozone alone. Because the extent to which free radicals were formed from ozone was not considered, and because system parameters that affect the rate and degree of free radical formation were not adequately characterized, the results of such studies are often not transferable to systems other than the exact one studied.

Ozone AOPs

Because the kinetics of molecular ozone oxidation of most hazardous organic compounds are generally not adequate to economically utilize this process for waste destruction, modifications of the conventional ozone-oxidation process have been pursued. These modifications promote the formation of strongly oxidizing free radical compounds, which exhibit oxidation kinetics that are often orders of magnitude faster than those exhibited by molecular ozone. Processes that are designed to utilize free radicals to increase the effectiveness of the oxidation process have been termed Advanced Oxidation Processes, or *AOP*s (Glaze et al., 1987). Although the mechanisms of formation of these radicals is not yet completely understood, the hydroxyl radical (OH•) has been found to be the most important of these radicals.

As previously noted, the kinetics of oxidation by hydroxyl radicals is generally much faster than the kinetics of oxidation by molecular ozone. Like molecular ozone, the hydroxyl radical has generally been observed to oxidize organic compounds following a bimolecular reaction mechanism, and kinetic studies have verified the corresponding second-order reaction kinetics. However, while molecular ozone is a very selective oxidant, with second-order rate constants varying over 12 orders of magnitude, the hydroxyl radical is a very unselective oxidant, with second-order rate constants varying over only two to three orders of magnitude; its typical rate constants are, approximately, $10^8 - 10^{10}$ $M^{-1}s^{-1}$ (Farhataziz and Ross, 1977), or one million times higher than second-order rate constants for molecular ozone oxidation. The hydroxyl radical is also a stronger oxidant than molecular ozone, with a standard oxidation potential of 2.80 volts at a temperature of 25°C. Because

TABLE 5-7 SECOND-ORDER RATE CONSTANTS FOR MOLECULAR OZONE OXIDATION

Compound	Rate constant ($M^{-1}*s^{-1}$)	Temperature (°C)
Benzene sulfonate ion	0.23 ± 0.05 (1)	23
Benzoate ion	1.2 ± 0.2 (1)	23
Benzene	2 ± 0.4 (1)	23
Toluene	14 ± 0.4 (1)	23
o-Xylene	90 ± 20 (1)	23
Anisole	290 ± 50 (1)	23
Phenol	1,300 ± 300 (1)	23
Tetrachloroethylene	<0.1 (1)	20
Trichloroethylene	17 ± 4 (1)	20
1,1 Dichloroethylene	110 ± 20 (1)	20
t 1,2 Dichloroethylene	5,700 ± 1,000 (1)	20
Ethanol	0.37 ± 0.04 (1)	20
Acetone	0.032 ± 0.006 (1)	20
2-Butanone	0.07 ± 0.02 (1)	20
2-Chlorophenol	1,100 ± 300 (2)	23
4-Chlorophenol	600 ± 100 (2)	23

(1) Hoigné and Bader, 1983-1.

(2) Hoigné and Bader, 1983-2.

of its faster kinetics and stronger oxidation potential, the hydroxyl radical can oxidize certain organic compounds which are recalcitrant to molecular ozone oxidation. Tables 5-7 and 5-8 provide second-order rate constants for the oxidation of various organic compounds by molecular ozone and hydroxyl radicals, respectively. Comparison of the rate constants provided in Tables 5-7 and 5-8 supports the observation that the hydroxyl radical oxidizes most compounds significantly faster than molecular ozone.

The chemistry of ozone decomposition in "real" aqueous systems is not yet completely understood. Even in "pure" water, the chemistry of ozone is quite complicated. Figure 5-17 provides a description of the fate of ozone in pure water, along with the rate constants of the various reactions (Staehelin and Hoigné, 1985). Further discussions on the kinetics and mechanisms of ozone decomposition are provided by Grasso and Weber (1989), Chelkowska et al. (1992), and Tomiyasu et al. (1985).

As can be seen in Figure 5-17, ozone can react with the hydroxide ion to form superoxide anion (O_2^-), which, in a sequence of reactions, reacts with ozone to eventually form the hydroxyl radical (OH•). The hydroxyl radical can in turn react with ozone to form the superoxide anion, resulting in a cyclical chain-reaction sequence which consumes ozone and forms the hydroxyl radical.

Other solutes can serve, like the hydroxide ion, as initiators of ozone decomposition, also resulting in the formation of the hydroxyl radical. Typical initiators of ozone decomposition are OH•, H_2O_2, Fe^{+2}, ultraviolet light, formate, and humic acids (Staehelin and Hoigné, 1985). Scavengers such as the carbonate and bicarbonate ions, react with the hydroxyl radical to form secondary radicals that do not promote the further decomposition of ozone, and that tend to quench the chain reaction. A single initiation step can cause the decomposition of hundreds of ozone molecules before chain termination, producing numerous highly oxidizing hydroxyl radicals. The net result is a complicated sequence of cyclical reactions that consume ozone, produce hydroxyl radicals, and oxidize reduced aqueous species. A simpli-

TABLE 5-8 SECOND-ORDER RATE CONSTANTS FOR HYDROXYL RADICAL OXIDATION

Compound	Rate constant ($M^{-1}*s^{-1}$)	Temperature (°C)
Benzene	670×10^7 (1)	22–25
Benzoate ion	560×10^7 (1)	22–25
Chlorobenzene	620×10^7 (1)	22–25
Ethanol	185×10^7 (1)	22–25
Acetone	7×10^7 (1)	22–25
Hydrogen peroxide	3.5×10^7 (1)	22–25
Bicarbonate ion	1.5×10^7 (1)	22–25
Carbonate ion	20×10^7 (1)	22–25
Tetrachloroethylene	230×10^7 (2)	25
Trichloroethylene	400×10^7 (2)	25

(1) Hoigné and Bader, 1976.

(2) Farhataziz and Ross, 1977.

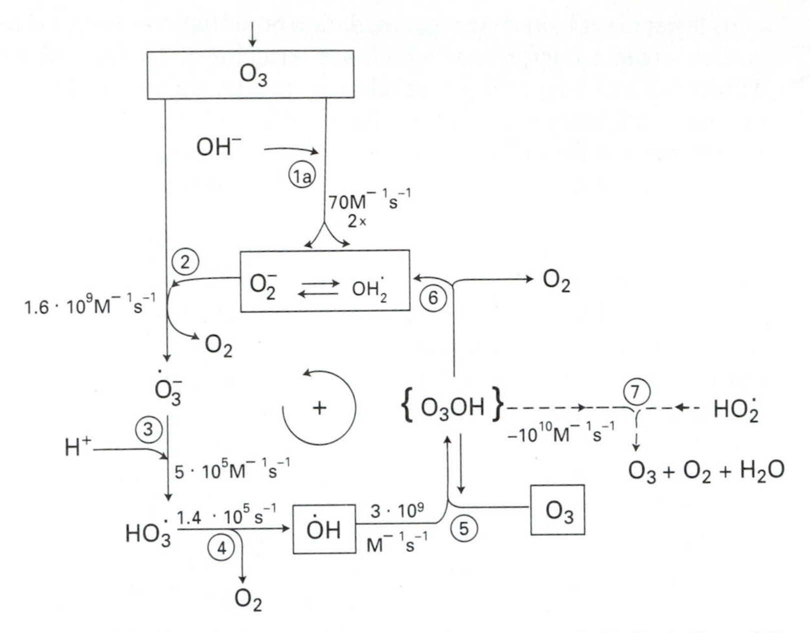

Figure 5-17 Reactions of aqueous ozone in "pure water." (J. Staehelin and J. Hoigne, "Decomposition of Ozone in Water in the Presence of Organic Solutes Acting as Promoters and Initiators of Radical Chain Reactions," *Environmental Science and Technology,* Vol. 19, No. 12, 1985, p. 1206. Reprinted with permission from the American Chemical Society. Copyright 1985, American Chemical Society.)

fied description of the reactions involving ozone, hydroxide ion, and the hydroxyl radical is provided in Figure 5-18 (Aieta et al., 1988).

The carbonate ion quenches the hydroxyl radical more quickly than the bicarbonate ion, and therefore, one would expect that an increase in solution pH would reduce the kinetics of the oxidation processes involving the hydroxyl radical. However, because the hydroxide ion also serves to initiate the decomposition of molecular ozone and the formation of the hydroxyl radical, increasing the solution pH may in some cases increase the overall oxidation kinetics (Glaze and Kang, 1988). The effect of pH on the kinetics of an ozone *AOP* is therefore dependent upon the solution carbonate alkalinity.

Hydrogen peroxide, or more accurately, the hydrogen peroxide anion (HO_2^-) has been identified as an initiator of ozone decomposition (Glaze and Kang, 1988), and processes utilizing ozone and hydrogen peroxide have been proposed as alternative technologies for the treatment of water and wastewater contaminated with synthetic organic compounds (Aieta et al., 1988). Figure 5-19 summarizes the reactions involving ozone and hydrogen peroxide that produce the hydroxyl radical (Aieta et al., 1988).

Other *AOP*s that utilize ozone are processes that use ultraviolet light together with ozone, and sometimes also hydrogen peroxide, to promote the formation of hydroxyl radicals and the oxidation of organic compounds (Lewis et al., 1990; U.S. EPA, 1990; Glaze, 1987; Buckley et al., 1988; Jones et al., 1985). Figure 5-20 pro-

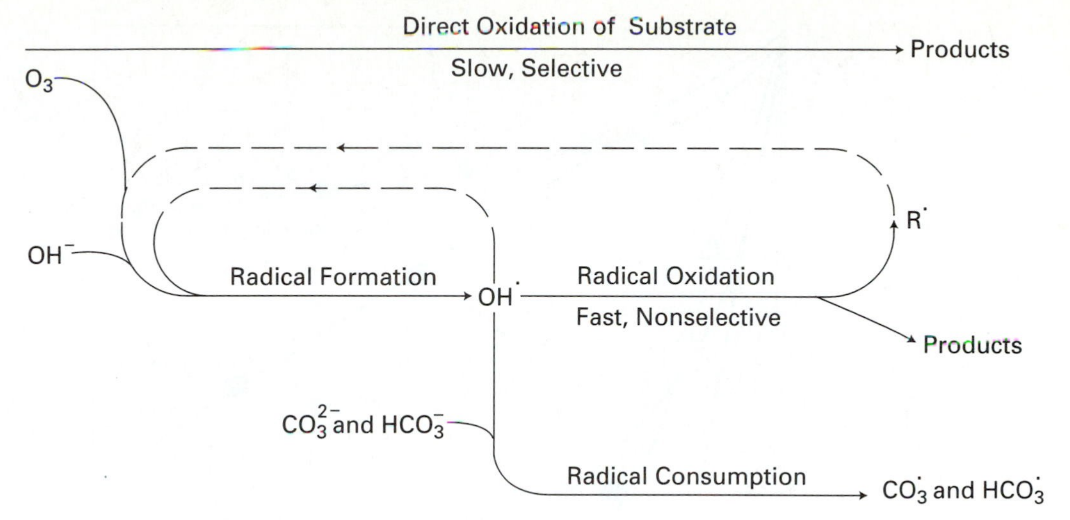

Figure 5-18 Pathways of reaction of ozone. (Aieta et al., "Advanced Oxidation Processes for Treating Groundwater Contaminated with TCE and PCE: Pilot Scale Evaluations," *Journal AWWA*, Vol. 80, No. 5, 1988, p. 64. Reprinted by permission of the American Water Works Association.)

vides a comparison of the rate of oxidation of various chlorinated organic compounds by ozone alone, and ozone with UV light, demonstrating that the effectiveness of the oxidation reaction is in general enhanced by the inclusion of UV light in the process (Glaze, 1987). The reactions of UV light with ozone and hydrogen peroxide may be summarized as follows (Topudurti et al., 1993):

$$O_3 + h\nu + H_2O \rightleftharpoons H_2O_2 + O_2 \tag{5.32}$$

$$H_2O_2 + h\nu \rightleftharpoons 2OH\bullet \tag{5.33}$$

Because in *AOPs* utilizing UV light, the hydroxyl radical is actually formed by reaction of hydrogen peroxide with the UV radiation, UV/Ozone *AOPs* are sometimes enhanced by the addition of hydrogen peroxide. The UV/Ozone/Hydrogen Peroxide process has been patented by Ultrox International of Santa Ana,

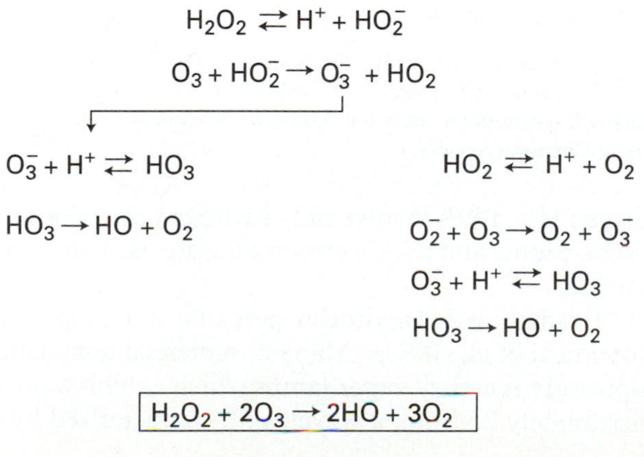

$$H_2O_2 \rightleftharpoons H^+ + HO_2^-$$

$$O_3 + HO_2^- \rightarrow O_3^- + HO_2$$

$$O_3^- + H^+ \rightleftharpoons HO_3 \qquad HO_2 \rightleftharpoons H^+ + O_2^-$$

$$HO_3 \rightarrow HO + O_2 \qquad O_2^- + O_3 \rightarrow O_2 + O_3^-$$

$$O_3^- + H^+ \rightleftharpoons HO_3$$

$$HO_3 \rightarrow HO + O_2$$

$$\boxed{H_2O_2 + 2O_3 \rightarrow 2HO + 3O_2}$$

Figure 5-19 Expected stoichiometry for hydroxyl radical formation from ozone and hydrogen peroxide. Aieta et al., "Advanced Oxidation Processes for Treating Groundwater Contaminated with TCE and PCE: Pilot Scale Evaluations," *Journal AWWA*, Vol. 80, No. 5, 1988, p. 67. Reprinted by permission of the American Water Works Association.)

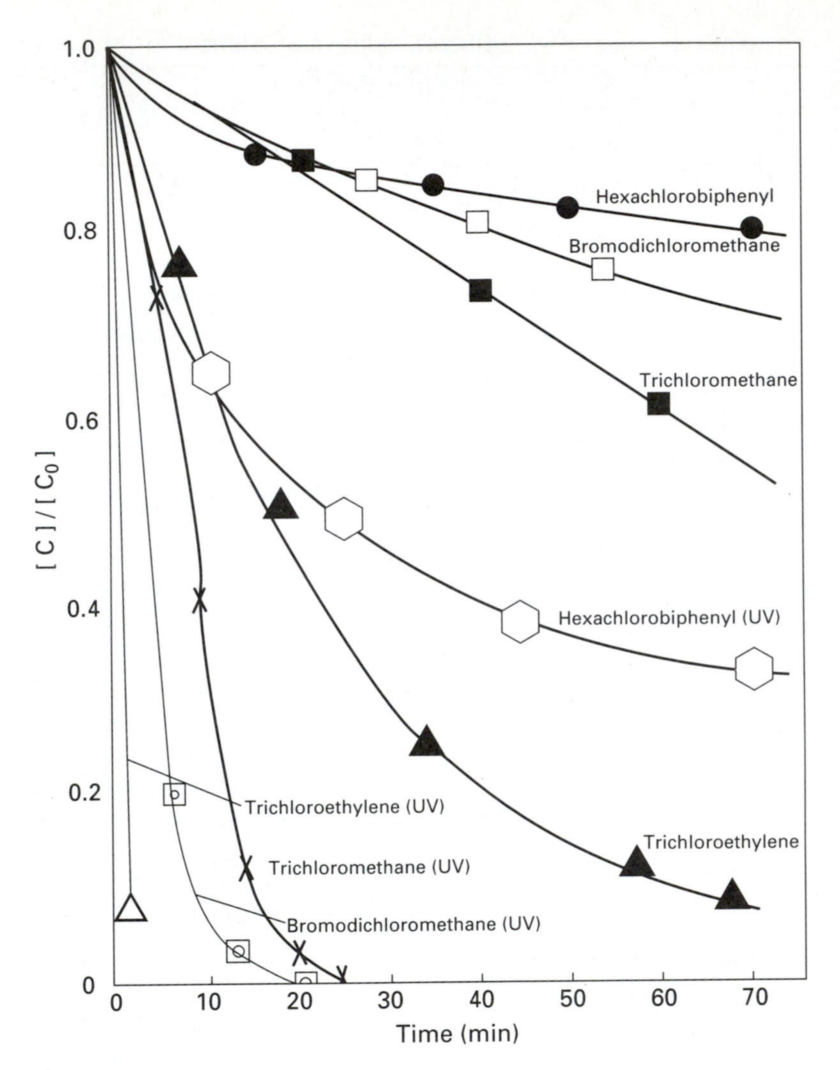

*At pH 6–7, ozone dose rate = 1.0–1.4 mg/L min; UV frequency is 254 nm from a low-pressure mercury lamp, with a flux of 0.42 W/L.

Figure 5-20 Destruction of chlorinated organics. (W. H. Glaze, "Drinking Water Treatment with Ozone," *Environmental Science and Technology*, Vol. 21, No. 3, 1987, p. 229. Reprinted with permission from the American Chemical Society. Copyright 1987, American Chemical Society.)

California. Alternatively, some UV *AOP*s involve only hydrogen peroxide and UV light. In *AOP*s where UV light, ozone, and hydrogen peroxide are used, the potential reaction pathways are numerous.

The absorbence of UV radiation by hydrogen peroxide is maximized at a wavelength of 220 nm (Topudurti et al., 1993). Many commercial applications of UV photolysis utilize low-pressure mercury vapor lamps which exhibit a predominant UV emission at approximately 254 nm, a wavelength characterized by a low molar extinction coefficient for hydrogen peroxide. However, recently developed medium-pressure lamps produce UV emissions at predominant wavelengths closer to the optimal wavelength of 220 nm, and UV lamps using Xenon can be designed to produce a spectral emission with the maximum intensity near 220 nm (Topudurti et al., 1993). A schematic of one commercial UV/Ozone/Hydrogen Peroxide oxidation system is provided in Figure 5-21 (Lewis et al., 1990).

Another oxidation process that involves the hydroxyl radical as the principal oxidant is the process utilizing ferrous iron (Fe^{+2}) and hydrogen peroxide, the so-called Fenton's reagent, to produce the hydroxyl radical (Sedlak and Andren, 1991 a, b). Although this process is promising conceptually, little pilot or full scale data utilizing the Fenton's reagent for hazardous waste treatment is available.

Ozone AOPs Process Design

Important factors in the design of ozone oxidation processes and the various *AOP*s include the following:

- contaminant concentration
- type and concentrations of constituents capable of exerting oxidant demand
- applied and transferred ozone dosage (ozone and ozone *AOP*s)
- reactor residence time
- reactor mixing characteristics
- temperature
- gas to liquid ratio (ozone and ozone *AOP*s)
- applied H_2O_2 dosage (ozone/H_2O_2, UV/H_2O_2, UV/ozone/H_2O_2)
- ratio of applied H_2O_2 to ozone (ozone/H_2O_2, UV/ozone/H_2O_2)
- applied and absorbed UV intensity; watts/cm^2 (UV processes)
- turbidity
- pH
- alkalinity

Because of the complicated kinetics of ozone oxidation processes and *AOP*s, design of such processes generally requires bench and/or pilot scale studies. Only through such preliminary studies can the effects on the process of the design parameters be fully quantified. However, some generalizations can be made. For oxidation by both molecular ozone and hydroxyl radical, the kinetics of the oxidation process has been generally observed to be first-order to both the oxidant (ozone or hydroxyl radical) and the contaminant of interest. However, this important fact cannot easily be applied to process design because the oxidant concentration is not easily related to the applied oxidant dosage due to the complicated and numerous reaction sequences involving ozone and the hydroxyl radical. In systems utilizing ozone, applied ozone dosages are typically on the order of 10 ppm to 100 ppm, with ozone applied at a ratio of 1.5 lb to 3.0 lb of ozone per lb of contaminant (Spivey et al., 1986; Lewis et al., 1990). *AOP* reaction residence times typically range from one minute to 60 minutes.

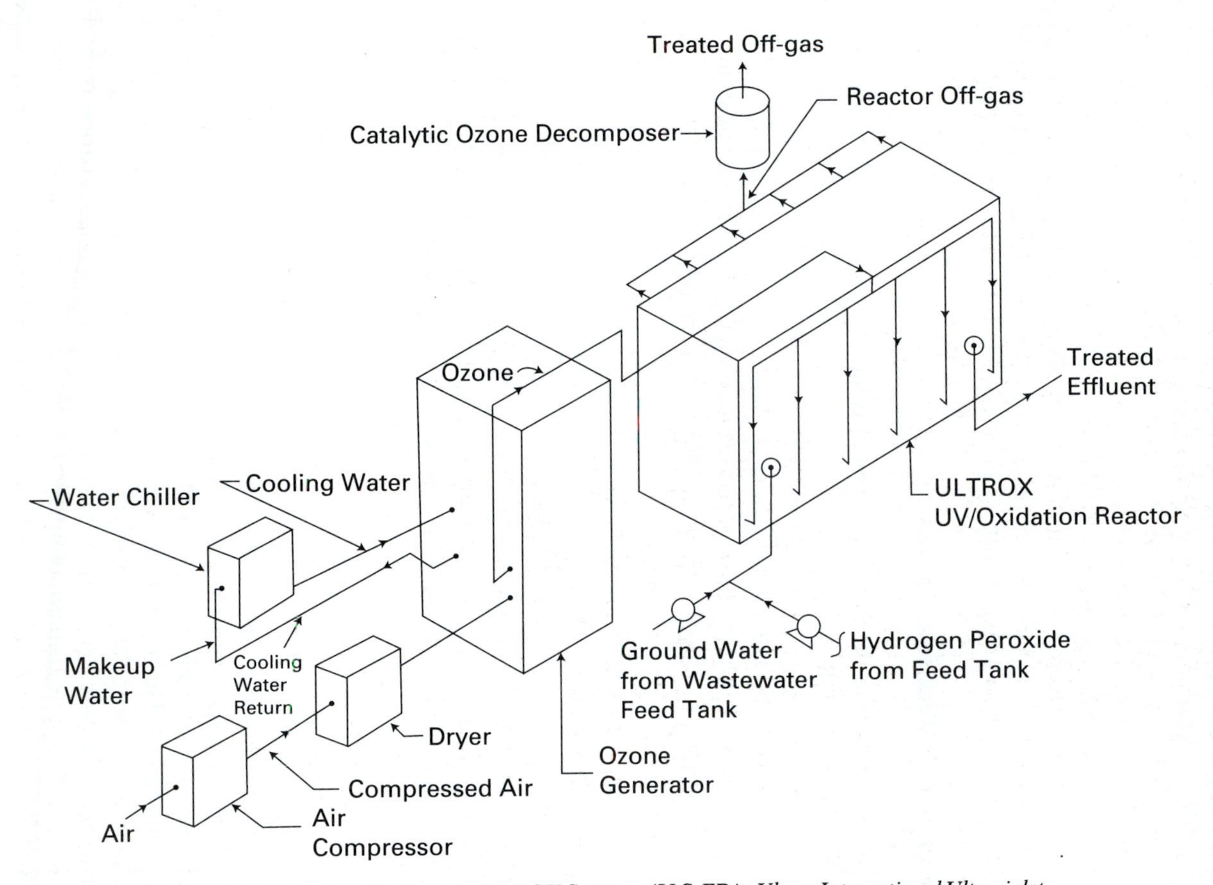

Figure 5-21 Isometric view of ULTROX System. (U.S. EPA, *Ultrox International Ultraviolet Radiation/Oxidation Technology: Applications Analysis Report,* EPA/540/A5-89/012, September, 1990, p. 7.)

Because the kinetics of ozone oxidation reactions and reactions induced by other *AOPs* have generally exhibited positive reaction orders to both the oxidant concentration and the contaminant concentration, the oxidation kinetics will generally increase with increasing contaminant concentrations and increasing oxidant concentration (and applied dose). It has also generally been observed that oxidation rates increase with increasing pH, and decreasing alkalinity. However, this effect of pH is not always observed, and is dependent upon the alkalinity of the solution, and the effect that solution pH has on the speciation of the contaminant. As in many chemical reactions, the rate of oxidation reactions involving ozone and hydroxyl radicals is increased with increasing temperature. Oxidation kinetics in systems involving UV light have generally been observed to increase with an increase in the UV intensity (Borup and Middlebrooks, 1988). High turbidity can reduce the effectiveness of systems utilizing UV light, because the turbidity absorbs some of the UV light.

AOP Treatment Costs

Estimating the treatment costs for a given *AOP* is difficult because the oxidant dosage will vary depending upon the nature and concentration of the pollutant of interest, the concentration of background oxidizable constituents, and the required degree of treatment. A perusal of the literature indicated a wide range of treatment costs associated with the various potential *AOP*s. As compared to processes involving ozone alone, processes using UV light have the additional power costs associated with the UV light production. In fact, the power costs associated with the UV light can be greater than the power costs associated with the ozone production (Arisman and Musick, 1980). The power consumption costs associated with the ozone production process can be quite high, making treatment processes using ozone relatively expensive. Typical power requirements for ozone production range 16 and 24 watt-hr per gram of ozone produced for oxygen and air-fed systems, respectively (Glaze, 1987). Processes using hydrogen peroxide have the additional chemical costs associated with the hydrogen peroxide. The use of UV light and/or hydrogen peroxide, in conjunction with ozone, is often justified by the resultant order of magnitude increase in the kinetics of the oxidation reaction, and a subsequent reduction in the required ozone dosage (Buckley et al., 1988). Fletcher (1987) reported that treatment costs using UV/Ozone ranged from $0.25/1,000 gallons for a slightly contaminated groundwater to $40/1,000 gallons for a highly contaminated industrial wastewater. For treatment of chlorinated VOC contaminated groundwater using a UV/ozone/hydrogen peroxide process, Fletcher (1991) reported that the system energy consumption was 11 kwh/h of operation. Zeff and Harris (1983) give the costs for the treatment of an agrichemical plant wastewater (TOC of 2,500 mg/l) using the UV/Ozone ULTROX process as $29/1,000 gallons. Buckley et al. (1988), reported that the typical treatment costs associated with the RAYOX process (UV/ozone/peroxide) ranged from $0.75 to $6/1,000 gallons. Arisman and Musick (1980) state that the treatment costs for the destruction of a PCB contaminated wastewater (50 mg/l) using the ULTROX process were $4.35/1,000 gallons. Aieta et al. (1988) compared the costs associated with air stripping to ozone/peroxide for the treatment of a PCE (15 mg/l) and TCE (140 mg/l) contaminated groundwater. The treatment costs were estimated as $0.094/1,000 gallons for ozone/peroxide, $0.075/1,000 gallons for air stripping, and $0.27/1,000 gallons for air stripping with gas phase GAC.

Applications of Ozone and Other AOPs

Applications of ozone oxidation and other various *AOPs* for hazardous waste treatment have generally been for the oxidation of dissolved organic contaminants in groundwaters or industrial wastewaters. It is not likely that these processes are feasible for the oxidation of wastes containing significant solid fractions, because of the excessively high oxidant demand imposed by the presence of the solids. Ozone oxidation processes and other *AOP*s can achieve high degrees of contaminant oxidation (greater than 99%) when optimized. Although relatively expensive compared to such conventional technologies as air stripping and carbon adsorption, the oxidation processes have the advantage of (potentially) destroying the contaminants, instead of simply inducing a change of phase with subsequent concentrating of the contaminants prior to ultimate disposal. However, it is not certain that oxidation of the contaminants of interest occurs through complete contaminant mineralization. Care must be taken that the oxidation process does not produce partially oxidized compounds that are also hazardous. Because of increasing required oxidant dosages with increased contaminant concentrations and total organic carbon concentrations, ozone oxidation and the other *AOP*s are generally not cost effective at high contaminant concentrations or high total organic carbon concentrations.

Although many bench-scale and some pilot-scale studies on the destruction of hazardous wastes by ozone oxidation and *AOP*s have been published, information on full-scale applications of these processes is scarce. Gurol et al. (1985) performed bench scale experiments on the destruction of cyanide in synthetic plating wastes using ozone. Although ozone effectively oxidized the free cyanide ion over a range of pH from 2.5 to 12, hydrogen cyanide was not effectively oxidized. Cyanide complexed with cuprous copper (Cu^{+1}) was quickly oxidized by the addition of ozone, at rates exceeding the oxidation of free cyanide. However, ozone was not effective in oxidizing cyanide complexed with ferric iron (Fe^{+3}). By adding cuprous copper to the ferric cyanide solution, the ozone oxidation kinetics were significantly enhanced. Cyanate was produced by the oxidation of the cyanide wastewaters. Although further addition of ozone eventually oxidized the cyanate, the kinetics of cyanate oxidation were much slower than the kinetics of free cyanide ion oxidation.

Whitlow and Roth (1988) analyzed the kinetic data available from the literature on the oxidation of a number of contaminants by ozonation. The majority of the data evaluated focused on the oxidation of phenols and cyanides. An empirical kinetic expression was evaluated, which included as variables the initial and steady-state contaminant concentration, the ozone application rate, and the pH (OH^- concentration). While phenol oxidation by ozone was found to be first-order to the steady-state phenol concentration, the oxidation of cyanide by ozone was found to be zero order to cyanide. The solution pH was not found to be a very significant kinetic parameter, while the ozone application rate was found to be a very significant kinetic parameter.

Much of the interest in *AOPs* is focused on applications for treatment of VOC contaminated groundwater. In a pilot-scale test of an ozone/hydrogen peroxide process, Aieta et al. (1988) studied the kinetics of oxidation of trichloroethylene (TCE) and tetrachloroethylene (PCE). The reactor was a continuous flow, sparged and baffled reactor. Reactor residence times were varied from 2.7 to 15 minutes, transferred ozone dosages were varied from 0.99 mg/l to 8.73 mg/l, applied hydrogen peroxide dosages were varied from 0.67 mg/l to 3.0 mg/l, and the pH was varied from 7.25 to 7.55. The concentrations of TCE ranged from 107 ppb to 143 ppb, and the concentrations of PCE ranged from 11 ppb to 16 ppb. Destruction efficiencies for TCE and PCE were observed to range from 50–96% and 41–88%, respectively. The kinetic data was analyzed assuming pseudo-first order (to the contaminant) reaction kinetics, with ozone and hydrogen peroxide at steady-state excess concentrations. The optimal mass ratio of applied hydrogen peroxide to ozone was determined to be 0.5, with removal efficiencies determined to be principally a function of the transferred ozone dosage. This dependency of the removal efficiency on the transferred ozone dosage indicated that the reactions were mass transfer limited by ozone phase transfer, with hydrogen peroxide serving to improve the mass transfer rate. Rate constants determined from the data analysis indicated that while an applied dosage of 9 mg/l of ozone would be required to achieve 95% TCE destruction in an ozone-only system at a residence time of 15 minutes, only 4 mg/l of transferred ozone would be required with the addition of 2 mg/l of transferred hydrogen peroxide dosage.

The Ultrox (UV/ozone/hydrogen peroxide) process was field-evaluated for the destruction of chlorinated VOCs in groundwater by Lewis et al., (1990). The groundwater pH and alkalinity were approximately 7.2 and 600 mg/l (as $CaCO_3$), respectively. Once optimized, removal efficiencies for TCE and total chlorinated VOCs as high as 99% respectively were obtained in a 150 gallon tank at a residence time of 40 minutes. The test unit contained 24 UV lamps, operating at 65 watts each. The optimized system parameters were an ozone dose of 110 mg/l (maximum in the study), a hydrogen peroxide dose of 13 mg/l, and all 24 of UV lamps operating. It was observed that for some of the contaminants (most principally 1, 1 dichloroethane and 1,1,1 trichloroethane), a significant fraction of the total removal was due to stripping (volatilization), as opposed to oxidation.

ION EXCHANGE

Ion exchange is a chemical treatment process used to remove dissolved ionic species from contaminated aqueous streams. Treatment for both anionic and cationic contaminants can be effected by ion-exchange processes. Although some ion exchange materials are used to remove organic constituents, such as humic and fulvic acids from aqueous streams (Fu and Symons, 1990) or phenol recovery from industrial process streams (Bolto and Pawlowski, 1987), ion exchangers are generally used to remove dissolved inorganic species from aqueous streams. Some common applications of ion exchange are the removal of hardness from drinking water supplies, demineralization of water for laboratory or industrial use, treatment of nuclear power plant cooling waters, and the removal of nitrates from potable groundwater (Dorfner, 1977). Also, ion exchange has been demonstrated to be a suitable technology for the removal of radium from potable groundwater supplies (Mangelson and Lauch, 1990; Subrammonian et al., 1990). Ion exchange is used in industry for the treatment of boiler water and wastewaters, and in various different production processes (Dorfner, 1977; Bolto and Pawlowski, 1987; Grammont et al., 1986). Common applications of ion exchange for the treatment of hazardous wastes are the removal and/or recovery of metal ions and cyanide from industrial rinse waters and wastewaters, treatment of contaminated groundwater, and treatment of landfill leachate.

Historically, the first known ion-exchange materials were natural inorganic minerals such as clays and aluminosilicates (Dorfner, 1977). For instance, clinoptilite, a sodium–calcium aluminum silicate, is still used today for ammonia removal because the mineral has a strong affinity for ammonium ions. However, most applications of ion exchange currently utilize synthetic ion exchange materials. These synthetic ion exchangers generally consist of polymers that have been designed to exhibit ion-exchange capacity.

Ion-exchange processes are effected by contacting an ion-exchange material with a solution containing ions that are to be exchanged, that is, removed from solution. Both cationic and anionic exchangers are available. Ion exchangers are insoluble high-molecular weight polyelectrolytes that have fixed ionic groups attached to a solid matrix. Figure 5-22 provides a schematic diagram of one type of synthetic cationic exchange resin. The fixed ionic groups are electrostatically balanced by mobile counterions of opposite charge, which are free to move within the solvated pore spaces within the exchanger. The fixed groups are surrounded by a membrane that allows the passage of solvent and dissolved ions. The counterions can pass through the membrane and be exchanged with other ions of similar charge in the solution. Electroneutrality is maintained within the exchanger by the charges of the fixed ionic groups being balanced electrostatically by close proximity of the required number of counterions. For example, a fixed ionic group of charge -1 could be balanced by one counterion of valency $+1$. If the mobile counterion has a charge of $+2$, then each counterion would be balanced by two fixed groups of charge -1. Thus, due to the electroneutrality requirements within the exchanger, the exchange of counterions is stoichiometric, where the stoichiometry of the reaction is dependent upon the charges of the exchanging counterions. This phenomenon is known as the equivalence of exchange. To a large degree, ions of the same charge as the fixed group, known as co-ions, are excluded from the ion exchanger due to electrostatic repulsion. Cationic exchangers, whose fixed ionic groups are all anions, will exchange cations. Conversely, anionic exchangers, which have fixed cationic groups, will exchange anions.

Consider a cationic ion-exchange resin with fixed anionic charges of R^{-1} which initially are balanced by counterions A^{+1}. Batch contacting of this resin with a solution containing the cation B^{+1} will allow the following ion exchange reaction to occur:

$$(R^{-1})A^{+1} + B^{+1}_{(aq)} \rightleftharpoons (R^{-1})B^{+1} + A^{+1}_{(aq)} \tag{5.34}$$

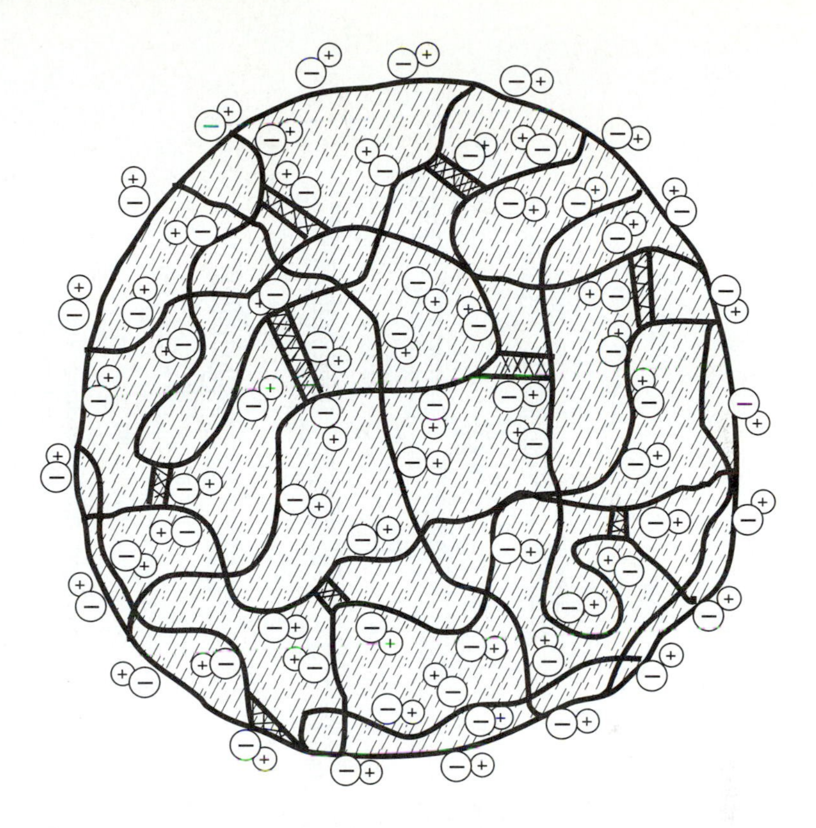

⊖ Fixed negatively charged exchange site, i.e., SO_3^-

⊕ Mobile positively charged exchangeable cation, i.e., Na^+

Polystyrene chain

Divinylbenzene crosslink

Water of hydration

Figure 5-22 Schematic picture of a gel-type strong acid cation exchanger. (R. Kunin, "Ion Exchange for the Metal Products Finisher," *Products Finishing,* April 1969, p. 67. Reprinted by permission of Gardner Publications, Inc.)

In general, ion-exchange reactions are reversible, and a mass action interpretation can be given for the reaction described in Equation 5.34, resulting in the following equilibrium expression:

$$K_B^A = \frac{\{RB\}\{A^{+1}\}_{\text{aq}}}{\{RA\}\{B^{+1}\}_{\text{aq}}} \tag{5.35}$$

where {} indicate the activities of the species enclosed by the brackets. Consider again the reaction described in Equation 5.34. As the exchanger is contacted with the solution, the gradient in concentration of the A^+ and B^+ cations between the solution and the exchanger causes an exchange reaction to occur. Initially, as the exchanger

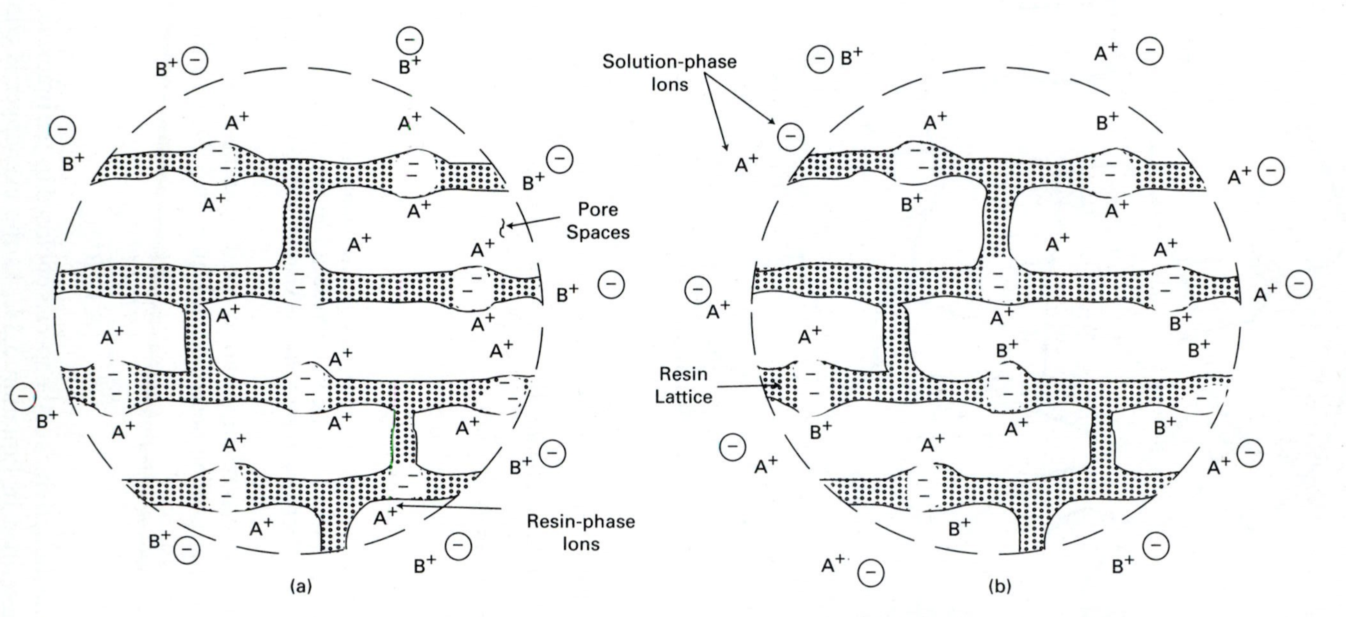

Figure 5-23 Schematic diagrams of a cation exchange resin framework with fixed exchange sites prior to and following an exchange reaction. (a) Initial state prior to exchange reaction with cation B^+; (b) equilibrium state after exchange reaction with cation B^+. (W. J. Weber, *Physicochemical Processes for Water Quality Control,* Copyright 1972, p. 263. Reprinted by permission of John Wiley & Sons, Inc.)

contains no B^+ cations, the reaction is driven strongly to the right, with B^+ cations being taken up by the resin, and A^+ cations being released to the solution. However, as more and more exchange sites of the exchanger are taken up by B^+ ions, the driving force for exchange decreases until an equilibrium is established between the two phases (Figure 5-23), with the relative concentrations of exchangeable ions related by some expression analogous to Equation 5.35. It has generally been observed that ion exchangers exhibit characteristic preferences, or selectivities, for ions (Helfferich, 1962). Therefore, the ratio of concentrations of the various ions in solution will likely be different from the concentration ratio in the resin phase at equilibrium. Ion-exchange equilibrium constants are dependent upon a number of factors, including the temperature, the nature of the exchanger fixed (functional group), the relative charges of the exchangeable ions, and the effective relative size of the exchangeable ions.

Ion exchangers are generally utilized in column reactors so that a high degree of exchanger utilization is achieved. Figure 5-24 provides a schematic view of an ion-exchange column where a waste stream containing the contaminant B^+ is removed from solution by passage through a column of ion exchange resins initially saturated to the A^+ ion. As the waste stream passes through the column, the B^+ cations are quickly exchanged with A^+ cations, with the exchange reaction occurring in a relatively narrow zone of the column. With continued passage of wastewater through the column, more and more B^+ cations are removed from the solution, and the active zone of exchange moves down the column. Eventually the entire column becomes saturated with B^+ ions, and the column is said to be exhausted. A useful characteristic of ion exchanger is that the exchange reaction is generally reversible. Therefore, once an exchanger has been exhausted, that is, there is no longer a sufficient concentration driving force for exchange to occur, the exchanger can be regenerated back to its original form, or other forms, by contacting it with a solution containing an excess of the desired counterion. In the example reactor provided in Figure 5-24, the column would be regenerated by a regenerant solution containing a high concentration of A^+ cations. The high concentration of the regenerant is needed to drive the equilibrium backwards so that the exchanger is again saturated to the A^+ counterion. In this way, ion exchangers can be used multiple times by sequential service and regeneration runs. The spent regenerant will contain a concentrated solution of both A^+ and B^+, presumably requiring additional treatment. However, this volume of waste regenerant is small relative to the volume of wastewater treated. Still, it is important to realize that ion exchange is ultimately a volume-reduction process. Regeneration and other operational aspects of ion exchange will be discussed in more detail later.

Ion-Exchange Types

As previously mentioned, both natural and synthetic ion exchangers are available. However, due to their greater stability, higher exchange capacity, and greater homogeneity of their exchange properties, synthetic ion exchange materials are predominantly used today. Because of their prevalence of use, much of the rest of the chapter

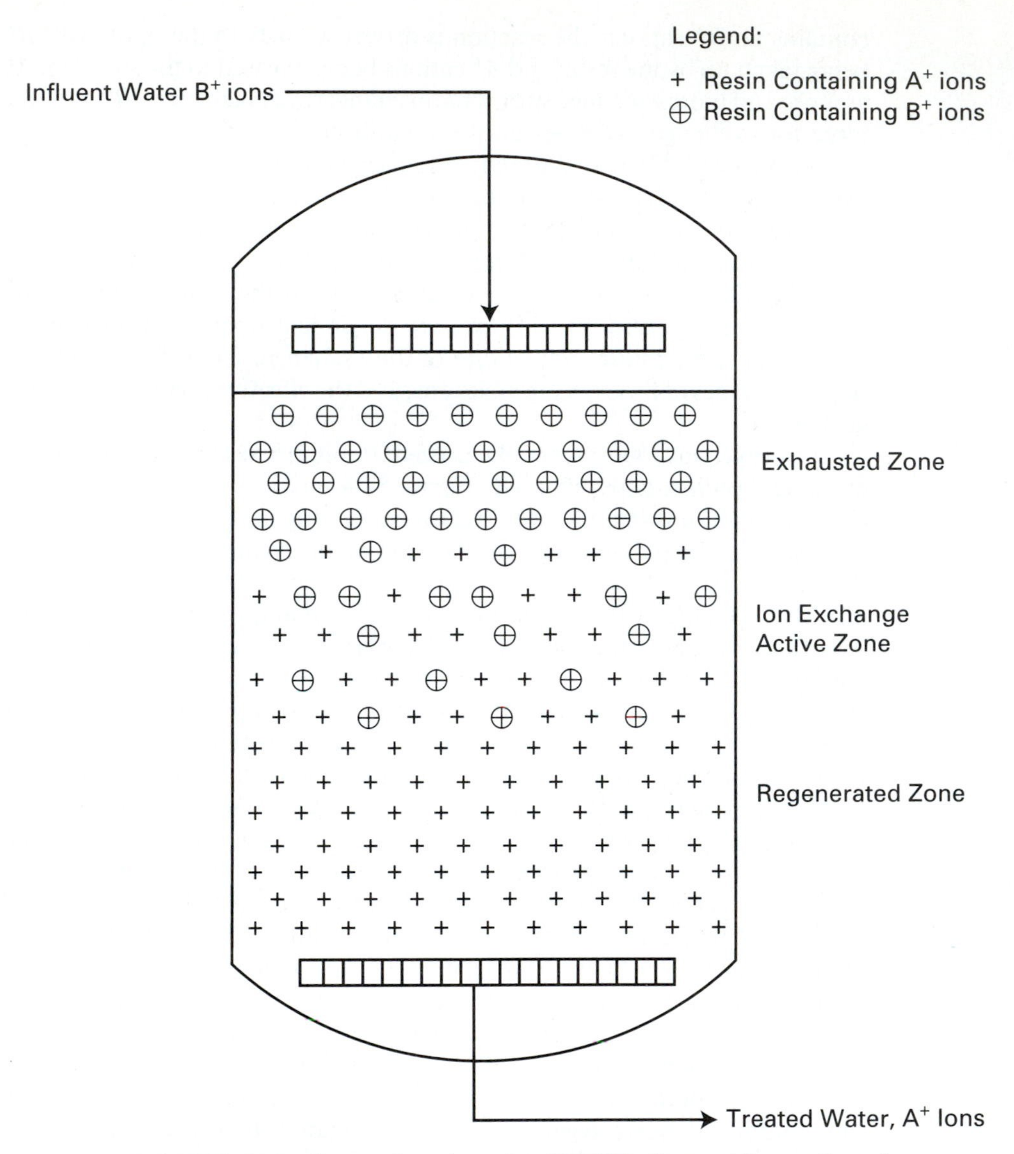

Figure 5-24 Ion exchange column in service. U.S. EPA, *Summary Report: Control and Treatment Technology for the Metal Finishing Industry, Ion Exchange,* EPA/625/8-81-007, June 1981, p. 32.)

will focus on synthetic ion exchanger resins. The reader interested in (natural) inorganic ion exchangers should consult Amphlett (1964).

Synthetic ion exchangers are generally polymeric materials (resins) that have been chemically treated to render them insoluble, and to exhibit ion exchange capacity. Most often, the exchanger is in the form of spherical resin beads. However, ion-exchange membranes are also available. The resins have active ionogenic groups which are covalently bonded to the polymer. These fixed groups are either permanently ionized or are capable of ionization or the acceptance of protons to form a charged site. Mobile ions of opposite charge to the fixed ionic groups (counterions) can be exchanged from the external solution by the resin. An ion with a charge that is the same sign as the charge of the fixed ionic group is known as a co-ion.

Various different polymers have been used as ion exchangers. Copolymers of styrene and divinylbenzene (DVB), are the most common synthetic ion-exchange materials. The DVB serves as a cross-linking agent, which renders the polymer insoluble, and provides desired mechanical stability and elasticity. The resultant material is an irregular, three-dimensional matrix of macromolecular hydrocarbon chains. The degree and distribution of the cross-linking agent is important in that it determines the internal pore structure. The size and spatial distribution of the pores have a large effect on the kinetics of the exchange process and the equilibrium distribution. While a more highly cross-linked resin is more stable, it is also more brittle. Furthermore, the rate of ion exchange is decreased with increasing degree of cross-linking. It has generally been observed that selectivity increases with increasing cross-linking.

Physical Properties of Ion Exchangers

Ion exchangers can be characterized by a number of physical properties including particle size, density, degree of cross-linking, resistance to oxidation, and thermal stability. Ion-exchange resins are commonly available commercially with particle diameters ranging from 0.04 mm to 1.2 mm, and with cross-linking percentages ranging from 5% to 25% DVB. Typical specific gravities of wet cationic and anionic exchange resins are 1.10 – 1.35 and 1.05 – 1.15, respectively, with typical bulk wet densities (volume includes void fraction) of ion-exchange resins ranging from 600 mg/L to 800 mg/l. Because ion-exchange resins imbibe water and swell to varying degrees depending on the exchanger and solution characteristics, other important physical properties of ion-exchange resins are their swelling characteristics and moisture content. The moisture content and degree of swelling of ion exchangers vary with the ionic form of the exchanger. For instance, a cationic exchanger with all exchangeable sites in the form of H^+ will contain more water, and occupy more volume, than the same exchanger in the K^+ form. Because ion-exchange resins are generally used in column reactors, the change in moisture content of the resins with the exchanger composition must be considered to insure that swelling does not stress and damage the reactor as the resins are utilized. Swelling characteristics of ion exchangers are also a function of the degree of cross-linking. More highly cross-linked resins are more resistant to swelling than ion exchange resins with low degrees of cross-linking.

Perhaps the most critical property of an ion exchanger is the quantity of counterions that can be exchanged by the exchanger. This quantity is generally characterized by the term *capacity,* which is expressed in units such as milliequivalents/gram. The capacity must be described in terms of equivalents because ion exchange reactions are balanced by charge. For example, two counterions of unit valency can be exchanged with one counterion that is divalent. Because the capacities of ion exchangers vary with the ionic form of the exchanger, exchange capacities should be referenced in terms of ionic form. Also, because the density of the exchanger changes with moisture content, ion exchanger capacities vary with the exchanger moisture content.

The dry-weight capacity of an ion-exchange resin is defined in units such as milliequivalents per gram of dry resin, and is a measurement of the number of functional groups available for exchange in a given mass of dry resin. Therefore, dry-weight capacities are basically constants for a given resin type. Fully sulfonated styrene-DVB resins have dry-weight capacities of approximately 5.0 meq/g. Strong-base resins exhibit a greater variability in dry-weight capacity (2.0 meq/g to 5.0 meq/g) because the number of functional groups that attach to the matrix benzene ring can vary depending upon the manufacturing process (Montgomery, 1985).

Ion-exchange capacities are also defined in terms of wet-volume capacity in units of equivalents per liter of wet-bed volume, with the bed volume defined as the volume of resins *plus* interparticle water contained in the void space of the bed. In other words, if a totally water-saturated column of ion-exchange resins, plus the water between the resins, occupies one liter and contains 2.0 equivalents of exchangeable sites, then the wet-volume capacity of the resin is operationally defined as 2 equivalents/liter. It is obvious that this wet-volume capacity (though more quickly applicable to process applications) is prone to greater variability than a dry-weight resin capacity. Typical wet-volume capacities of strong acid resins are approximately 2.0 eq/l (in Na^+ form), while capacities of strong-base resins vary from 1.0 eq/l to 1.4 eq/l (in the Cl^- form).

Ion-exchange capacities can be determined by titrations of the resins with bases (for cationic exchangers) or acids (for anionic exchangers). Consider a strong-acid ion exchanger totally in the hydrogen form and contained in a beaker of pure water being titrated with a sodium hydroxide solution. As the sodium hydroxide is contacted with the resin, the sodium ions are taken up by the exchanger with an equivalent release of hydrogen ions. The hydrogen ion is, of course, neutralized by the added hydroxide, with no significant change in solution pH. However, as more and more sodium hydroxide is added to the solution, the exchanger eventually becomes completely saturated to the sodium ion (i.e., all exchangeable sites within the exchanger are occupied by sodium ions). Once the exchanger is saturated to sodium, any further addition of sodium hydroxide results in a rapid increase in pH, because there are no more hydrogen ions being released from the exchanger to neutralize the added hydroxide ion. The point at which the solution pH exhibits a sharp rise is the equivalence point of the titration, which defines the ion-exchange capacity. Figure 5-25 includes a titration curve of a strong acid cationic exchange resin. The plot clearly shows that the capacity of the strong-acid ion exchanger is 5 meq/g. Also included in Figure 5-25 are titration curves of a weak-acid resin and a polyfunctional cationic-exchange resin. In contrast to the strong-acid resin, the capacities of weak-acid resins are a function of solution pH, and therefore are not as clearly defined as strong-acid resin capacities.

Example 5-13

a) The capacity of a strong-acid ion exchange resin is determined by titration with 2 N sodium hydroxide. The wet volume of the exchanger to be titrated is 10 ml,

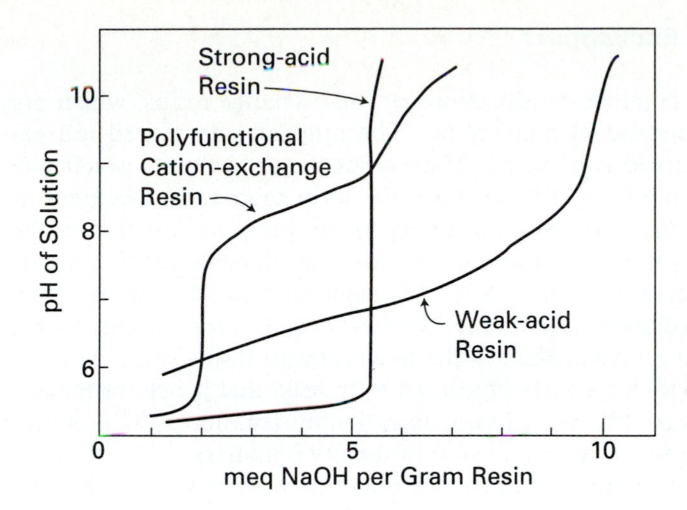

Figure 5-25 pH titration curves of various cation-exchange resins. F. G. Helfferich, *Ion Exchange,* 1962, p. 82, McGraw-Hill. Reprinted by permission of F. G. Helfferich.

and by pH titration, the capacity of the exchanger is achieved by the addition of 11 ml of the sodium hydroxide titrant. The resin is then dried and found to weigh 4.5 grams. Calculate the resin wet-volume capacity and dry-weight capacity.

b) Assume a column of this exchanger with a wet volume of one liter is to be used to remove divalent cadmium ions from a solution. If the concentration of the cadmium in the solution is 0.1 M, and if the process is 100% efficient, calculate the volume of the cadmium solution that can be treated through the column before column exhaustion.

Answer

a) The wet-volume capacity is simply the number of equivalents needed to equilibrate the resin in the titration, divided by the wet volume of the resin:

$$\text{Wet volume capacity} = \left(\frac{0.011 \text{ liter titrant}}{0.01 \text{ liter resin}}\right)\left(\frac{2 \text{ eq titrant}}{\text{liter}}\right) = 2.2\,\frac{\text{equivalents}}{\text{liter}}$$

The dry weight capacity is simply the number of equivalents needed to equilibrate the resin in the titration, divided by the dry weight of the resin:

$$\text{Dry weight capacity} = \left(\frac{0.011 \text{ liter titrant}}{4.5 \text{ grams resin}}\right)\left(\frac{2 \text{ eq titrant}}{\text{liter}}\right)\left(\frac{1000 \text{ meq}}{\text{equivalent}}\right)$$

$$= 4.89\,\frac{\text{meq}}{\text{liter}}$$

b) With a wet volume of one liter, the column is capable of exchanging 2.2 equivalents of (free) divalent cadmium from the solution, or 1.1 moles (moles calculated as the equivalents divided by the valency). Therefore, the volume of the cadmium solution that can be treated is calculated as:

$$\text{Volume} = 1.1 \text{ moles}\left(\frac{1 \text{ liter}}{0.1 \text{ mole}}\right) = 11 \text{ liters}$$

Classification of Ion Exchangers

There are a number of different classifications of ion-exchange resins, which are distinguished by their method of preparation. The originally developed ion exchange resins can be described as gel resins. These types of resins exhibit a practically homogeneous distribution of water throughout the resin matrix. Imbibement of water causes swelling of the resin, with the ability to swell being a function of the degree of cross-linking. Polystyrene resins are typically available at DVB contents ranging from overall averages of 2% to 20%. Although gel resins do not exhibit a completely uniform distribution of DVB, the distribution of the cross-linking chains are more uniform in the gel resins than in the macroporous resins. Macroporous resins are synthesized in such a way to produce a resin bead that is heterogeneous. The macroporous resins contain two phases: homogenous regions of DVB matrix and water and water filled void regions devoid of the DVB matrix.

Ion-exchange materials are characterized as either cationic or anionic, depending upon whether the material exchanges cations or anions. Because cationic exchangers generally contain a fixed ionogenic group, which can behave similar to an acid (donate protons), cationic exchangers are often termed acidic exchangers. Similarly, anionic ion exchangers are often termed basic ion exchangers. Typical terminology used in describing ion-exchange reactions is defined in Table 5-9. Ion exchangers can be further classified as a strong or weak acid (cationic) or base (anionic) ion exchanger, depending upon the degree to which the ionizable group of the exchanger will dissociate. Strong-acid and strong-base ion exchangers completely dissociate over all pH values of practical interest. In other words, under normal operating conditions, the hydrogen ion does not compete with other cations for exchange sites of strong-acid cationic exchangers, and therefore, strong-acid cationic exchangers are effective for nearly all pH values of practical interest. Likewise, the hydroxyl ion does not compete with other anions for exchange sites of strong-base ion exchangers. Therefore, the ion exchange capacities of strong-acid and strong-base ion exchangers do not vary significantly with pH.

Weak-acid and weak-base ion exchangers behave similar to weak acids and bases, and do not completely dissociate over the entire range of practical pH values. Therefore, the number of sites available for ion exchange, and the capacity of these exchangers, are a function of pH. Figure 5-26 (U.S. EPA, 1981) illustrates the pH dependency of the capacity of typical weak-acid and weak-base ion exchangers. The capacity of weak-acid ion exchange resins decreases with decreasing pH, while the capacity of weak-base ion exchange resins decreases with increasing pH. Some typical examples of each of the four broad classes of ion exchangers are provided below. Ion exchange resins are available from a number of commercial suppliers. Bolto and

TABLE 5-9 TERMINOLOGY OF ION-EXCHANGE IONIC SPECIES

Type of ion exchanger	Fixed charges	Counter-ions	Co-ions
Cation exchanger	(−)	(+)	(−)
Anion exchanger	(+)	(−)	(+)

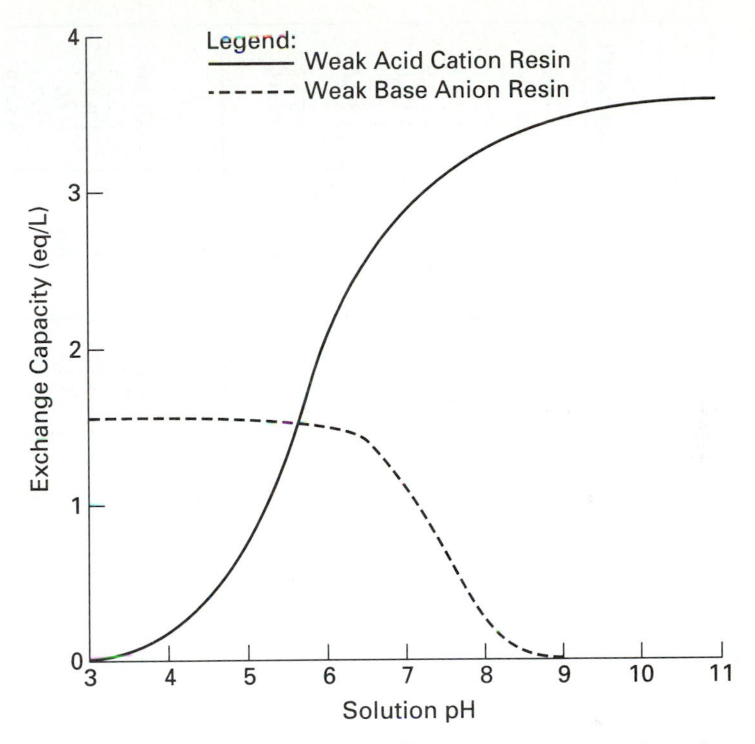

Figure 5-26 Exchange capacity of weak acid cation and weak base anion resins as a function of solution pH. (U.S. EPA, *Summary Report: Control and Treatment Technology for the Metal Finishing Industry, Ion Exchange,* EPA/625/8-81-007, June 1981, p. 6.)

Pawlowski (1987) provide a catalogue of commercial ion-exchange resins and suppliers. The physical properties of various typically used ion exchange materials are provided in Table 5-10 (Perry et al., 1984).

Strong-Acid Resins. The most common strong-acid ion exchanger resins are sulfonated polystyrene resins cross-linked with DVB, with sulfonic acid groups serving as the active ion-exchange sites. In the following descriptions, the term *R* refers to the polymeric structure of the resin to which the active exchange sites are bonded. The following describes the exchanger reaction between a sulfonic acid cationic exchanger in the hydrogen ion form, with a solution containing cadmium ions:

$$2R - SO_3H + Cd^{+2} \rightleftharpoons (R - SO_3)_2Cd + 2H^+ \quad (5.36)$$

Strong-acid ion exchangers are so named because they behave similarly to strong acids in that they readily give up a hydrogen ion (and subsequently associate with another cation). Because of this ability, the capabilities of strong-acid ion exchangers are basically independent of solution pH. Furthermore, because practically all other cations are more strongly attracted to the strong acid exchanger fixed ionic groups than is hydrogen, a strong-acid ion exchanger in the hydrogen form can be used to remove practically all other cations from solution. However, because of the relative

TABLE 5-10 PHYSICAL PROPERTIES OF ION-EXCHANGE MATERIALS

Material	Shape of particles[1]	Bulk wet density (drained), kg/L	Moisture content (drained), % by weight	Swelling due to exchange, %	Maximum operating temperature[2], °C	Operating pH range	Exchange capacity	
							Dry, equivalent/kg	Wet, equivalent/L
Cation exchangers: strongly acidic								
Polystyrene sulfonate								
Homogeneous (gel) resin	S				120-150	0-14		
4% cross-linked		0.75-0.85	64-70	10-12			5.0-5.5	1.2-1.6
6% cross-linked		0.76-0.86	58-65	8-10			4.8-5.4	1.3-1.8
8-10% cross-linked		0.77-0.87	48-60	6-8			4.6-5.2	1.4-1.9
12% cross-linked		0.78-0.88	44-48	5			4.4-4.9	1.5-2.0
16% cross-linked		0.79-0.89	42-46	4			4.2-4.6	1.7-2.1
20% cross-linked		0.80-0.90	40-45	3			3.9-4.2	1.8-2.0
Macroporous structure								
10-12% cross-linked	S	0.81	50-55	4-6	120-150	0-14	4.5-5.0	1.5-1.9
Sulfonated phenolic resin	G	0.74-0.85	50-60	7	50-90	0-14	2.0-2.5	0.7-0.9
Sulfonated coal	G							
Carbon exchangers: weakly acidic								
Acrylic (pK 5) or methacrylic (pK 6)								
Homogeneous (gel) resin	S	0.70-0.75	45-50	20-80	120	4-14	8.3-10	3.3-4.0
Macroporous	S	0.67-0.74	50-55	10-100	120		~8.0	2.5-3.5
Phenolic resin	G	0.70-0.80	~50	10-25	45-65	0-14	2.5	1.0-1.4
Polystyrene phosphonate	G,S	0.74	50-70	<40	120	3-14	6.6	3.0
Polystyrene aminodiacetate	S	0.75	68-75	<100	75	3-14	2.9	0.7
Polystyrene amidoxime	S	~0.75	58	10	50	1-11	2.8	0.8-0.9
Polystyrene thiol	S	~0.75	45-50		60	1-13	~5	2.0
Cellulose								
Phosphonate	F						~7.0	
Methylene carboxylate	F,P,G						~0.7	
Greensand (Fe silicate)	G	1.3	1-5	0	60	6-8	0.14	0.18
Zeolite (Al silicate)	G	0.85-0.95	40-45	0	60	6-8	1.4	0.75
Zirconium tungstate	G	1.15-1.25	~5	0	>150	2-10	1.2	1.0

Anion exchangers: strongly basic								
Polystyrene-based								
Trimethyl benzyl ammonium (type I)								
Homogeneous, 8% CL	S	0.70	46-50	~20	60-80	0-14	3.4-3.8	1.3-1.5
Macroporous, 11% CL	S	0.67	57-60	15-20	60-80	0-14	3.4	1.0
Dimethyl hydroxyethyl ammonium (type II)								
Homogeneous, 8% CL	S	0.71	~42	15-20	40-80	0-14	3.8-4.0	1.2
Macroporous, 10% CL	S	0.67	~55	12-15	40-80	0-14	3.8	1.1
Acrylic-based								
Homogeneous (gel)	S	0.72	~70	~15	40-80	0-14	~5.0	1.0-1.2
Macroporous	S	0.67	~60	~12	40-80	0-14	3.0-3.3	0.8-0.9
Cellulose-based								
Ethyl trimethyl ammonium	F				100	4-10	0.62	
Triethyl hydroxypropyl ammonium					100	4-10	0.57	
Anion exchangers: intermediately basic (pK 11)								
Polystyrene-based	S	0.75	~50	15-25	65	0-10	4.8	1.8
Epoxy-polyamine	S	0.72	~64	8-10	75	0-7	6.5	1.7
Anion exchangers: weakly basic (pK 9)								
Aminopolystyrene								
Homogeneous (gel)	S	0.67	~45	8-12	100	0-7	5.5	1.8
Macroporous	S	0.61	55-60	~25	100	0-9	4.9	1.2
Acrylic-based amine								
Homogeneous (gel)	S	0.72	~63	8-10	80	0-7	6.5	1.7
Macroporous	S	0.72	~68	12-15	60	0-9	5.0	1.1
Cellulose-based								
Aminoethyl	P						1.0	
Diethyl aminoethyl	P						~0.9	

[1] Shapes: C–cylindrical pellets; G–granules; P–powder; S–spheres.

[2] When two temperatures are shown, the first applies to H^+ form for cation, or OH^- form for anion, exchanger; the second to salt ion.

From R. H. Perry, D. W. Green, and J. O. Maloney, eds., *Perry's Chemical Engineers' Handbook*, p. 16–10, 1984. Reprinted by permission of McGraw-Hill.

selectivity of the exchanger for the hydrogen ion, regeneration of a strong-acid exchanger to the hydrogen form requires a great excess of hydrogen ions (very low pH) in the regenerant. For applications other than de-ionization (water softening, heavy metal removal), strong-acid exchangers are often used in the sodium form, where they can be regenerated cost-effectively using a concentrated solution of sodium chloride.

Weak-Acid Resins. Weak-acid ion exchanger resins generally possess carboxylic acid groups as the active exchange sites. Weak-acid exchangers are so named because the fixed carboxylic group behaves as a weak acid, dissociating to varying degrees depending upon the solution pH. At low pH values (<3), the hydrogen ion competes effectively with other cations, thereby resulting in low effective cationic exchange capacities at low pH. Conversely, at high solution pH values (>10), very few carboxyl groups will be protonated, and therefore, the effective cationic exchange capacity of the weak-acid resins will be at the maximum at high solution pH values. Although weak-acid exchangers cannot be used to deionize solution because of the pH dependent capacity and selectivity, an advantage of the affinity of these exchangers for the hydrogen ion is that they can be regenerated to the hydrogen form using significantly less acid than is required to regenerate a strong acid exchanger. Weak-acid ion-exchange resins are often copolymers of acrylic or methacrylic acid with DVB. A diagram of the typical structure of weak-acid ion-exchange resins is provided in Figure 5-27 (Bolto and Pawlowski, 1987).

An example of an exchange reaction involving free cadmium ions and a weak-acid resin is provided below.

$$2R - COOH + Cd^{+2} \rightleftharpoons (R - COO)_2Cd + 2H^+ \tag{5.37}$$

Strong-Base Resins. Strong-base anion exchange resins generally have quaternary ammonium groups ($R - R_3N^+OH^-$) connected to aromatic rings of styrene-DVB copolymers via a methylene group. This type of exchanger is a strong-base exchanger because the quaternary ammonium groups remain ionized at all pH values. Two classes of strong-base resins are commonly used. Strong-base resins where the R_3 (aminating agent) component of the structure is a trimethyl group, $(CH_3)_3$, are known as Type I strong-base resins, while resins with a dimethylethanolamine group as the aminating agent are termed Type II strong-base resins. Type II

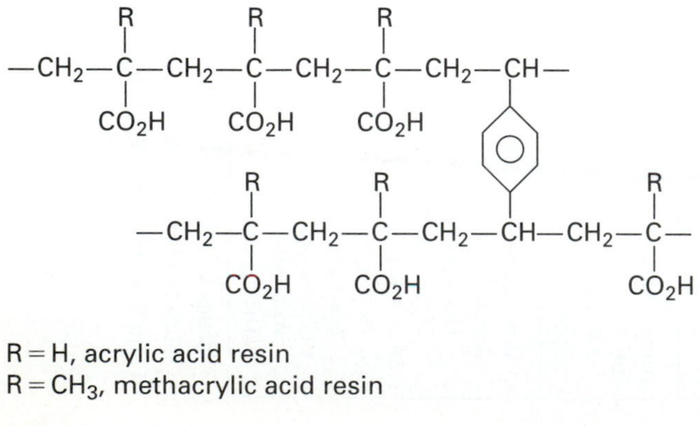

R = H, acrylic acid resin
R = CH_3, methacrylic acid resin

Figure 5-27 Typical structure of weak acid ion exchange resin. B. A. Bolto and L. Pawlowski, *Wastewater Treatment by Ion Exchange,* E & F. N. Spon, New York N.Y. (1987).)

resins are less stable chemically, have a lower concentration of exchange sites, lower basicity, and lower regenerant requirements than Type I resins. Strong-base resins can also be produced using polyacrylamide instead of polystyrene as the polymer matrix, with DVB still used as the cross-linking agent. Analogous to strong-acid exchangers, strong-base exchangers are highly ionizable and can be used at practically all rangers of pH. Regeneration of strong-base resins is generally achieved using concentrated sodium hydroxide solutions.

Weak-Base Resins. Weak-base resins contain functional groups that are derived from weak-base amines, such as tertiary ($-NR_2$), secondary ($-NHR$), or primary ($-NH_2$) amino groups. Weak base resins can be produced with the matrix/cross-linking agent groups as polystyrene/DVB or polyacrylamide/DVB. Weak-base resins can also be produced by step growth or condensation polymerizations, to produce the phenol-formaldehyde resins containing amino groups, or epoxy polyamines. Because the functional groups are derived from weak-acid groups, the hydroxyl ion will compete with other anions for the exchange sites at higher levels of solution pH. Because regeneration of weak-base resins requires less regenerant than required for strong-base resins, the use of weak-base resins is generally preferred. Weak-base resins are often used in combination with a strong-acid exchange column in demineralization systems. The solution to be demineralized is first passed through a strong-acid exchanger where all cations are removed and replaced in the solution with hydrogen ions. The resultant acidic solution then passes through the weak-base exchanger. Because the pH of the solution is low, the weak-base exchanger is nearly completely ionizable, exhibits maximum exchange capacity, and can remove nearly all the anions (other than OH^-) from the solution.

Other Exchange Resins. Ion-exchange resins that have been specifically designed to selectively remove specific ionic species from solutions are available. The most prominent of the ion-specific exchange resins are the so-called chelating resins, which contain imidodiacetic acid groups attached to a cross-linked polystyrene matrix. The imidodiacetic acid group is similar to the soluble phase chelating agent EDTA (ethylenediaminetetraacetic acid), which is a weak acid and which forms strong complexes with heavy metal cations. Therefore, chelating acidic ion-exchange resins exhibit high selectivities toward heavy metals, and being weakly acidic, are easily regenerated under acidic conditions. Table 5-11 provides a listing of ion-specific exchangers and the ions to which they exhibit specificity (Bolto and Pawlowski, 1987).

Ion-Exchange Equilibria

A number of different approaches have been used to model ion-exchange equilibria (Helfferich, 1962). The principal difference among the various approaches to describing ion-exchange equilibria is the method used to characterize the exchanger phase activities. In general, the exchanger phase cannot be treated like the solution phase because of the high concentrations present within the ion exchanger. For example, a conventional cross-linked ion-exchange resin (10% DVB cross-linking) will exhibit an internal concentration of five to six molal (Weber, 1972). One of the

TABLE 5-11 ION EXCHANGERS CONTAINING SPECIFIC GROUPS SELECTIVE FOR PARTICULAR IONS

Ion	Specific group or exchanger
Ammonium	Clinoptilite (inorganic matrix)
Arsenic	Fluorone
Beryllium	Diallyl phosphate
Bismuth	Pyrogallol
Boron	N-Methylglucamine (Rohm and Haas Amberlite XE-243)
Calcium	Gallic acid, imidodiacetic acid, and diallyl phosphate
Caesium	Methylene sulphonic acid (Bayer-Lewatit DN and Diamond Shamrock Duolite C-3)
Cobalt	8-Hydroxyquinoline β-Diketone Ethylenediaminetetra-acetic acid
Copper	Diphenylthiourea 8-Hydroxyquinoline Anthranilic acid β-Diketone Ethylenediaminetetra-acetic acid N-(Hydroxyalkyl)picolylamine (Dow XFS-9195, XFS-4196) Phosphonic acid (Duolite ES-63) Amidoxime (Duolite CD-346)
Cyanide	Anion exchanger (Rohm and Haas Amberlite XE-275 for ferrocyanide)
Germanium	Fluorone Polyhydric phenols
Gold	Polyisothiouronium (Ayalon SRAFION-NMRR)
Iron	Alginic acid Diallyl phosphate Hydroxamic acid m-Phenylglycine Thiol (AKZO Chemical Co.) Anthranilic acid
Lead	Pyrogallol (AKZO Chemical Co. - IMAC T-73)
Magnesium	Alginic acid Phenyldiaminoacetic acid
Mercury	Thiol (AKZO Chemical Co. - IMAC TMR; Diamond Shamrock Duolite ES-465) Polyisothiouronium (Ayalon SRAFION-NMRR) Pyrogallol Dithiocarbamate (Nippon Soda - NISSO Alm)

TABLE 5-11 *(Continued)*

Ion	Specific group or exchanger
Nickel	Dimethylglyoxime Thiol (AKZO Chemical Co.) 8-Hydroxyquinoline β-Diketone Ethylenediaminetetra-acetic acid
Nitrate	Alkylated amidines
Palladium	Aminophenol and nitro groups Guanidine
Platinum	Polyisothiouronium (Ayalon SRAFION-NMRR) Guanidine
Potassium	Dipicrylamine
Silver	Thiol Aminocarboxylic acid
Strontium	Dially phosphate
Thorium	Arsonic acid
Titanium	Chromotropic acid
Uranyl oxide	Schiff base–dinitrophenol
Zinc	Anthranilic acid Phosphonic acid
Zirconium	Phosphonic acid

Source: B. A. Bolto and L. Pawlowski, *Wastewater Treatment by Ion Exchange,* pp. 18–19 (1987). Reprinted by permission of E. & F. N. Spon.

more common approaches to modeling ion-exchange equilibria, originally proposed by Ekedahl et al. (1950), and Argersinger et al. (1950), is provided below.

Consider the following general binary ion-exchange reaction between soluble phase species A and B, with generalized valencies, z_A and z_B, and an ion-exchange resin (Res):

$$z_B ARes + z_A B^{z_B} \rightleftharpoons z_A BRes + z_B A^{z_A} \qquad (5.38)$$

An equilibrium constant can be defined for Equation 5.41 as:

$$K_{eq} = e^{-\Delta G/RT} = \frac{f_B^{z_A} N_B^{z_A} \gamma_A^{z_B} [A^{z_A}]^{z_B}}{f_A^{z_B} N_A^{z_B} \gamma_B^{z_A} [B^{z_B}]^{z_A}} \qquad (5.39)$$

where γ_i is the solution phase activity coefficient for component i, N_i is the mole fraction of component i in the resin phase, and f_i is the resin phase activity coefficient for component i. The selectivity coefficient, K_a is often defined as:

$$K_a = \frac{N_B^{z_A} \gamma_A^{z_B} [A^{z_A}]^{z_B}}{N_A^{z_B} \gamma_B^{z_A} [B^{z_B}]^{z_A}} \tag{5.40}$$

with the selectivity coefficient, and the equilibrium constant, related by

$$K_{eq} = K_a \frac{f_B^{z_A}}{f_A^{z_B}} \tag{5.41}$$

The selectivity coefficient expresses the relative distribution of ions between the exchanger phase and the solution phase. The relative selectivities of common ionic constituents with various types of ion-exchange resins are provided in Table 5-12 (U.S. EPA, 1981; Patterson, 1985).

Example 5-14

Consider the following ion-exchange systems and provided selectivity coefficients. Assuming that at equilibrium, the solution phase molar concentrations are equal, calculate the ratio of mole fractions in the resin phase at equilibrium, assuming that resin phase activity coefficients are unity (i.e., $K_a = K_{eq}$).

Exchangeable ions (terminology from Equation 5.38)		
A	*B*	K_a
H^+	Na^+	1.2
Cd^{++}	Ba^{++}	2.9
Na^+	Cd^{++}	3.0

Answer

a) For the hydrogen/sodium system, the solution is as follows:

$$K_a = \frac{N_{Na}[H^+]}{N_H[Na^+]} = 1.2 \qquad [H^+] = [Na^+] \qquad \frac{N_{Na}}{N_H} = 1.2$$

b) For the cadmium/barium system, the equilibrium composition of the resin phase is calculated as:

$$K_a = \frac{N_{Ba}[Cd^{+2}]}{N_{Cd}[Ba^{+2}]} = 2.9 \qquad [Ba^{+2}] = [Cd^{+2}] \qquad \frac{N_{Ba}}{N_{Cd}} = 2.9$$

c) For the sodium/cadmium binary system, the selectivity coefficient is defined as follows:

$$K_a = \frac{N_{Cd}[Na^{+1}]^2}{N_{Na}^2[Cd^{+2}]} = 3.0 \qquad [Cd^{+2}] = [Na^{+1}] \qquad \frac{N_{Cd}}{N_{Na}} = ?$$

TABLE 5-12 SELECTIVITY OF ION-EXCHANGE RESINS TO VARIOUS IONS IN ORDER OF DECREASING PREFERENCE

Strong acid cation exchange resin	Weak acid (carboxylic) cation exchange resin	Strong base anion exchange resin (type I)	Strong base anion exchange resin (type II)	Weak base anion exchange resin
Barium (+2)	Hydrogen (+1)	Iodide (−1)	Iodide (−1)	Hydroxide (−1)
Lead (+2)	Copper (+2)	Bisulfate (−1)	Bisulfate (−1)	Sulfate (−3)
Calcium (+2)	Calcium (+2)	Nitrate (−1)	Nitrate (−1)	Chromate (−2)
Nickel (+2)	Magnesium (+2)	Bromide (−1)	Bromide (−1)	Nitrate (−1)
Cadmium (+2)	Potassium (+1)	Cyanide (−1)	Cyanide (−1)	Phosphate (−3)
Copper (+2)	Sodium (+1)	Bisulfite (−1)	Bisulfite (−1)	Iodide (−1)
Zinc (+2)		Nitrite (−1)	Nitrite (−1)	Bromide (−1)
Magnesium (+2)		Chloride (−1)	Chloride (−1)	Chloride (−1)
Cesium (+1)		Bicarbonate (−1)	Bicarbonate (−1)	Fluoride (−1)
Rubidium (+1)		Hydroxide (−1)	Hydroxide (−1)	
Potassium (+1)		Dihydrogen Phosphate (−1)	Dihydrogen Phosphate (−1)	
Ammonia (+1)		Fluoride (−1)	Fluoride (−1)	
Sodium (+1)		Sulfate (−2)		
Hydrogen (+1)				
Lithium (+1)				

In other words, the information provided is not adequate to solve the problem, because the distribution of ions between the exchanger phase and solution phase in binary systems of mixed valencies is dependent upon the total solution concentration. This can be explained as follows. Let Φ be the total solution normality (equivalents/liter). It follows then that the solution phase molar concentrations of sodium and cadmium can be calculated as:

$$\Phi = [Na^{+1}] + 2[Cd^{+2}]$$

and it was given that $[Na^{+1}] = [Cd^{+2}]$. Thus $[Na^{+1}] = [Cd^{+2}] = \frac{\Phi}{3}$. Furthermore, $N_{Cd} = 1 - N_{Na}$

Substituting into K_a definition,

$$K_a = \frac{N_{Cd}\left[\frac{\Phi}{3}\right]^2}{(1 - N_{Cd})^2\left[\frac{\Phi}{3}\right]} = 3.0$$

which can be solved at a given Φ to determine N_{Cd} and N_{Na} for the described conditions of equivalent solution phase molar concentrations of the two counterions at equilibrium. The following summarizes the effect that total solution phase normality has on the relative distribution of counterions at equilibrium under the described conditions. Increasing the total solution phase concentration tends to reduce the preference of the exchanger for counterions of higher valency. Note that in ion-exchange system where binary exchange reactions involve only the exchange of counterions with the same valencies, this effect of solution phase concentration is not observed.

Total solution normality	Equilibrium mole fraction Cd
0.20	0.862
0.05	0.923
0.01	0.967

Although early attempts at quantifying ion-exchange equilibria relied upon an assumption of a constant selectivity coefficient, it was subsequently discovered that selectivity coefficients vary with changing resin phase composition. Figure 5-28 provides a plot of the selectivity coefficient with varying resin phase composition for the sodium-hydrogen binary ion-exchange system on a strong-acid synthetic ion-exchange resin (Davidson and Argersinger, 1953).

The solution phase activity coefficients are defined using the conventional standard and reference states whereby the component solution phase activity coefficient approaches unity as the component molar concentration approaches zero. The exchanger phase standard and reference states are defined as the mono-ion exchanger in equilibrium with an infinitely dilute solution of that ion, such that for exchangeable component i:

$$\begin{aligned} \text{Fugacity}_i &\longrightarrow 1 \quad \text{as} \quad N_i \longrightarrow 1 \\ a_i &\longrightarrow 1 \quad \text{as} \quad N_i \longrightarrow 1 \end{aligned} \tag{5.42}$$

This definition is analogous to activity and activity coefficient definitions for solid solutions, and therefore, one can view the exchanger phase as a solid solution of the various resinate species. With the above definitions, the binary equilibrium constant and exchanger phase activity coefficients can be determined by applying the Gibbs–Duhem equation to the exchanger phase. Practically, the equilibrium constant can only be determined by a series of observations of selectivity coefficients with changing resin phase composition (such as Figure 5-28), where the equilibrium constant is defined by the following equation:

$$K_{eq} = \int_0^1 K_a d\, N_B \tag{5.43}$$

Therefore, under the thermodynamic approach described above, the ion-exchange equilibrium constant is an integrated average of the selectivity coefficient as it varies with resin phase composition. The resin phase activity coefficients (fs), which will generally vary with resin phase composition, can also be determined from the same ion exchange isotherm used to calculate the equilibrium constant (Argersinger et al., 1950).

An obvious drawback of the abstract thermodynamic approach to ion exchange equilibria presented above is that it does not provide a means of predicting equilibria distributions based upon the physical and chemical parameters of the exchanging species and the ion exchanger. Although physical models of ion exchange equilibria have been developed and can explain ion exchange phenomenon qualitatively (Gregor, 1948; Donnan and Guggenheim, 1932), the available physical models have not been found to be quantitatively accurate. Some qualitative generalizations regarding

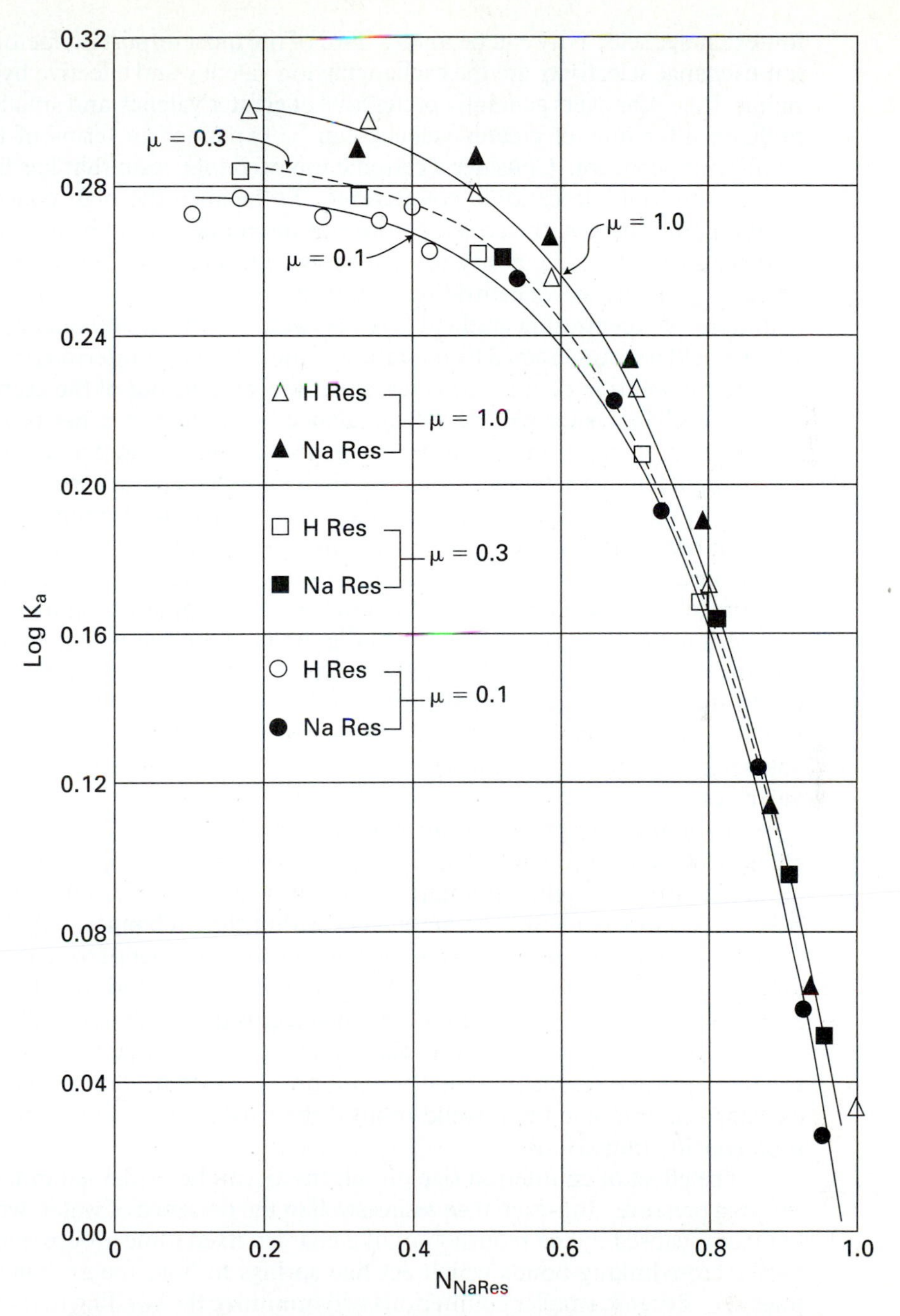

Figure 5-28 Sodium-hydrogen exchange at three ionic strengths. (A. W. Davidson and W. J. Argersinger, "Equilibrium Constants of Cation Exchange Processes," *Annals N.Y. Academy of Sciences,* Vol. 57, No. 3, 1957, p. 109. Reprinted by permission of the New York Academy of Sciences.)

ion-exchange selectivity can be made. Two of the most important factors governing ion-exchange selectivity are the exchanging ion valency and effective hydrated ionic radius. Ion exchangers generally prefer ions of greater valency and smaller size. The preference for ions of greater valency can be explained in terms of the Donnan equilibrium concept. Consider a cationic ion exchange resin that has fixed anionic charges, and mobile cationic counterions. To balance the high concentration of fixed anionic charges, the concentrations of the mobile counterions within the exchanger are on the order of five to eight equivalents per liter. However, the solution phase is generally at concentrations significantly less than this. Therefore, a great counterion concentration gradient exists between the inside and outside of the resin, which could be counteracted by migration of the cationic counterions from inside the resin to the solution phase. However, as cations migrate out of the exchanger, fixed anionic exchange sites will be left unbalanced, resulting in a net negative charge within the exchanger, a net positive charge in the solution, and a potential gradient (Donnan potential) between the phases. This potential causes a repulsion of co-ions (anions in this case) from the exchanger, and attraction of counterions from the solution back into the exchanger. Thus an equilibrium is established whereby the high counterion concentration gradient between the exchanger phase and the solution phase is balanced by a potential gradient that would become excessive if the counterion concentrations were to equilibrate between the phases by migration of counterions from within the exchanger to the solution phase. Because the force of electrostatic attraction is proportional to the ionic charge, counterions of greater valency are attracted more strongly to the exchanger than counterions of smaller valencies. Therefore, ion exchangers are selective toward counterions of higher valencies.

A simpler interpretation of the preference for higher valency ions in the exchanger phase is as follows. The fixed ions within an exchanger must be balanced electrostatically by mobile counterions. Because, for example, a divalent cation can balance two monovalent fixed anion sites within the exchanger, only half as many divalent counterions are required to balance a given number of fixed ionic sites than an electrostatically equivalent number of monovalent counterions. The fewer number of divalent counterions within the exchanger is thermodynamically more favorable than the electrostatically equivalent (and greater) number of monovalent counterions because it results in a smaller equilibrium concentration gradient across the exchanger membrane than would result if the fixed anionic sites were balanced by monovalent counterions.

The effect of counterion size on selectivity can be explained in terms of resin swelling pressure. Ion-exchange resins swell in the presence of water, with the swelling that is caused by the repulsion of like-charged fixed ionic groups being balanced by the cross-linking bonds which act like springs to hold the exchanger structure together. Because smaller counterions will minimize the swelling that is opposed by the cross-linking, smaller counterions of a given valency are preferred to larger counterions of a given valency. The preference of ion exchangers for higher valency counterions can also be explained by this physical description, in that a single divalent counterion will be smaller than the net size of two monovalent cations. Thus, a divalent counterion will occupy less room inside an exchanger than two monovalent counterions, and thus will be preferred. It should be noted that in this qualitative explanation of ion exchange selectivity, the effective "size" of the counterion must be taken as the hydrated ionic radius, and not simply the ionic radius. This is important in that for a given sequence of ions, the ions of largest ionic radius are generally the ions of smallest hydrated ionic radius, and vice versa.

Typically observed selectivity orders generally follow selectivity orders predicted by the above-described physical models for ion exchange. For example, the predicted preference of ion exchangers for counterions of greater valency matches the following observed selectivity orders:

$$PO_4^{-3} > SO_4^{-2} > Cl^{-1} \quad \text{and} \quad Th^{+4} > Nd^{+3} > Ca^{+2} > Na^{+1} \qquad \text{(valency effect)}$$

The effect of the hydrated counterion radius is illustrated in the following observed selectivity orders:

$$\text{Small} \longrightarrow \longrightarrow \longrightarrow \longrightarrow \longrightarrow \longrightarrow \text{Large}$$

$$Cs^{+} > Rb^{+} > K^{+} > Na^{+} > Li^{+}$$

$$Ba^{+2} > Sr^{+2} > Ca^{+2} > Mg^{+2} > Be^{+2}$$

Note that within a given series of ions, the hydrated ionic radius is generally inversely proportional to the unhydrated ionic radius, and therefore, the ion with the largest *ionic radius* will be the most preferred, having the smallest *hydrated ionic radius.* Thus, among the monovalent alkali metals, Cs^{+} is the most preferred (having the smallest hydrated ionic radius), while Li^{+} is the least preferred (having the largest hydrated ionic radius). Likewise, for the divalent alkali earth-metals, Ba^{+2} is the smallest and most selected for, while Be^{+2} is the largest and least selected for. There are, however, exceptions to the above heuristics for ion-exchange selectivity under certain conditions where the selectivity sequences cross over. For example, the following selectivity sequence for anion exchange includes an exception where the divalent chromate anion is less preferred than monovalent anions nitrate and iodate:

$$SO_4^{-2} > I^{-1} > NO_3^{-1} > CrO_4^{-2} > Br^{-1}$$

Also, at high solution phase concentrations (> 1,000 mg/l TDS), the effect of valency is diminished to the point that some monovalent counterions may be preferred to some divalent counterions, as is indicated in Figure 5-29. This effect of solution phase concentration on selectivity is illustrated in the following example.

Example 5-15

Again, let Φ be the total solution phase normality in an ion-exchange system involving the sodium and cadmium counterions. In the previous example, it was demonstrated that the resin phase mole fraction of cadmium could be calculated through the following relationship:

$$K_a = \frac{N_{Cd}\left[\dfrac{\Phi}{3}\right]^2}{(1 - N_{Cd})^2\left[\dfrac{\Phi}{3}\right]} = 3.0$$

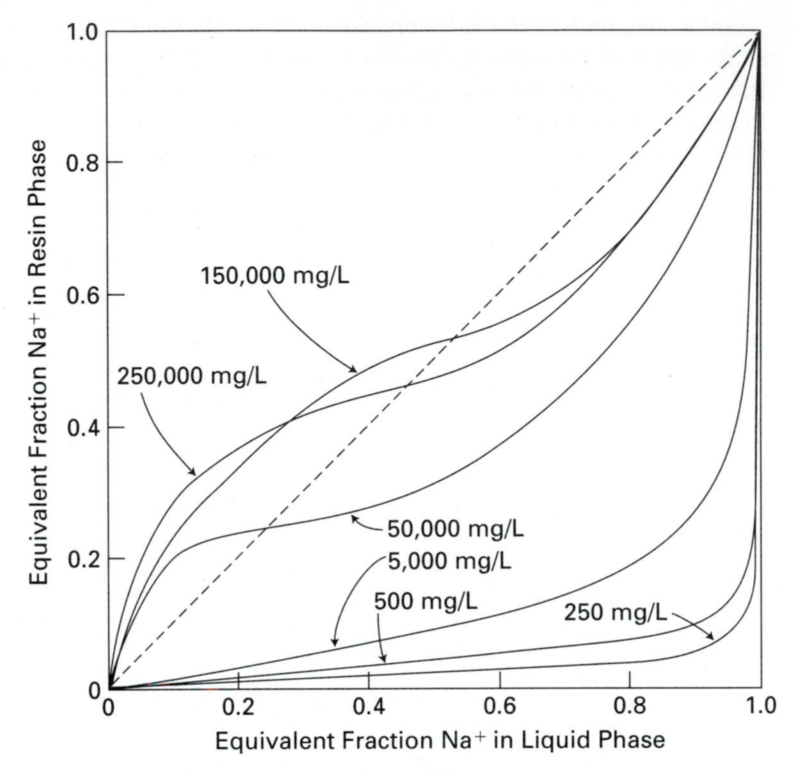

Figure 5-29 Na^{+}–Ca^{+2} equilibria for sulfonic acid cation exchange resin. (Rohm and Haas Company, *Amberlite Ion Exchange Resins,* Vols. 1 and 2, 1973. Reprinted by permission of the Rohm and Haas Company.)

With varying solution phase total normality, the equilibrium mole fraction of cadmium is given in the following table.

Total solution normality	Equilibrium mole fraction Cd	Equilibrium mole fraction Na
0.2	0.862	0.138
2.0	0.627	0.373
5.0	0.483	0.517

Therefore, as the solution phase normality is increased from 2.0 to 5.0, the counterion preference of the ion exchanger changes from the divalent cadmium ion to the monovalent sodium ion. *[The above calculation is not strictly valid for a number of reasons, including the likelihood of complexation and precipitation reactions of cadmium at such high solution phase normalities. However, the calculation illustrates the point that typical selectivity orders based on valency can cross over at high solution phase concentrations.]*

Other factors than hydrated ionic radius and ionic valency can affect relative selectivities. For example, large ions such as organic ions or inorganic complexes may be too large to penetrate the exchanger matrix, and are effectively screened from the exchanger. This effect is more important for highly cross-linked resins, because the cross-linking results in smaller openings into the exchanger matrix. It follows that ionic series for which selectivity orders follow the relative counterion hydrated ionic radius exhibit greater differences in selectivity as the cross-linking increases. At very low degrees of cross-linking, the effect of ionic size becomes negligible.

Interactions of the counterion with the fixed ionic group, or with solution phase species, can also affect selectivity. For example, if a ligand is present in the solution phase that strongly associates with the counterion, the preference of the exchanger for the counterion is diminished. An example of this would be a solution containing EDTA, which strongly complexes heavy metals. Such effects can be accounted for if the speciation of the metal ion in the solution phase is accurately determined, and the free (exchanging) metal ion concentration in the solution phase is used in the ion exchange equilibrium calculation, and not the total metal ion concentration. The ability of some resins to strongly attract certain counterions is enhanced when the resin's fixed ionic group strongly associates with the exchangeable counterion (e.g., a fixed EDTA group and the Fe^{+2} counterion). Also, a resin with a fixed carboxylic group (that has a strong affinity for hydrogen ions) will prefer the hydrogen ion to many other cations.

Further discussions on the causes of ion exchange selectivity, and the typically observed selectivity orders, can be found in Helfferich (1962) and Reichenberg (1966). A comparison of selectivity coefficients for various ions and exchangers is provided in Table 5-13 (Weber, 1972). A more complete list of ion exchange equilibria constants can be found in Marcus and Howery (1975).

Ion Exchange Kinetics

Conceptually, the kinetics of ion exchange are similar to the kinetics of other adsorption processes and can be viewed as the overall rate associated with the following reactions:

1. Transport of the exchanging ion from the bulk of the external solution to the surface of the exchanger or the boundary layer surrounding the exchanger;
2. Transport of the exchanging ion through the boundary layer;
3. Pore transport of the ion within the exchanger to the active ion exchange site;
4. The exchange process at the active exchange site;
5. Pore transport of the exchanged counterion that was formerly associated with the fixed exchanger ionic group;
6. Transport through the exchanger boundary layer at the exchanger particle surface, and;
7. Transport from the external surface of the exchanger to the bulk of the solution.

TABLE 5-13 ION EXCHANGE SELECTIVITIES AT 25°C

Exchanging ion	Degree of cross-linking 4%	8%	16%
Monovalent cations			
Li	1.00	1.00	1.00
H	1.32	1.27	1.47
Na	1.58	1.98	2.37
K	2.27	2.90	4.50
Rb	2.46	3.16	4.62
Cs	2.67	3.25	4.66
Ag	4.73	8.51	22.9
Tl	6.71	12.4	28.5
Divalent cations			
Mg	2.95	3.29	3.51
Ca	4.15	5.16	7.27
Sr	4.70	6.51	10.1
Ba	7.47	11.5	20.8
Pb	6.56	9.91	18.0
Monovalent anions	(2% c.l.)		
OH	0.80	0.50	
F		0.08	
Cl	1.0	1.0	
Br	2.7	3.5	
I	9.0	18.5	
NO_3		3.0	
SCN	6.0	4.3	
ClO	9.0	10.0	

Adapted from Weber (1972); original source is Meites (1963). Data are for polystyrene base resins, sulfonic acid cation exchanger, and type 2 quaternary base anion exchanger. Data of Bonner and Smith (1957) for cation exchanges; Gregor et al. (1955) for anion exchanges. A value greater than 1 indicates preferential adsorption of ion named against the reference ion (Li for anions).

It can generally be assumed that the kinetics of the exchange process at the active ion-exchange site (step 4) is instantaneous. Therefore, the overall kinetics of the exchange reaction are dependent upon the rate at which the transport processes (steps 1 – 3 and 5 – 7) occur. In column reactors or stirred batch reactors, the rates of steps 1 and 7 are also generally fast, and therefore not rate limiting. Therefore, the kinetics of the overall exchange reaction are in most applications either controlled by the rate of transport across the hypothetical film surrounding the outside (and inside) of the exchanger, or the intraparticle pore diffusion transport rate. The overall kinetics may be film-diffusion controlled or pore-diffusion controlled, depending upon a number of exchanger, solution, and process variables.

The rate of film diffusion is proportional to the concentration gradient across the film, and inversely proportional to the size of the film boundary layer thickness. Therefore, the rate of film diffusion is increased with increase solution phase concentration (increasing concentration gradient), increased mixing or agitation (which reduces the boundary layer thickness), and decreasing exchanger particle diameter.

The rate of pore diffusion is unaffected by external solution mixing, increases with decreasing exchanger particle size, and decreases with an increasing degree of exchanger cross-linking (which increases the resistance to pore diffusion).

Depending upon whether the exchange reaction is film or pore diffusion controlled, a change in the exchanger system variables may have no effect or a very significant effect. For instance, if the kinetics of the exchanger reaction are pore diffusion controlled, increasing the concentration gradient between the solution phase and the exchanger phase (i.e., increase the concentration of the solution phase) will have little or no effect on the overall rate of reaction. Likewise, if the kinetics of the exchange reaction are film-diffusion controlled, switching from a highly cross-linked resin to a resin with a lower cross-linking would not likely increase the overall rate of the reaction.

Practically, the solution phase concentration generally determines whether the exchange reaction is pore-diffusion controlled or film-diffusion controlled. When the solution-phase concentration is high (such as when regenerating an exchanger), the reaction is generally limited by the rate of pore diffusion. Conversely, at low solution phase concentrations, the overall rate of the reaction is generally controlled by the rate of film diffusion.

Understanding and modeling the kinetics of ion exchange processes are particularly important in predicting the behavior of ion exchange column performance. Specifically, the kinetics of the exchange reactions will affect the shape of the breakthrough curve. For further information on ion exchange kinetics and prediction of column behavior, the interested reader should consult Helfferich (1962).

Ion Exchange Processes

Ion exchangers can be used in batch or continuous-treatment applications. In a batch treatment scenario, the exchanger and a solution are mixed in a container and the exchange reaction between the solution and the exchanger are allowed to come to equilibrium. At equilibrium, the contaminant concentration in the solution phase will be dependent upon the selectivity of the exchanger for the various system counterions, the initial concentrations in solution of exchangeable counterions, the capacity of the exchanger, and the relative quantity of exchanger and solution. Because the equilibrium distribution of the counterions will result in contaminants remaining in solution, batch ion exchange treatment processes are not very effective. Furthermore, batch regeneration of the exchanger is not efficient because some quantity of the contaminant will be present in the regenerant solution upon contact with the used exchanger, and therefore, complete regeneration is not possible by batch regeneration. For these reasons, ion exchangers are rarely used in batch treatment systems.

Example 5-16

Silver is to be removed from a solution by batch ion exchange with a strong-acid ion exchange resin initially saturated in the sodium form. The applicable exchanger reaction is given below.

$$\text{NaRes} + \text{Ag}^+ \rightleftharpoons \text{AgRes} + \text{Na}^+$$

The equilibrium constant for the reaction is 3.0 and the exchanger capacity of the resin is 5 meq/g. If 100 ml of solution is contacted with one gram of the resin, calculate the equilibrium composition of the exchanger if the solution initially contains ionic silver at a concentration of:

a) 0.1 Molar
b) 0.001 Molar

Assume that the solution contains no exchangeable cations other than silver.

Answer

a) There are four equilibrium system unknowns: $[Na^+]$, $[Ag^+]$, N_{Na}, N_{Ag}. Therefore, four independent equations relating the four system variables must be derived to determine the equilibrium compositions. First, perform system mass balance on silver and sodium ions. With a capacity of 5 meq/g and one gram of resin, the total quantity of sodium in the system is 5 meq, or 0.005 mole. The total quantity of silver in the system is simply the number of moles of silver initially in solution, or 0.01 mole. Equilibrium is achieved by redistribution of the silver and sodium ions between the solution phase and the exchanger. However, the total number of moles of silver and sodium initially, and at equilibrium, are the same. Therefore, the following two equations can be derived:

$$0.005 \text{ mole Na} = (0.1 \text{ liter})[Na^+] + 1 \text{ gram}\left(0.005\,\frac{\text{equivalents}}{\text{gram}}\right)\left(1\,\frac{\text{mole total}}{1 \text{ equivalent}}\right)\left(\frac{N_{Na} \text{ moles Na}}{1 \text{ mole total}}\right)$$

$$0.01 \text{ mole Ag} = (0.1 \text{ liter})[Ag^+] + 1 \text{ gram}\left(0.005\,\frac{\text{equivalents}}{\text{gram}}\right)\left(1\,\frac{\text{mole total}}{1 \text{ equivalent}}\right)\left(\frac{N_{Ag} \text{ moles Ag}}{1 \text{ mole total}}\right)$$

The remaining two equations are the definition of the equilibrium constant and the relationship between the equilibrium exchanger mole fractions:

$$N_{Na} + N_{Ag} = 1$$

$$K_{eq} = \frac{N_{Ag}[Na^+]}{N_{Na}[Ag^+]} = 3.0$$

The above four equations can be solved simultaneously to give the following equilibrium compositions:

$$[Na^+] = 0.041 \text{ M}$$

$$[Ag^+] = 0.059 \text{ M}$$

$$N_{Na} = 0.186$$

$$N_{Ag} = 0.814$$

The results clearly indicate the limitations of a batch ion-exchange treatment process. Even though the resin selectivity for silver is significantly greater than for sodium (evidenced by the dominance of silver in the exchanger phase at equilibrium), the solution-phase equilibrium silver concentration is still quite high. To remove silver to a low equilibrium solution-phase concentration, a large relative mass of exchanger initially saturated to sodium would have to be contacted with the solution.

b) Similarly, for an initial solution-phase silver ion concentration of 0.001 molar, the following equilibrium composition can be calculated:

$$[Na^+] = 9.933 \times 10^{-4} \text{ M}$$

$$[Ag^+] = 6.718 \times 10^{-6} \text{ M}$$

$$N_{Na} = 0.980$$

$$N_{Ag} = 0.020$$

Results for (b) are markedly different from those for (a) because the initial solution phase concentration of silver represents a small number of moles relative to the number of exchangeable sites available within the ion exchanger. In this case a small equilibrium concentration of silver in the solution phase is achievable by batch treatment. However, on an industrial scale, the quantity of ion exchanger required to produce acceptable solution phase equilibrium contaminant concentrations will generally be so large that the process would not be practical or economically viable.

Columnar Ion Exchange Processes

It is clear from the above example and discussion that batch processes are not an efficient means of utilizing ion exchangers. Practical applications of ion exchange processes generally involve utilizing the ion exchanger in a column. The most common columnar ion exchange process is a fixed bed process, whereby the aqueous stream to be treated is passed through a stationary bed of ion exchange materials. A properly operated fixed ion exchange column can achieve nearly complete separation of the aqueous contaminant from the solution, and is well suited to operating as a continuous treatment unit. Furthermore, using an ion-exchange column nearly completely utilizes the exchange capacity of the ion exchanger and allows for efficient and effective exchanger regeneration.

The operation of an ion exchange column is depicted in Figure 5-30, where the hypothetical monovalent counterion B is being exchanged in a column initially saturated with the hypothetical monovalent counterion A. At the start of the process, as the solution enters the column, an exchange reaction between the solution phase and the exchanger phase occurs causing the transfer of counterion B from solution to the exchanger, along with an equivalent exchange of counterion A from the exchanger to the solution. As one follows the initial parcel of the aqueous phase as it passes further through the column, the concentration of the B counterion is observed to decrease until eventually, all B counterions have been removed from the aqueous parcel, no further exchange occurs, and the exchanger downstream from the exchange front remains saturated to the A counterion.

As more and more solution is passed through the column, some length of the column at the column entrance eventually becomes saturated to B counterions, and the exchange process does not begin in the column until some distance downstream from the column entrance. However, as long as the entire column is not saturated,

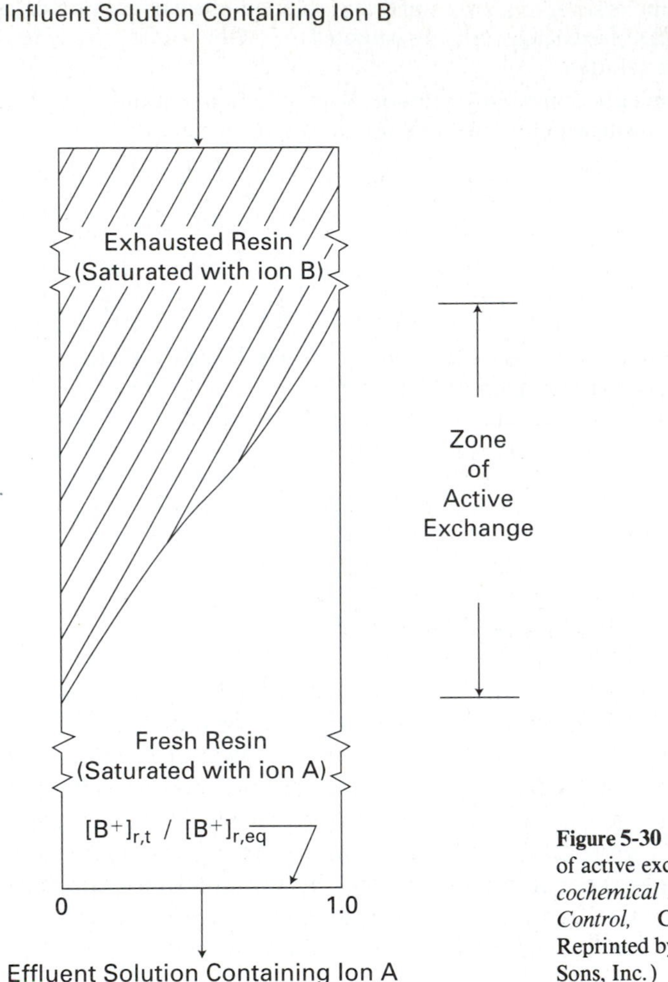

Figure 5-30 Longitudinal profile of zone of active exchange. (W. J. Weber, *Physicochemical Processes for Water Quality Control,* Copyright 1972, p. 284. Reprinted by permission of John Wiley & Sons, Inc.)

there will be a portion of the column toward the column exit that remains saturated to the *A* counterion. Figure 5-31 demonstrates this behavior of ion exchange columns. The zone of the column near the column inlet that is saturated to the feed solution is known as the saturated zone, the zone of the column near the column exit that retains exchanger at the initial exchanger composition is known as the solute-free zone, while the zone between the saturated zone and the solute-free zone is termed the active adsorption, or mass transfer zone. The exchange reaction in the column is effectively confined to the length of the mass transfer zone. The mass transfer zone will over time (and volume of column throughput) move through the column until *breakthrough* of the counterion *B* is observed at the column effluent. Prediction of column behavior requires prediction of the shape, size, and movement of the mass transfer zone from start of the treatment process until breakthrough.

Upon breakthrough, the ion-exchange column must be regenerated, which in this hypothetical case would be achieved by passing a concentrated solution of the *A*

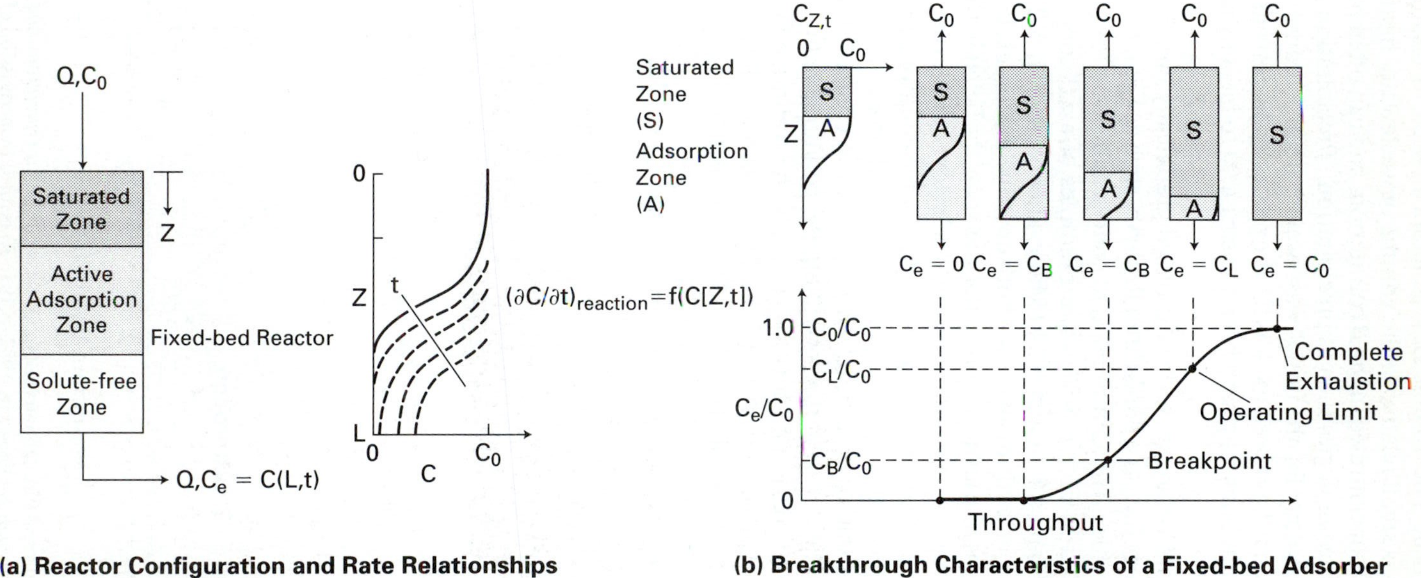

C_0 ; C_e = column influent and effluent concentration of solute
C_B = breakpoint concentration of solute
C_L = operating limit concentration of solute
Q = liquid flow rate through column
Z = longitudinal coordinate
t = time

Figure 5-31 Typical fixed-bed column behavior. (W. J. Weber and E. H. Smith, "Removing Dissolved Organic Contaminants from Water," *Environmental Science and Technology,* Vol. 20, No. 10, 1986, p. 976. Reprinted with permission from the American Chemical Society, Copyright 1986, American Chemical Society.)

counterion through the bed. The volume of regenerant needed to resaturate the column to A counterions is small relative to the volume of wastewater treated. The spent regenerant is highly concentrated in the B counterion, and can often be treated using more conventional technologies such as precipitation. Precipitation may not have been a feasible technology for the initial wastewater stream because the contaminant concentrations may have been too low for the process to be cost effective, or the required discharge limitations may have been too stringent to meet by precipitation.

Ion exchange column behavior can be expressed graphically either as an effluent concentration history (i.e., as a breakthrough curve, such as Figure 5-32) or as an axial concentration profile. An example effluent concentration history (or breakthrough curve) is provided in Figure 5-32 while examples of axial concentration profiles are provided in Figures 5-33 and 5-34 (Maneval et al., 1985). While axial concentration profiles and breakthrough curves both describe the ion-exchange column behavior, the two types of graphical representations show the behavior from two different frames of reference. An effluent concentration history displays the solution phase concentration at one point in the column (generally the column effluent) as a function of effluent volume. In contrast, an axial concentration profile holds constant the volume of feed solution (or time) and displays the composition of the exchanger phase as a function of axial distance along the column length.

The ordinate of the breakthrough curve is generally the column effluent concentration of the contaminant of interest, while the abscissa is generally the number of bed volumes passed through the column. The bed volume is defined as the volume of the unpacked column. In the provided figure, contaminant concentrations in the effluent of the exchanger are very low until the breakthrough point, which has in this

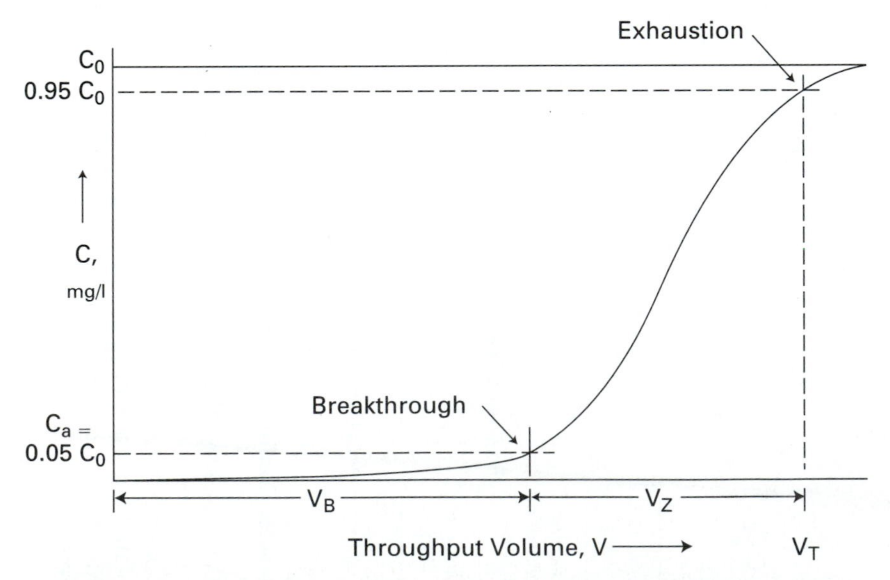

Figure 5-32 Typical breakthrough curve. (T. D. Reynolds, *Unit Operations and Processes in Environmental Engineering,* 1982, p.197. Reprinted by permission of PWS Publishing Company.)

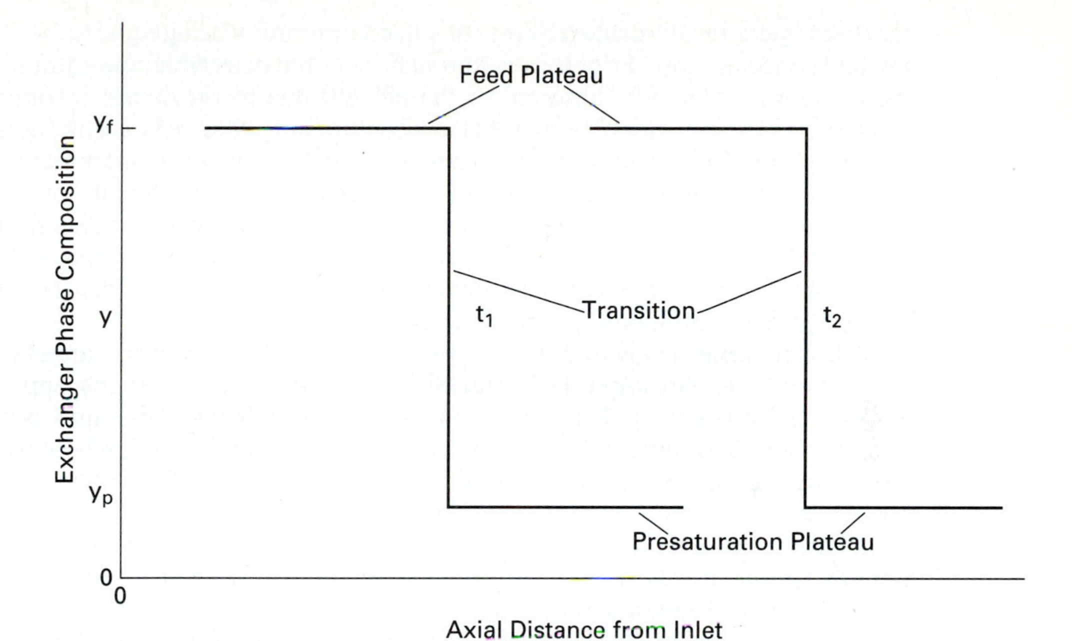

Figure 5-33 Movement of abrupt transition (shown at times t_1 and $t_2 > t_1$) under premise of local equilibrium. (J. E. Maneval, G. Klein, and J. Sinkovic, *Selenium Removal from Drinking Water by Ion Exchange,* U.S. EPA, Water Engineering Research Laboratory, 1985.)

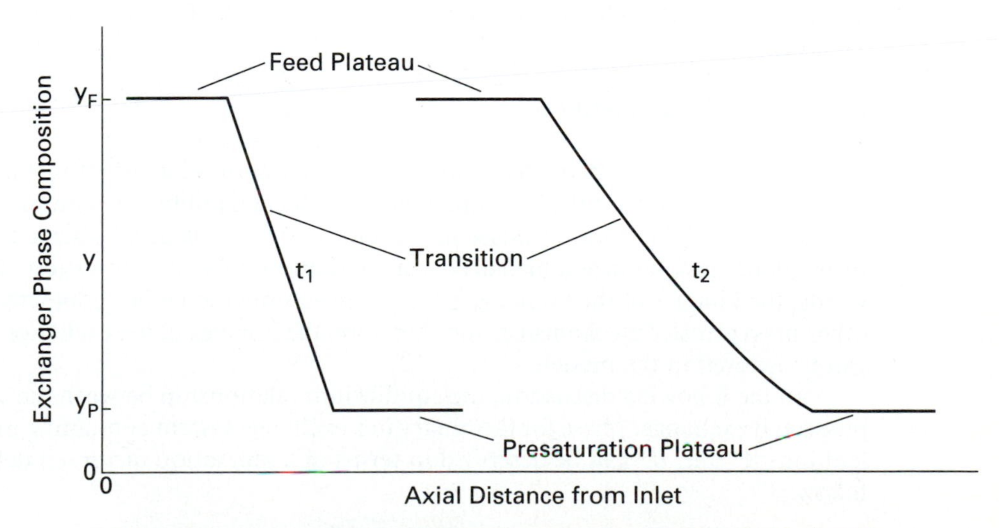

Figure 5-34 Movement of gradual transition (shown at times t_1 and $t_2 > t_1$) under premise of local equilibrium. (Maneval, Klein, and Sinkovic, *Selenium Removal from Drinking Water by Ion Exchange,* U.S. EPA, Water Engineering Research Laboratory, 1985.)

case been operationally defined by the effluent concentration being equal to 5% of the influent concentration. Prior to breakthrough, a low but detectable concentration of the contaminant may be observed in the effluent due to previously incomplete regeneration cycles. This observance of low contaminant effluent concentrations in the effluent prior to breakthrough is termed *leakage.* Because the contaminant concentration in the ion-exchange effluent can increase relatively rapidly once any breakthrough occurs, breakthrough is often defined at a concentration significantly less than the allowable maximum discharge concentration. This precaution allows adequate time to observe the breakthrough and to regenerate the exchanger before the effluent concentrations become excessive.

Also indicated in Figure 5-32 is the *exhaustion* point of the ion exchange bed, at which point the exchanger bed influent and effluent concentrations approach equality. At ion exchange bed exhaustion, the influent stream and the entire bed are at or very near equilibrium, and therefore, no further ion exchange can be achieved (assuming the influent stream composition does not change). Practically, ion exchange beds are rarely operated to the point of exhaustion, because bed regeneration is performed upon breakthrough, which is generally defined at effluent concentrations that occur prior to exhaustion. Regeneration of an ion exchange bed is typically required after utilization of only 60%–75% of the column saturation capacity.

Although ion exchange column behavior can be predicted, developing and solving a predictive model for multicomponent systems with inclusion of kinetic limitations and dispersion are very challenging mathematically. Furthermore, such a model requires accurate exchange equilibrium constants, dispersion parameters, and mass transfer rate constants, which are not always available. Therefore, full scale column design is often done based upon bench scale column testing of the exchange process using samples of the waste to be treated. However, for purposes of illustration, it is worthwhile discussing a simple predictive model for ion exchange column behavior.

Local Equilibrium Model

The simplest approach in modeling ion exchange column behavior is to assume local equilibrium throughout the column. Under the local equilibrium assumption, the composition of the solution phase and exchanger phase is at all points and time throughout the column assumed to be related by an equilibrium expression. In other words, the kinetics of the exchange process are assumed to be fast compared to the other mass transfer mechanisms, and therefore, the kinetics of the exchange process can be ignored in the model.

In the following discussion, the equilibrium relationship between the solution phase and exchanger phase for the binary ion-exchange system containing monovalent ions A^+ and B^+ can be described in terms of a separation factor (α) defined as follows:

$$\alpha_B^A = \frac{y_A x_B}{y_B x_A} \tag{5.44}$$

where y_i is equivalent fraction in resin phase and x_i is equivalent fraction in solution phase. Note that the separation factor is identical to a selectivity coefficient as defined in Equation 5.41 for the case of monovalent counterions and negligible solution phase and resin phase nonidealities.

The local equilibrium model is developed by first performing a differential mass balance over a slice of column length dz.

$$\epsilon \frac{\partial C}{\partial t} + (1 - \epsilon) \frac{\partial \overline{C}}{\partial t} + u \frac{\partial C}{\partial z} = \frac{\partial}{\partial z}\left(D \frac{\partial C}{\partial z}\right) \quad (5.45)$$

where ϵ is the packed column void fraction, C and $\overline{C}$ are the solution phase and exchanger phase normality for the component of interest, t is time, z is axial distance along column, and D is a dispersion coefficient. The term on the right side of Equation 5.45 describes dispersion. Formally, local equilibrium assumption does not preclude dispersion. However, the convention in the literature is that so-called local equilibrium sorption or exchange models generally also assume the dispersion is negligible. The following discussions will assume that dispersion is negligible, and therefore, the generalized mass balance can be written for each exchangeable component as:

$$\epsilon \frac{\partial C}{\partial t} + (1 - \epsilon) \frac{\partial \overline{C}}{\partial t} + u \frac{\partial C}{\partial z} = 0 \quad (5.46)$$

where C and $\overline{C}$ are the solution- and exchange-phase normalities of any exchangeable component. In the case of a binary ion-exchange system, $\partial \overline{C}/\partial t$ can be eliminated from Equation 5.46 through the following relationship:

$$\frac{\partial \overline{C}}{\partial t} = \frac{d\overline{C}}{dC} \frac{\partial C}{\partial t} \quad (5.47)$$

Substituting Equation 5.47 into Equation 5.46 results in the following relationship:

$$\frac{\partial C}{\partial t} + u_C \frac{\partial C}{\partial z} = 0$$

$$u_C = \frac{\dfrac{u}{\epsilon}}{1 + \dfrac{1 - \epsilon}{\epsilon} \dfrac{d\overline{C}}{dC}} \quad (5.48)$$

The local equilibrium assumption is applied by calculating $d\overline{C}/dC$ through the definition of the separation factor (Equation 5.44). Equation 5.48 can be solved by rewriting z and t in terms of dimensionless variables (DeVault, 1943; Walter, 1945, Tondeur and Bailly, 1986). The solution of Equation 5.48 is dependent upon whether the exchange reaction in the column is "dispersive" or "compressive." Compressive behavior, which results in an abrupt transition of exchanger phase composition as evidenced in an axial concentration profile (Figure 5-33), occurs when a column is operated so that a weakly retained ionic species originally retained by the exchanger is replaced by a more strongly retained ionic species that is being removed from the solution phase by passage through the column. Alternatively, dispersive column behavior (Figure 5-34) results in a gradual transition; it occurs when the column is operated so that a more strongly preferred ionic species that is originally retained by the exchanger is replaced by a less preferred ionic species that is removed from solution through the column operation. For example, the monovalent sodium ion is more preferred by most synthetic ion exchanger resins than the hydrogen ion. Removal of sodium ions from a solution by passage through a strong-acid resin packed ion exchange column originally saturated to hydrogen ions would result in compressive behavior and an abrupt transition in the concentration profile as the front of sodium moves through the bed. Conversely, dispersive behavior and a gradual transition occurs when a column initially saturated to the sodium ion is used to remove hydrogen ions from solution.

Gradual transition. For the case of dispersive behavior (Figure 5-34) and a gradual transition front, with the exchanger initially saturated to the more preferred counterion A, and a step input of feed containing only the less preferred counterion B, the solution phase and exchanger phase composition within the mass transfer zone of the ion-exchange column is:

$$\begin{aligned} x_A &= \frac{1}{\alpha - 1}\left(\sqrt{\frac{\alpha}{T}} - 1\right) \\ x_B &= 1 - x_A \\ y_A &= \frac{1}{\alpha - 1}\left(\sqrt{\frac{\alpha}{T}} - 1\right) \\ y_B &= 1 - y_A \end{aligned} \tag{5.49}$$

with the dimensionless throughput parameter T, which can be defined as the ratio of the number of equivalents that have exited the column to the number of equivalents of exchange capacity retained by the column, is given as:

$$T = \frac{C_0(V - \epsilon v)}{\rho_b Q v} \tag{5.50}$$

where C_0 is the feed solution total normality, V is the volume of feed solution that has exited the column, v is the bed volume, Q is the capacity of the exchanger (equivalents/mass), and ρ_b is the bulk density of the exchanger (Klein et al., 1967). The throughput parameter T corresponds to a dimensionless time variable for a bed of constant length. Outside of the mass transfer zone, the solution-phase composition is equal to the feed composition upstream of the mass transfer zone (B at concentration C_0), and downstream of the mass transfer zone, the solution phase contains only A at a concentration of C_0.

Example 5-17

Consider an ion exchange column containing strong acid ion exchanger resins saturated to the sodium ion. The ion exchange column is being used to treat a solution of lithium at a concentration of 0.05 M. Assuming local equilibrium behavior, calculate the volume of solution that can be treated before the effluent lithium concentration is equal to 0.1, 0.5, and 1.0 of feed. The following additional information is given:

Bed diameter = 1 ft
Bed length = 6 ft
Packed bed void fraction = 0.35
Resin capacity = 5 meq/g
Resin bulk density = 800 g/L
Separation factor (α) = 1.6 (lithium preferred)

Answer

Plot X_B versus T using Equation 5.49 and determine values of the throughput parameter T where $X_B = 0.1, 0.5, 1.0$. From Equation 5.50, calculate V for the three respective T values. The results are summarized below.

X_{Li}	T	V (liters)
0.1	0.675	7,247
0.5	0.974	10,148
1.0	1.6	17,114

Abrupt transition. For the case of compressive behavior in a binary ion-exchange system (see Figure 5-33), the transition point is separated by the presaturation plateau (downstream of the transition front) and the feed plateau (upstream of transition front). The point at which the abrupt transition occurs can be calculated as (Tondeur and Bailly, 1986):

$$T = \frac{y_{i_F} - y_{i_p}}{x_{i_F} - x_{i_p}} \tag{5.51}$$

For the case where the feed solution contains only component i, and with the exchanger initially saturated to component j (and $y_i = x_i = 0$ in the presaturation plateau), the transition point is given by:

$$T = \frac{y_{i_F}}{x_{i_F}} \tag{5.52}$$

Limitations of Binary Local Equilibrium Model

Because of the many simplifying assumptions made in the development of the binary ion exchange local equilibrium model for column behavior, application of the model must be made with care. Many ion exchange column processes are designed to treat wastes containing multiple exchangeable ions. In such cases, the relative selectivities of all exchangeable ions must be considered in determining the feasibility of ion exchange treatment of the wastewater. For instance, an ion-exchange process proposed for removing chromium as the chromate anion (CrO_4^{-2}) from a wastewater that also contains the sulfate ion in significant concentration must consider the relative selectivities of the exchanger for the chromate and sulfate anions. If the sulfate ion is preferred by the exchanger (which typically is the case for weak-base exchange resins), then, depending upon the relative concentrations of sulfate and chromate, the process may not be efficient because the capacity of the exchanger to remove chromate from solution may be impaired due to sulfate exchange, resulting in breakthrough of chromate well prior to what would be expected based upon the theoretical capacity of the column, assuming that the chromate was the only exchangeable anion. The local equilibrium approach that has been presented can be generalized to predict ion exchange column behavior for multicomponent exchange systems (Klein et al., 1967; Tondeur and Klein, 1967; Helfferich, 1967). However, the mathematical complexities incurred by consideration of multiple exchangeable components are significant. The mathematic complexities are further increased when constant separation factors are not assumed.

Additional complexity in the modeling of ion exchange columns arises with the inclusion of kinetic limitations and dispersion (Helfferich, 1962). Even if the necessary dispersion coefficients and rate constants are available to input to a full multicomponent ion-exchange column model, the resulting model can generally only be solved using numerical methods. For these reasons, design of fixed bed ion exchange processes often involves bench or pilot studies that can generate data used to scale-up to a full system design.

Column Design and Operation

Waste Characterization. The first step in the design of ion exchange treatment systems is to characterize the waste to be treated. This characterization should include a determination of the concentration of ionic species, and the solution pH, alkalinity, and acidity. Also, the presence and concentration of organic compounds, especially those that can serve as complexing agents with ionic species of interest, should be determined, as well as the presence of any oxidants or reductants that could alter ionic valencies, or attach the ion exchange resins. Finally, the turbidity and suspended solids concentration should be considered to insure that solids loading of the exchanger will not impair column performance. Pretreatment for solids removal is recommended for wastes containing filterable solids.

Determine Treatment Objectives. The objectives of the treatment process must be determined prior to final process design. The design engineer should verify that ion exchange is capable of achieving the desired treatment objectives and whether the process would be reasonably cost-effective. Specifically, these considerations should include whether the ionic composition and relative ionic selectivities are amenable to achieving the treatment objective via ion exchange. For example, a solution with high concentrations of divalent and trivalent cations that is to be treated to remove a relatively unselected for monovalent cation may not be a good candidate to be treated by ion exchange.

Select Media and Size. The next step in the process design is to select the ion-exchange media and media size. Obviously, one must select an ion exchanger that can effectively remove the ion(s) of interest from the waste. Review of published selectivities and guidance from resin manufacturers and suppliers can provide the necessary information to select the appropriate media. Here again, a good characterization of the waste allows one to correctly select the appropriate media, and insure that exchange of ions other than the contaminants of interest does not exhaust exchanger capacity and diminish process effectiveness. Theoretical maximum exchanger capacities for the ions of interest can be used to evaluate process cost-effectiveness under optimal conditions. Regeneration of the exchanger should be considered because the cost of regeneration and spent regenerant treatment/disposal is a significant fraction of the overall process cost. The exchanger particle size should be considered as it affects process kinetics, column solids entrainment, and column pressure drop.

Prediction of Column Performance. Because of the mathematical complexities associated with predicting ion exchange column performance in cases where multiple exchangeable ions are present, especially when kinetic limitations and counterion dispersion must be considered, the design of many ion-exchange processes is accomplished by bench or pilot scale treatability studies.

Treatability Studies. After selecting a medium or a number of different ion exchangers for further consideration, treatability studies are often performed using the waste of interest. Bench scale columns 1 inch to 3 inches in diameter and 2–3 feet long are adequate to obtain quality process design data. The purpose of the treatability study is to determine the service flow rate and length of service run, and the regenerant dose and concentration and length of regeneration run. These are the primary design parameters to be optimized. Other factors such as bed pressure drop and resin stability can also be assessed through the treatability study.

The principal treatability data obtained from such small-scale pilot tests are saturation loading curves (breakthrough curves) and elution curves. The loading curve is obtained by passing the waste through the column and periodically collecting samples of the effluent for chemical analysis. This process is continued until the column is nearly saturated (and at equilibrium with the influent), and the effluent of the column is at approximately the same composition as the influent of the column. The saturation loading curve is then plotted as the effluent concentration as a function of the number of (empty) bed volumes treated. Often, loading curves for a number of different flow rates are obtained to assess the effect of loading rate and optimize the process variable. An example loading curve illustrating the effect of volumetric loading rate, is provided in Figure 5-35 (U.S. EPA, 1981). As the figure illustrates, operating at a lower loading rate will generally allow one to obtain a greater capacity per unit volume of resin. However, for a given volumetric flow rate, an increase in column diameter (at greater capital cost) is required to reduce the loading rate. Therefore, there is a trade-off between (increased) capital costs and (decreased) operating costs.

Service loading rate can be expressed in various units. The volumetric loading rate is generally expressed as the ratio of the volumetric loading rate to the empty bed volume of the column. Volumetric loading rate is often expressed in units such as gpm/ft^3 or bed-volumes/hr, and is inversely proportional to bed contact time. The surface area loading rate (or sometimes termed hydraulic loading rate) is the ratio of the volumetric flow rate to the bed cross-sectional area, and is often in units such as gpm/ft^2, or m/hr. The surface area loading rate is directly related to the fluid linear velocity through the column, and must be maintained within acceptable ranges to

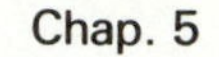

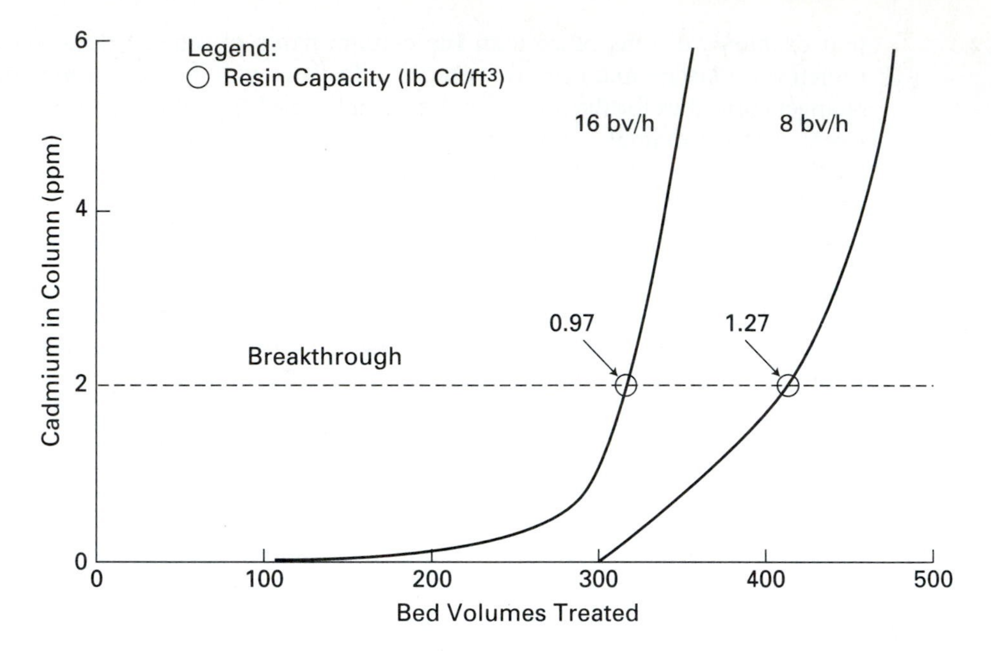

Note.—Feed Solution: 50 ppm Cd^{+2}, 1,000 ppm $CaCl_2$, pH = 4.0.

Figure 5-35 Influence of flowrate on chelating resin capacity. (U.S. EPA, *Summary Report: Control and Treatment Technology for the Metal Finishing Industry, Ion Exchange,* EPA/625/8-81-007, June 1981, p. 25)

avoid excessive pressure drops or damage to the exchanger media. Typical ranges for volumetric loading rate and surface area loading rate are 1 to 10 gpm/ft^3 and 4 to 15 gpm/ft^2, respectively. The results of runs at various loading rates can be used to determine the optimal values in the trade-off between capital and operating costs. Figure 5-36 (Kunin, 1958) illustrates the results of various loading runs, and shows the reduction of exchanger utilization with increasing service flow rate.

After each service run, the exchanger must be regenerated to the desired initial ionic form. Various regenerants and regenerant concentrations and flow rates can be assessed to optimize the regeneration cycle. These regeneration runs are performed in the treatability study in a manner similar to the service runs, with samples of the effluent periodically collected and analyzed to develop an elution curve. The elution curves can be used to choose the optimum regeneration protocol with respect to resultant operation capacity and regeneration efficiency. Although the exchanger can theoretically be fully regenerated by excessive contacting with regenerant, full regeneration is rarely necessary or economically practical. An example elution curve is provided in Figure 5-37 (Montgomery, 1985). The variables of interest during regeneration are the regenerant concentration and flow rate, and the regenerant dosage expressed in units of mass of regenerant per empty bed volume of resin. Typical regenerant volumetric loading rates are 2 to 5 bed volumes/hr. By choosing the regenerant volumetric loading rate regenerant within this range, and varying the regenerant concentration, the optimum regenerant concentration can be

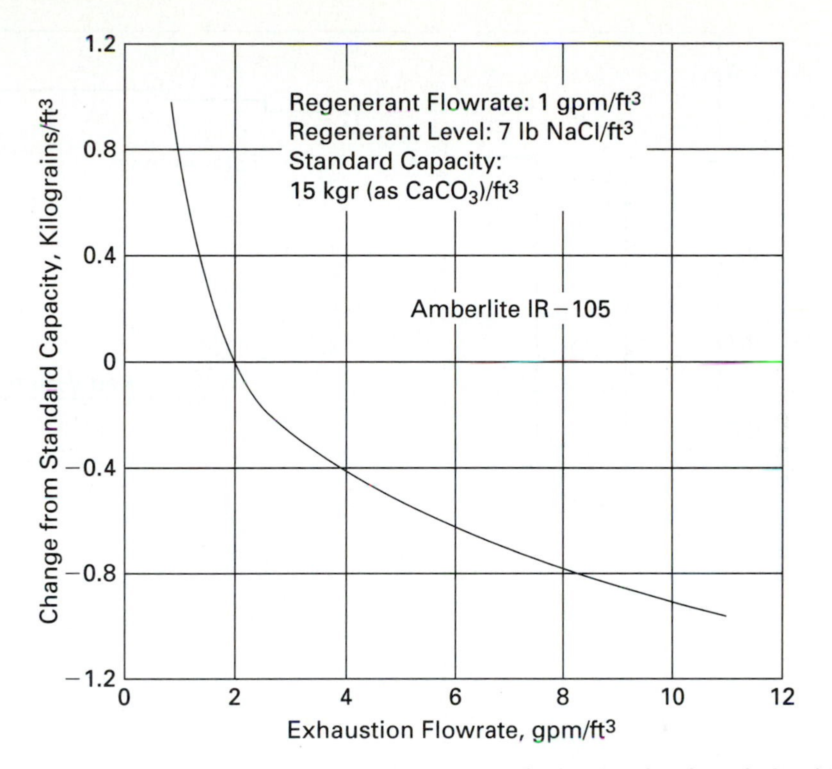

Figure 5-36 Effect of exhaustion flowrate on softening capacity of a sulfonic acid cationic exchange resin. (R. Kunin, *Ion Exchange Resins,* 2nd ed., John Wiley & Sons, 1958, p. 134.)

determined. The regenerant volumetric loading rate can then be optimized by varying this parameter with the regenerant concentration held constant at the optimized value. The regenerant dose is calculated as the product of the regenerant concentration and volumetric loading rate. The various regeneration runs can be used to optimize the regenerant dose in terms of column utilization and regeneration efficiency. Column utilization refers to the fraction of the exchange capacity of the column that is recovered by the regeneration step, while regeneration efficiency is the ratio of the number of equivalents of capacity recovered to the number of equivalents of regenerating ion applied during the regeneration step. By plotting the regeneration efficiency, column utilization, and the product of these two parameters, as a function of regenerant dosage, one can optimize the regeneration process. Figure 5-38 provides an example of this approach to optimizing the regeneration process, with the optimal point defined by the set of conditions whereby the product of the regeneration efficiency and column utilization is at a maximum.

Upon optimizing the service and regeneration cycle, a number of successive service and regeneration cycles should be performed to determine the steady state breakthrough point and leakage level. Leakage is the result of incomplete column regeneration, which produces low concentrations of the contaminant at the effluent prior to breakthrough. Leakage must be considered to insure that the optimizing process does not result in leakage concentrations that exceed acceptable levels.

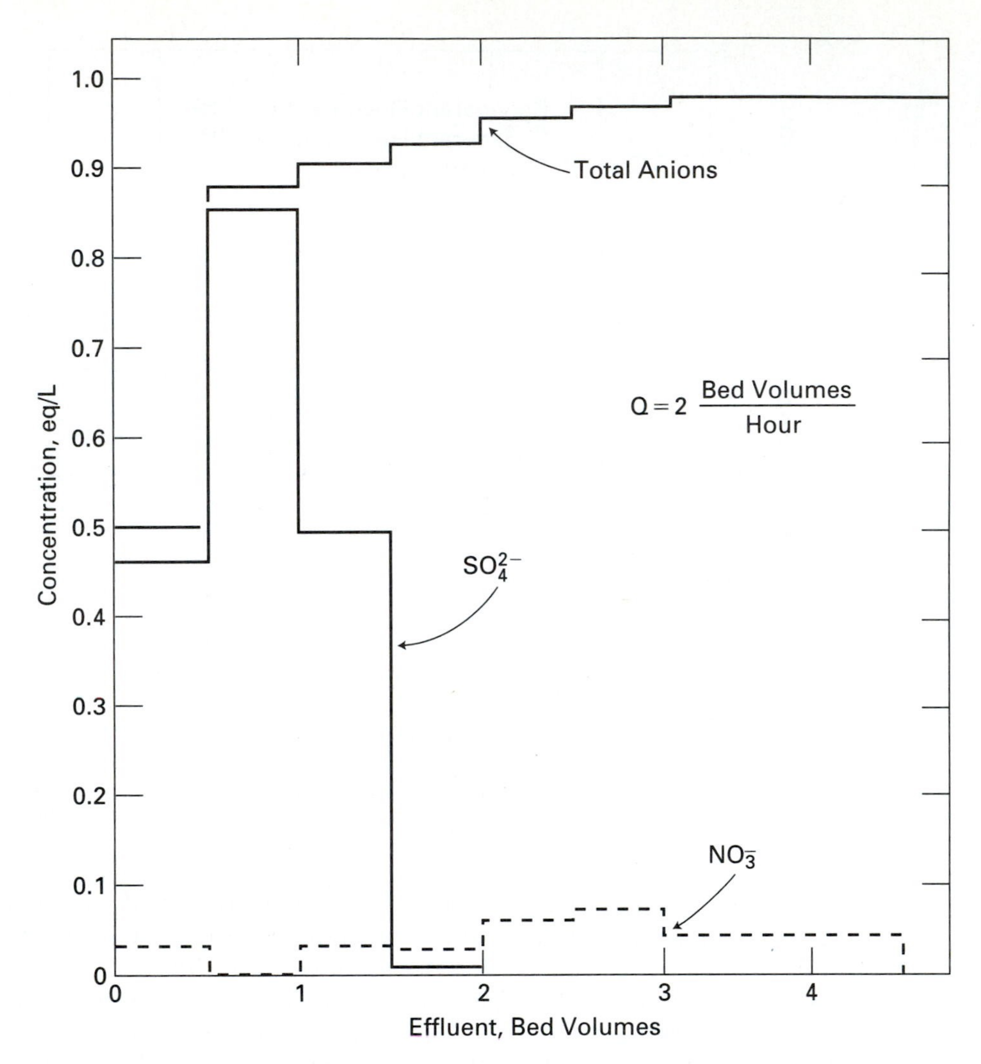

Figure 5-37 Elution of strong base resin in sulfate-nitrate form with 1.0 N NaCl.

Scale-up from the treatability study to full scale involves increasing the column diameter to maintain the surface area loading rate within acceptable range of 4 to 15 gpm/ft^2. The optimized volumetric loading rate is maintained in scale-up by increasing the column diameter and (sometimes) the bed depth. Again, typical volumetric loading rates range from 1 to 10 gpm/ft^3. Typical full-scale column depths range from 2.5 ft to 9 ft, and column diameters range up to 15 ft. If necessary, multiple columns in series can be used to reach the design volumetric loading rate. The required resin volume (empty bed volume) can be calculated as the ratio of the volumetric flow rate (i.e., gpm) to the volumetric loading rate (gpm/ft^3).

Process Configuration. The most common process configuration involves a fixed column bed of a single ion-exchange medium operating in a downflow service mode, with regeneration conducted either cocurrently (same flow direction as service

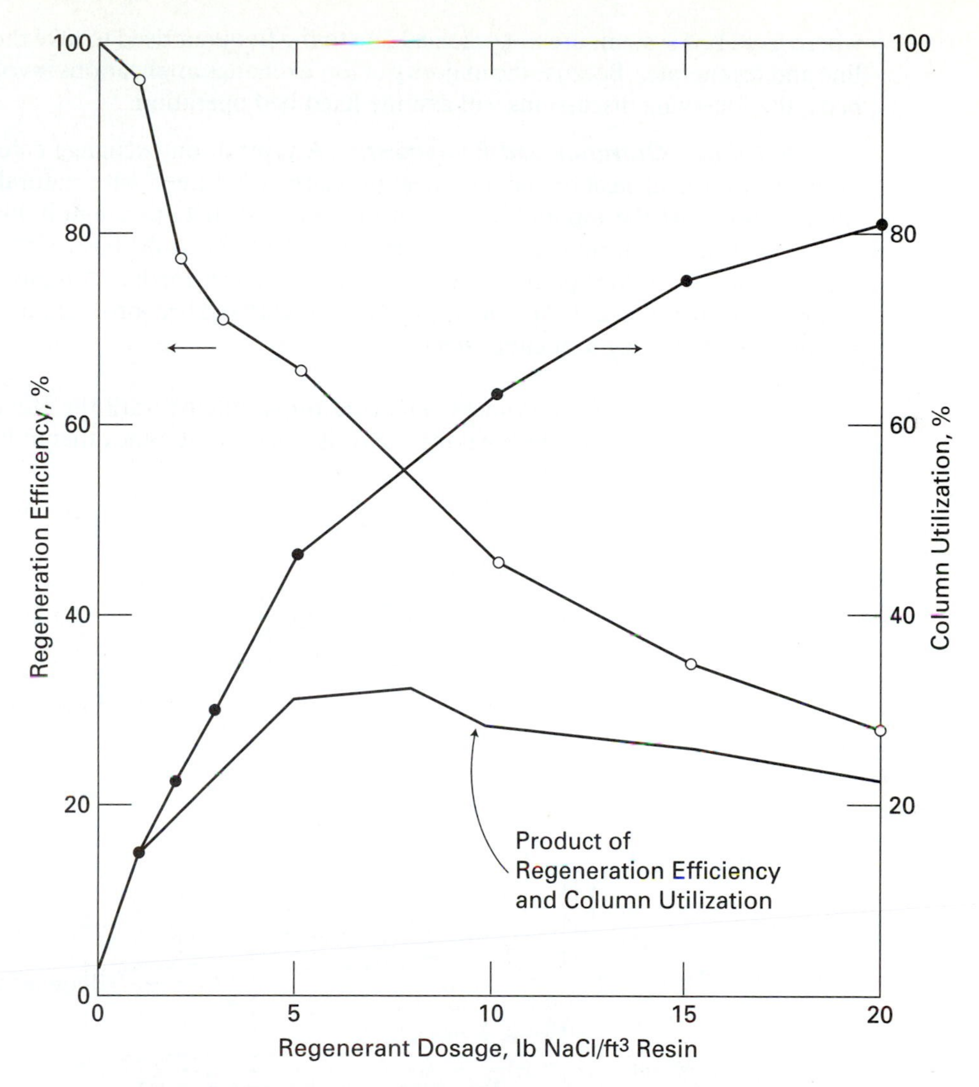

Figure 5-38 Efficiency and column utilization as a function of regeneration level for strong-acid resin used for softening. Conditions: influent hardness, 500 mg/L as $CaCO_3$; service flowrate, 2 gpm/ft^3; regeneration rate, 0.5 gpm/ft^3, 10% NaCl. (J. M. Montgomery, *Water Treatment Principles and Design,* Copyright 1985, p. 220. Reprinted by permission of John Wiley & Sons, Inc.)

flow) or countercurrently. However, many other configurations have been applied (Bolto and Pawlowski, 1987). Mixed beds containing more than one type of ion exchanger maintained in distinct layers have been used for water de-ionization. Continuous ion-exchange contactors have also been designed and operated. In these continuous ion-exchange systems, spent exchanger at the top of the column is continuously removed and regenerated, with regenerated exchanger returned to the bottom of the column. Movement of the exchanger is therefore countercurrent to the direction of flow of the process stream. Applications of such continuous-exchange systems are generally limited to treatment of highly concentrated wastes, where fixed bed systems are not practical due to the frequent need to take the bed off line and regenerate. Because the majority of ion-exchange applications involve fixed beds, the following discussions will assume fixed bed operation.

Full-Scale Operation and Equipment. A typical ion-exchange column is a vertical cylindrical steel or stainless steel pressure vessel lined with natural or synthetic rubber. At the top and bottom of the vessel, distribution manifolds are installed to insure uniform hydraulic distribution. A screen at the base of the bottom distribution manifold is present to support the exchanger media. A typical ion-exchange column is illustrated in Figure 5-39. A typical fixed bed ion-exchange process involves the following sequential steps:

- *Service run to breakthrough:* The column is utilized until the effluent concentrations exceed the breakthrough criteria, or until some other indication is

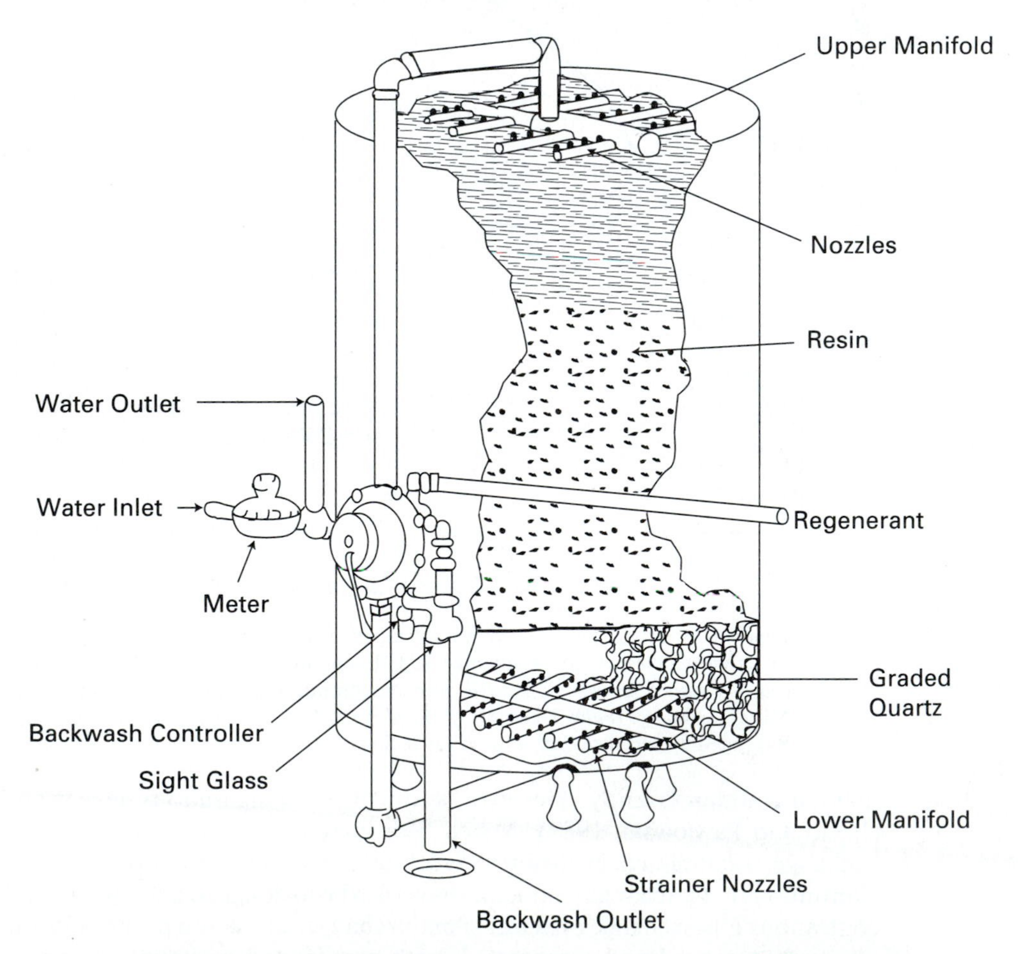

Figure 5-39 Typical ion exchange resin column. (R. Kunin, "Ion Exchange for the Metal Products Finisher," *Products Finishing,* May 1969, p. 74. Reprinted by permission of Gardner Publications, Inc.)

given that regeneration is required. Service runs are generally performed by down-flow through the column.

- *Backwash:* The column is then taken off-line and backwashed to remove suspended solids and to eliminate channels which may have formed during the service run. Backwash is generally done by upflowing clean water at a high flow rate through the column. Backwashing fluidizes the bed, releases any trapped suspended solids, and redistributes the exchanger particles according to particle size. Bed expansion is typically 50% to 100%, and adequate freeboard must be available to allow for this expansion. Backwashing is typically performed for 5 to 30 minutes, at rates of 2 to 10 gpm/ft^2. During backwashing, the larger particles will accumulate at the base of the column, while smaller particles will be lifted to the top section of the column. The resultant particle size distribution yields a uniform hydraulic flow pattern and serves to resist bed fouling by suspended solids.
- *Regeneration:* The column is regenerated by passing the regenerant through the column at the pre-determined volumetric flow rate. Typical regeneration flow rates range from 1 to 2 gpm/ft^2. Both cocurrent (regenerant flow direction same as service flow direction) and countercurrent regeneration is used, with cocurrent being more commonly applied. A disadvantage of cocurrent regeneration is that the exchanger media near the (service run) effluent of the column will be the most poorly regenerated media in the column, because the regenerant must pass through the entire column depth before reaching this media. As a result, residual contaminant in the media can result in contaminant leakage once the column is brought back on line. Countercurrent regeneration results in contacting the exchanger media near the service run effluent with fresh regenerant. This results in the service run effluent end of the column containing the most completely regenerated media after regeneration. Significant reductions in required regenerant can be achieved by performing countercurrent regeneration as opposed to cocurrent regeneration. Typically, 125% to 150% of the theoretical (stoichiometric column equivalent) mass of regenerant is required under countercurrent regeneration, while cocurrent regeneration often requires 200% to 400% of the theoretically required mass of regenerant. However, the reduced chemical costs of countercurrent regeneration are offset by increased capital costs. Regeneration time typically ranges from one to two hours.
- *Water Rinse:* After regeneration, a slow water rinse is first used to remove occluded regenerant. Typically, one to two bed volumes is passed through the column at a flowrate significantly less than the service flowrate. Following the slow water rinse, a fast water rinse at a flowrate of 1 to 1.5-gpm/ft^2 is applied to remove any traces of regenerant and to insure good media distribution. The fast rinse is typically conducted for a period of 10 minutes to 30 minutes, and total rinse water volume ranges from 30 to 100 gal/ft^3.
- *Return to Service:* Following the fast water rinse, the column is generally brought back into service. However, for resins which exhibit significant

swelling or shrinking during regeneration, a second backwash may be necessary to eliminate channeling or resin compression.

Applications of Ion Exchange in Hazardous Waste Treatment

Soluble hazardous constituents which are amenable to treatment by ion exchange include arsenic, barium, cadmium, chromium, cyanide (free and complexed), lead, mercury, selenium, and silver. Because ion exchange process costs are generally prohibitive for treatment of highly concentrated waste streams, ion exchange is typically used as a polishing step after precipitation. Stringent discharge limitations

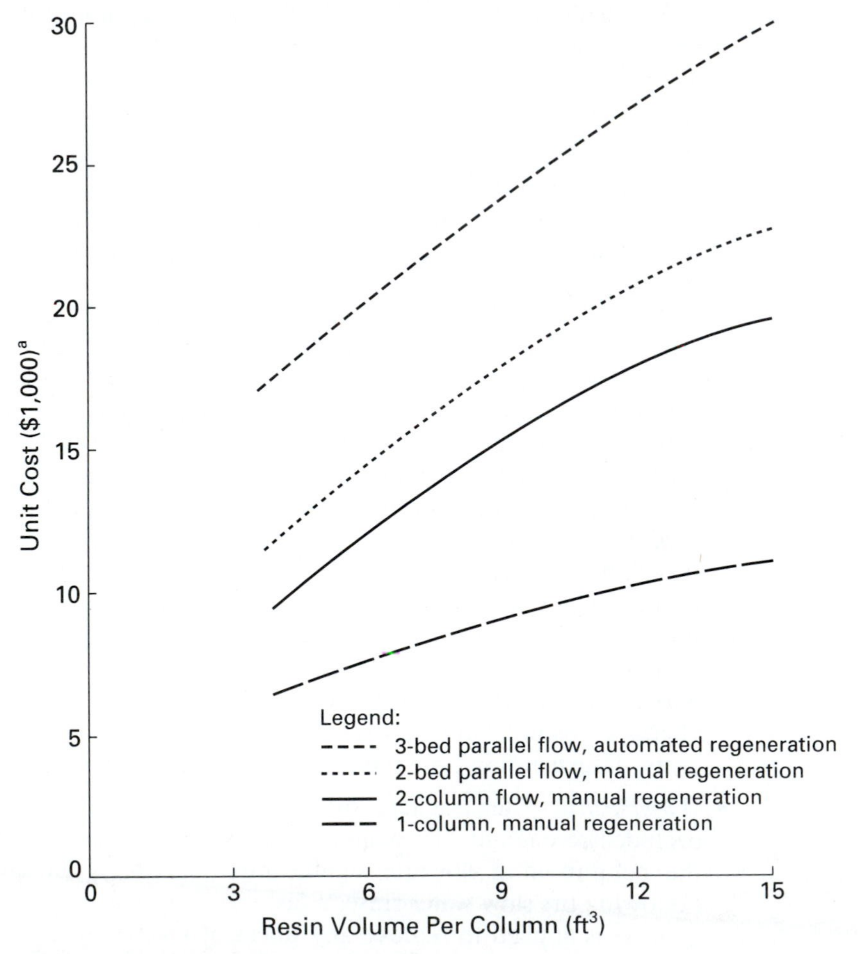

[a]1980 dollars. Add $200/ft³/column for chelating resin.

Note.—Skid-mounted unit with weak acid cation resin, acid and base regenerant, storage, and all internal piping and valves.

Figure 5-40 Unit costs of example ion exchange systems. (U.S. EPA, *Summary Report: Control and Treatment Technology for the Metal Finishing Industry. Ion Exchange,* EPA/625/8-81-007, June 1981, p. 33.)

which cannot be met by other conventional technologies can be met using ion exchange. Ion exchange is also sometimes applied when metal recovery by ion exchange (i.e., for precious metals) off-sets the higher ion exchange process costs.

Some of the most common applications of ion exchange processes for the treatment of hazardous contaminants are in the treatment of metal finishing wastes and wastewaters (U.S. EPA, 1981; Patterson, 1985; Calmon et al., 1979). Heavy metal ions typically present in these waste streams (cadmium, chromium, nickel, zinc, copper) can be removed by ion exchange processes that are used as a final polishing step after precipitation. Strong-acid, weak-acid, or chelating ion exchange resins can be used to remove soluble heavy metals. Reuse of treated electroplating rinsewaters after ion exchange treatment is a strong economic driving force even in cases when discharge limitations can be met by precipitation (North et al., 1985).

Spent chromic acid plating and anodizing baths can be purified for reuse by ion exchange (Calmon et al., 1980; Patterson, 1985, U.S. EPA, 1981). Chromium wastes containing chromate and dichromate can be removed by strong-base anion exchange. Because most strong-base ion exchangers are more highly selective to chromate, it is advantageous to adjust the pH of the waste to below six to insure that chromate is the predominant chromium species (as opposed to dichromate which predominates above pH of six). However, the pH must be maintained above a pH of four to avoid oxidation of the ion exchange resin. Regeneration of the spent exchanger with sodium hydroxide produces a concentrated solution of sodium chromate, which can then be passed through a strong-acid ion exchanger operating on the hydrogen cycle to recover chromic acid.

Studies have demonstrated that cyanides in coke plant wastes can be treated by first converting the cyanides to $Fe(CN)_6^{-4}$ by the addition of ferrous sulfate, followed by ion exchange with a strong-base anion-exchange resin operating on chloride cycle (Bessent et al., 1979). Conversion of the free cyanides to the ferro-cyanide complex is necessary for efficient removal. Also, free cyanides have been found to deteriorate ion exchange resins. Additional studies have demonstrated the effectiveness of strong-base ion exchange for removal of cyanide in synthesized metal plating wastes

TABLE 5-14 RESIN CAPACITY BASED ON TEST RESULTS

	Feed		Product		
Constituent	g/L	eq/L	g/L	eq/L	Change (eq/L)
Calcium	0.39	0.0195	0.3	0.015	0.0045
Magnesium	0.35	0.0292	0.27	0.0225	0.0067
Zinc	0.09	0.0028	[a]	[a]	0.0028
Sodium	0.06	0.0026	0.32	0.0139	−0.0113
Manganese	0.03	0.001	0.02	0.0006	0.0004
Cadmium	0.001	[a]	[a]	[a]	0.0001

[a] Negligible.

Note: Exchange requirements: per liter of feed, 0.0145 eq/L; per 175 bv of feed, 2.53 eq/L of resin.

TABLE 5-15 COLUMN SIZE DETERMINATION FOR THREE-COLUMN PARALLEL FLOW UNIT

Item	Factor
Flowrate:	
Per column	25 gal/min
Total	50 gal/min
Column cycle	4 hr
Exchange capacity per liter of feed	0.0145 eq
Capacity needed per column	$4 \times 60 \times 25 \times 3.79 \times 0.0145 = 330$ eq
Resin volume needed:	
Per column	$[330/(2 \text{ eq/L})] \times [1/(3.79)(7.48)] = 5.8 \text{ ft}^3$
With safety factor	$5.8/0.75 = 7.8 \text{ ft}^3$
Regenerant consumption per column per cycle:	
HCL (based on 100%)	45 lb
NaOH (based on 100%)	50 lb
Wash water	390 gal
Cost per cycle[1]	$15.09
Waste concentration factor	(6,000 gal wastewater per cycle)/(400 gal purge per cycle) = 15

[1] 1980 dollars

(Hassan et al., 1991). Again, the cyanide removal was effected by first converting free cyanide to ferro-cyanide complexes, followed by anionic ion exchange.

The feasibility of removal of selenium in contaminated groundwater by anionic ion exchange was demonstrated in bench scale experiments conducted by Boegel and Clifford (1986). The process evaluated involved first converting selenium in the form of SeO_3^{-2} (Se^{+4}) to SeO_4^{-2} (Se^{+6}) by chlorine oxidation.

TABLE 5-16 ANNUAL COST OF ION EXCHANGE TREATMENT SYSTEM[1]

Item	Cost
Investment ($)	23,000
Operating cost ($/yr):	
Labor, ½ hr/shift at $8/hr	2,000
Maintenance, 6% of investment	1,400
Regenerant chemicals, 4,000 hr at 2 hr/cycle	30,180
Total operating cost	33,580
Fixed cost ($/yr):	
Depreciation	2,300
Taxes and insurance	230
Total fixed cost	2,530
Total annual cost	36,110

[1] 1980 dollars

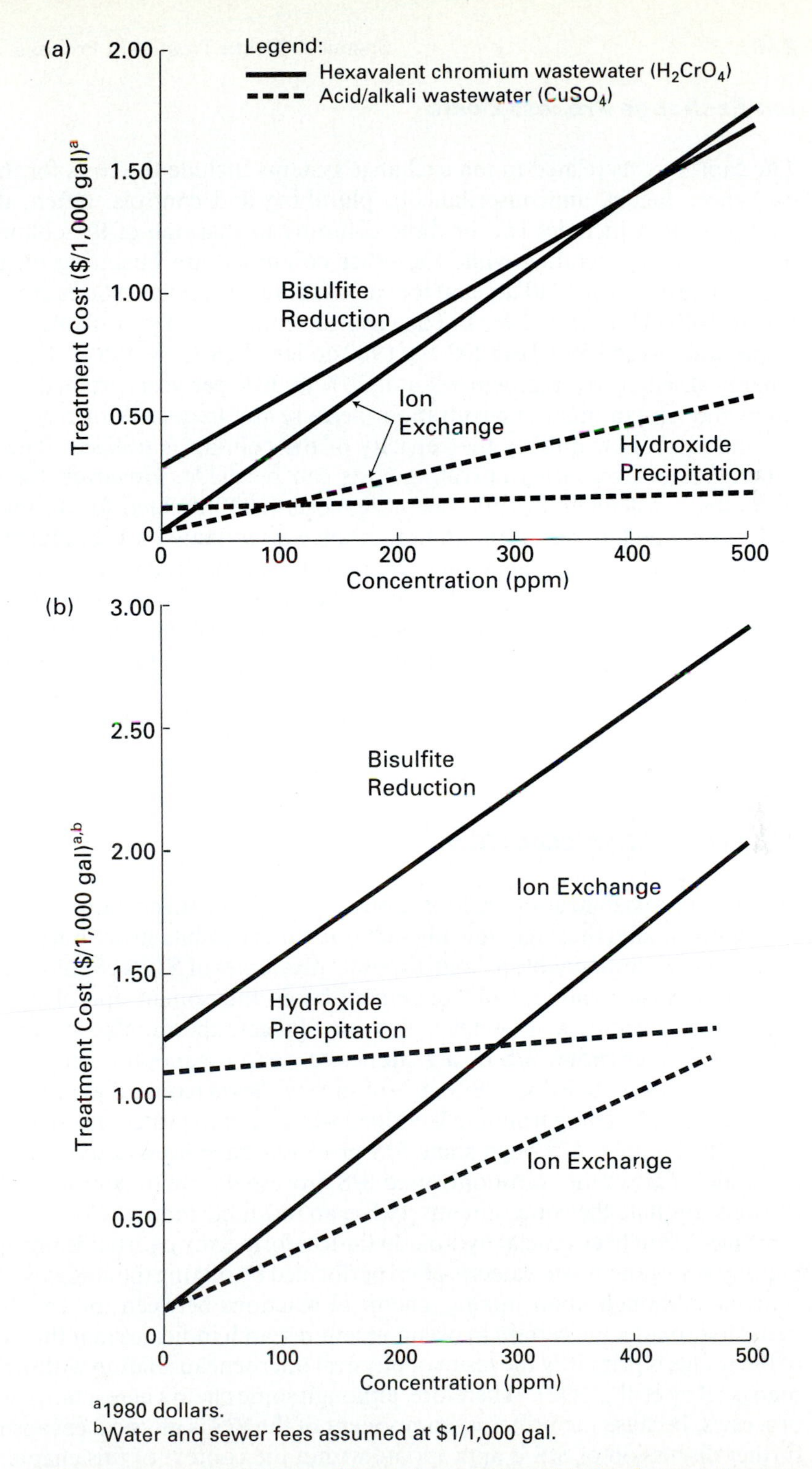

Figure 5-41 Comparative costs of ion exchange and precipitation for example wastewaters. (U.S. EPA, *Summary Report: Control and Treatment Technology for the Metal Finishing Industry, Ion Exchange,* EPA/625/8-81-007, June 1981, p. 17.)

Ion Exchange Process Costs

The capital costs related to ion exchange systems include the costs for the column, exchanger media, and miscellaneous plumbing and controls. Often, the ion exchange system includes two or three columns so that one of the columns can be off-line and regenerating while the other column(s) are operating on the service cycle. The costs (in 1980 dollars) for various column configurations are provided in Figure 5-40 (U.S. EPA, 1981). The costs for commercially available ion exchange resins range from $50/ft^3 to $200/ft^3 (1980 dollars, U.S. EPA, 1981). Operating costs (chemical, labor, replacement resin of 20% to 30% per year) are highly dependent upon the contaminant concentration, because the frequency of regeneration will depend upon how quickly the capacity of the column is utilized. Therefore, few generalizations regarding operating costs can be made. However, for illustrative purposes, a calculation of the operating costs (1980 dollars) for an ion exchange system designed to treat a zinc and cadmium contaminated wastewater at a flow rate of 50 gpm using a weak-acid ion exchanger, is summarized in Tables 5-14 through 5-16 (U.S. EPA, 1981). Figure 5-41 compares the overall costs of treating a hexavalent chromium wastewater and an acid/alkali wastewater containing copper, using ion exchange or the alternative conventional technology of precipitation (U.S. EPA, 1981).

SOLIDIFICATION/STABILIZATION

Solidification/stabilization (S/S) is a process used to convert liquid and semi-solid wastes into a solid that is chemically stable and that exhibits structural properties that render the waste amenable to land disposal. Examples of S/S technologies utilized in hazardous waste treatment include cement-based and cement-pozzolan-based S/S of heavy metal wastes and metal hydroxide sludges, thermoplastic S/S of organic wastes, polymerization S/S of organic wastes, and organophilic clay-based S/S of organic wastes (U.S. EPA, 1989). S/S of such wastes is required to reduce the risk of contaminants leaching from the landfilled waste, and to protect the structural integrity of the landfill. Although some S/S processes have been described as chemical processes, most of the commonly used S/S processes serve to dewater and/or physically encapsulate the contaminants within an insoluble matrix (Conner, 1990). For example, S/S of heavy metal hydroxide sludges formed by hydroxide precipitation of plating wastes and wastewaters is often performed by mixing the sludge with portland cement. Although upon mixing, chemical reactions between the cement and the metal hydroxides are certain to occur, recent research indicates that the stabilization of the sludge is primarily the result of physical microencapsulation within the cement matrix (Roy et al., 1992). Therefore, although some chemical reactions occur in S/S processes, because the principal component of the S/S is physical encapsulation, no further discussion of S/S is appropriate within the context of this chapter.

PROBLEMS

5.1. A waste hydrochloric acid solution (10^{-3} M HCl) is to be neutralized by the addition of a base. The waste solution is to be continuously treated at a flow rate of 10 gal/min. Assuming unit activity coefficients and a temperature of 25°C, calculate the rate of base addition, in lb/hr, that must be added to neutralize the waste to a pH of 7.0, for the following bases:

a) Sodium carbonate (assume as 100% Na_2CO_3)

b) Sodium hydroxide (assume as 50% NaOH)

5.2. A titration of a strong-base waste solution indicates that 1 liter of the waste can be neutralized to a pH of 7.0 with 2 ml of 4 N hydrochloric acid. However, it is desired to neutralize the basic waste solution using a 0.5 N waste sodium acetate solution. Determine the relative volumes of the waste acid solution and the waste sodium acetate solution that must be mixed to achieve a final pH of 6.0. Assume unit activity coefficients and a temperature of 25°C.

5.3. a) Calculate the equilibrium pH of a 10^{-4} M cadmium nitrate solution in pure water at 25°C. Account for solution phase nonidealities using the Debye–Hückel expression.

b) It is desired to precipitate cadmium as cadmium hydroxide, from the solution described above, to a total soluble phase cadmium concentration of 10^{-6} M. Determine the required dosage of 2 N NaOH (in ml added/liter of cadmium solution) required to achieve the desired treatment objective. Account for solution phase nonidealities using the Debye–Hückel expression, and assume a temperature of 25°C.

5.4. a) Divalent lead (Pb^{+2}) at a concentration of 10^{-2} M is to be precipitated from solution by the addition of sodium sulfide (Na_2S). If the solution is buffered at a pH of 7.0, and assuming a constant temperature of 25°C, determine the number of moles of sodium sulfide that must be added per liter of lead solution to bring the final Pb^{+2} concentration to 10^{-7} M. Neglect activity coefficient corrections.

b) Determine whether ignoring the formation of precipitated lead hydroxide is justified.

c) Assume the lead solution is initially at a pH of 7.0, but is not buffered. Determine the number of moles of sodium sulfide that must be added per liter of lead solution to bring the final Pb^{+2} concentration to 10^{-7} M. What is the final solution pH? [Ignore the formation of any lead–hydroxy complexes, but consider whether lead hydroxide will precipitate upon addition of the sodium sulfide solution.]

5.5. A waste sodium sulfide solution that is 10^{-3} M in total sulfide is to be used to precipitate cadmium from a waste cadmium nitrate solution that is 10^{-2} M in total cadmium.

a) Determine at 25°C the equilibrium pH of the two waste solutions (neglecting activity coefficient corrections).

b) Calculate the relative volumes of the two waste solutions that must be mixed to achieve a final equilibrium free cadmium concentration (Cd^{+2}) of 10^{-6} M.

5.6. a) Write the balanced redox equation for the reduction of hexavalent chromium ($Cr_2O_7^{-2}$) to trivalent chromium (Cr^{+3}) using ferrous (Fe^{+2}) iron.

b) If the initial dichromate concentration is 0.01 M, and the desired final dichromate concentration is 10^{-4} M, determine the required dosage of ferrous iron, and the equilibrium concentrations of all chromium and iron species, assuming a temperature of 25°C and an equilibrium pH of 4.5. Neglect activity coefficient corrections and the potential formation of trivalent chromium hydroxide precipitate.

5.7. Trichloroethylene (TCE) is to be oxidized in a continuous flow sparged reactor using ozone and hydrogen peroxide. The formation of the hydroxyl radical from ozone and hydrogen peroxide can be assumed to be instantaneous with respect to the mass transfer of ozone from the gas phase to the liquid phase, and with respect to the oxidation of TCE by hydroxyl radicals. The stoichiometry of the formation of the hydroxyl radical, and the oxidation of TCE, can be assumed as follows:

$$H_2O_2 + 2O_3 \longrightarrow 2OH\bullet + 3O_2$$

$$TCE + OH\bullet \longrightarrow \text{oxidized products}$$

a) Using the rate constant for TCE oxidation given in Table 5-8, determine the steady-state hydroxyl radical concentration needed in the reactor to reduce TCE from 5 mg/l to 0.005 mg/l if the reactor residence time is 30 minutes. Assume the reactor can be modeled as a complete mixed reactor.

b) If the reactor volume is 30 liters, what are the required applied dosages of ozone and hydrogen peroxide to achieve the desired degree of treatment for a residence time of 30 minutes. Assume the stoichiometry of the reactions are as above, with no other side reactions.

c) The mass transfer of ozone between the gas and aqueous phase can be modeled using Equation 5.30. [*This is not strictly valid, because Equation 5.30 is only applicable when aqueous phase ozone is unreactive. However, for illustration purposes, ignore this fact and use Equation 5.30.*] Determine the steady-state aqueous ozone concentration using Equation 5.30, given the following additional information:

$$K_L a = 1.8 \text{ min}^{-1}$$

$$H_{ozone} = 0.082 \text{ atm} \cdot \text{m}^3 \cdot \text{mole}^{-1}$$

$$[O_3]_{gas} = 3\% \text{ by weight (assume constant)}$$

d) At the aqueous phase ozone concentration calculated in c), determine whether oxidation of TCE in the reactor by molecular ozone oxidation is significant, using the oxidation rate constant provided in Table 5-7.

5.8. A strong-acid ion exchanger initially saturated to sodium is to be used to remove lead from a waste solution. The waste solution contains lead at a concentration of 100 mg/l, and the process is to be designed to meet a final solution phase lead concentration of 5 mg/l. Assume that no ions other than lead and significant to the ion-exchange process are present in the solution. The applicable ion exchange reaction is given below, and the equilibrium constant for the reaction is 4.5 (neglect activity coefficient corrections in both the solution and exchanger phase).

$$2NaRes + Pb^{+2} \rightleftharpoons PbRes + 2Na^{+1}$$

If the capacity of the exchanger (for both lead and sodium) is 4.8 meq/g (dry basis), and if 100 gallons of solution are to be treated, calculate the dry mass of exchanger required to achieve the desired effluent concentration.

5.9. An ion exchange column using a strong-acid ion exchanger (initially saturated to the sodium ion) is to be designed to remove divalent cadmium from a waste solution that contains only cadmium and magnesium in significant concentrations. The exchanger is selective to cadmium with respect to magnesium. Assume that the binary equilibrium constant for the exchange reaction involving cadmium and magnesium is 2.7, and that activity coefficients in the solution phase and the exchanger phase can be ignored. Other process information is provided below:

Waste solution flow rate	= 30 gpm
Initial cadmium concentration	= 200 mg/l (Cd^{+2})
Allowable effluent cadmium concentration	= 5 mg/l (Cd^{+2})
Initial magnesium concentration	= 400 mg/l
Resin wet-volume capacity	= 1.5 eq/l
Selectivity coefficient (Cd/Mg)	= 2.7

The column design should adhere to the following design parameters:

Surface area loading rate = 8 gpm/ft²
Volumetric loading rate = 2 gpm/ft³

a) Determine the required volume of resin, column diameter, exchanger bed depth, and total column depth. Total column depth (length) should allow for freeboard of 50% to account for expansion of the exchanger bed during regeneration and backwashing.
b) Determine the total volume of waste solution that can be treated through the column before the cadmium concentration in the effluent reaches 5 mg/l [*Hint: Assume at breakthrough that the entire column is in equilibrium with the influent composition. Calculate the mole fractions of cadmium and magnesium under this condition, and back-calculate the volume of solution that must pass through the column to equilibrate the entire column to this exchanger phase composition.*]
c) Determine the volume of backwash water generated per cycle assuming 10 minutes of backwash at 5 gpm/ft².
d) Assume that the exchanger will be regenerated using 4 N HCl at a flowrate of 1 gpm/ft², and that the total regenerant dosage will be 150% of the stoichiometric column equivalent capacity. Determine the regeneration time, and the total volume of spent regenerant produced.
e) Calculate the ratio of the volume of water treated to the sum of the backwash rinse water and waste regenerant. *This represents the waste volume reduction achieved by the exchanger process.*

REFERENCES

AIETA, E. M., REAGAN, K. M., LANG, J. S., MCREYNOLDS, L., KANG, J., AND GLAZE, W. H. "Advanced Oxidation Processes for Treating Groundwater Contaminated With TCE and PCE: Pilot-Scale Evaluations." *J. AWWA.* May (1988), pp. 64–72.

AMPHLETT, C. B. *Inorganic Ion Exchangers.* Elsevier, Amsterdam (1964).

ARGERSINGER, W. J., DAVIDSON, A. W., AND BONNER, O. D. "Thermodynamics and Ion Exchange Phenomena." *Trans. Kansas Acad. Sci.* 53:404–410 (1950).

ARISMAN, R. K., AND MUSICK, R. C. "Experience in Operation of a UV–Ozone Ultrox Pilot Plant for Destroying PCBs in Industrial Waste Effluent." *Proceedings of the 35th Industrial Waste Conference.* Lewis Publishers, Chelsea, Mich., pp. 802–810 (1980).

BALL, W. P., JONES, M. D., AND KAVANAUGH, M. C. "Mass Transfer of Volatile Organic Compounds in Packed Aeration." *J. Wat. Poll. Control Fed.* 56, 2: 127–136 (1984).

BESSENT, R. A., LUTHER, P. A., AND EKLUND, C. W. "Removal of Cyanides from Coke Plant Wastewaters by Selective Ion Exchange–Results of Pilot Testing Program." *Proceedings of the 34th Industrial Waste Conference, May 1979.* Lewis Publishers, Chelsea, Mich., pp. 47–62 (1980).

BHAKTA, D., SHUKLA, S. S., CHANDRASEKHARALAH, M. S., AND MARGRAVE, J. L. "A Novel Photocatalytic Method for Detoxification of Cyanide Wastes." *Environ. Sci. and Technol.* 26, 3: 625–626 (1992).

BHATTACHARYYA, D., AND CHEN, L. F. *Sulfide Precipitation of Nickel and Other Heavy Metals from Single and Multi-Metal Systems,* U.S. EPA, EPA/600/S2–86/051, July 1986.

BLACK, A. P., BUSWELL, A. M., EIDSNESS, F. A., AND BLACK, A. L. "Review of the Jar Test." *Jour. AWWA* 39, No. 11: 1414 (1957).

BLACK, A. P., AND HARRIS, R. J. "New Dimensions for the Old Jar Test." *Water & Wastes Eng.* December, 49 (1969).

BOEGEL, J. V., AND CLIFFORD, D. A. *Selenium Oxidation and Removal by Ion Exchange.* U.S. EPA, Cincinnati, Ohio, EPA/600/S2–86/031 (May 1986).

BOLTO, B. A., AND PAWLOWSKI, L. *Wastewater Treatment by Ion Exchange.* E. & F.N. Spon, New York, N.Y. (1987).

BONNER, O. D., AND L. L. SMITH. "A Selectivity Scale for Some Divalent Cations on Dowex 50." *J. Phys. Chem.* 61, No. 3, 326–329 (1957).

BORUP, M. B., AND MIDDLEBROOKS, E. J. "Photocatalyzed Oxidation of Organic Wastestreams." *Toxic and Hazardous Wastes: Proceedings of the 18th Mid-Atlantic Industrial Waste Conference,* G. D. Boardman, ed. Virginia Polytechnic Inst. and State University, pp. 554–568 (1988).

BRETON, M., ARIENTI, M., FRILLICI, P., KRAVETT, M., PALMER, S., SHAYER, A., AND SURPRENANT, N. *Technical Resource Document Treatment Technology for Solvent Containing Wastes,* EPA/600/S2–86/095, U.S. EPA, Cincinnati, Ohio, February, 1987.

BUCKLY, J. A., BROWN, P. M., CATER, S. R., LUONG, C. V., REED, D. W., STEVENS, D. S., AND TREMBLAY, L. S. "Rayox, A Cost-Effective Photooxidation Treatment for Wastewater." *Proceedings of the International Conference on Physiochemical and Biological Detoxification of Hazardous Wastes.* Technomic Publishing, Lancaster, U.K., Vol. I, pp. 369–393 (1988).

BUTLER, J. N. *Ionic Equilibrium.* Addison-Wesley, Reading, Mass., 1964.

CALMON, C., GOLD, H., AND PROBER, R., EDS. *Ion Exchange for Pollution Control.* Volume I, CRC Press, Boca Raton, Florida (1979).

CAMP, T. R. "Water Treatment." In *The Handbook of Applied Hydraulics."* C. V. Davis, ed. New York, NY, McGraw-Hill (1952).

CAMP, T. R. "Flocculation and Flocculation Basins." *Trans. ASCE* 120: 1 (1955).

CAMP, T. R. "Floc Volume Concentration." *Jour. AWWA* 60, No. 6: 656 (1968).

CAMP, DRESSER & MCKEE, INC. *Technical Assessment of Treatment Alternatives for Wastes Containing Corrosives.* Prepared for U.S. EPA under Contract No. 68-01-6403. Boston, Mass., September, 1984.

CAWLEY, W. A., ED. *Treatability Manual, vol III. Technologies for Control/Removal of Pollutants.* U.S. EPA 600/2-82-011c, September, 1981.

CHELKOWSKA, K., GRASSO, D., FABIAN, I., AND GORDON, G. "Numerical Simulations of Aqueous Ozone Decomposition." *Ozone Sci. & Eng.* 14: 1: 33–49 (1992).

CONNER, J. R. *Chemical Fixation and Solidification of Hazardous Wastes.* Van Nostrand Reinhold, New York (1990).

Crowle, V. A. "Effluent Problems as They Affect the Zinc Die-Casting and Plating-on-Plastics Industries." *Metal Fin. J.* 17, 194, pp. 51–54 (1971).

Davidson, A. W., and Argersinger, W. J. "Equilibrium Constants of Cation Exchange Processes." *Annals New York Academy of Sciences* 57: 105–115 (1953).

De Renzo, D. J. *Pollution Control Technology for Industrial Wastewater.* Noyes Data Corp., Park Ridge, New Jersey (1981).

DeVault, D., "The Theory of Chromatography." *J. Amer. Chem. Soc.* 65, 1943, pp. 532–540.

Donnan, F. G., and Guggenheim, E. A. "The Exact Thermodynamics of Membrane Equilibria." *Z. Physik. Chem.* 162A: 346–360 (1932).

Dorfner, K. D. *Ion Exchangers: Properties and Applications.* Ann Arbor Science, Ann Arbor, Mich. (1977).

Eckenfelder, W. W. *Industrial Pollution Control.* McGraw-Hill, New York, N.Y. (1966).

Eckenfelder, W. W. *Water Quality Engineering for Practicing Engineers.* Barnes and Noble, New York, N.Y. (1970).

Ekedahl, E., Hogfeldt, E., and Sillen, L. G. "Activities of the Components in Ion Exchangers." *Acta Chem. Scand.* 4: 556–558 (1950).

Evans, D. R., and Wilson, J. C. "Capital and Operating Costs-AWT." *J. Water Poll. Control. Fed.* 44, 1–13, (1972).

Farhataziz, P. C., and Ross, A. B. "Selective Specific Rates of Reactions of Transients in Water and Aqueous Solutions. P-III. Hydroxyl Radical and Perhydroxyl Radical and Their Radical Ions." *Natl. Stand. Ref. Data Ser.* 59, U.S. Natl. Bur. Stand. (1977).

Fletcher, D. B. *Waterworld News.* May/June, pp. 25–27 (1987).

Fletcher, D. B. "Successful UV/Oxidation of VOC-Contaminated Groundwater." *Remediation.* Summer 1991, pp. 353–357.

Fu, P. L. K., and Symons, J. M. "Removing Aquatic Organic Substances by Anion Exchange Resins." *J. Am. Water Works Assoc.* 82, 10, 70–77 (1990).

Glaze, W. H. "Chemistry of Ozone, By-Products and Their Health Effects." *Ozonation: Recent Advances and Research Needs.* AWWARF, Denver, Colo. (June 1986).

Glaze, W. H. "Drinking Water Treatment With Ozone." *Env. Sci. & Tech.* 21: 3: 224–230, (1987).

Glaze, W. H., Kang, J. W., and Chapin, D. H. "The Chemistry of Water Treatment Processes Involving Ozone, Hydrogen Peroxide and Ultraviolet Radiation." *Ozone Sci. and Engg.* 9: 4: 335 (1987).

Glaze, W. H., and Kang, J. "Advanced Oxidation Processes for Treating Groundwater Contaminated With TCE and PCE: Laboratory Studies." *J. AWWA.* May (1988), pp. 57–63.

Grammont, P., Rothschild, W., Sauer, C., and Katsahian, J. "Ion Exchange in Industry." In *Ion Exchange: Science and Technology,* edited by A. E. Rodrigues, Martinus Nijhoff. pp. 403–447, Boston, Mass. (1986).

Grasso, D., and Weber, W. J. "Mathematical Interpretation of Aqueous-Phase Ozone Decomposition Rates." *J. Env. Eng.* 115: 3: 541–559 (1989).

Grasso, D., Fujikawa, E., and Weber, W. J. "Ozone Mass Transfer in a Gas-Sparged Turbine Reactor." *Research Journal WPCF* 62: 3: 246–253 (1990).

Gregor, H. P. "A General Thermodynamic Theory of Ion Exchange Processes." *J. Am. Chem. Soc.* 70: 1293 (1948).

GREGOR, H. P., BELLE, J., AND MARCUS, R. A. *J. Am. Chem. Soc.* 77: 2731 (1955).

GUROL, M. D., AND NEKOUINAINI, S. "Effect of Organic Substances on Mass Transfer in Bubble Aeration." *J. WPCF* 57, 3: 235–240 (1985).

GUROL, M. D., BREMEN, W. M., AND HOLDEN, T. E. "Oxidation of Cyanides in Industrial Wastewaters by Ozone." *Environ. Progr.* 4, 1: 46–51 (1985).

HASSAN, S. Q., VITELLO, M. P., KUPFERLE, M. J., AND GROSSE, D. W. "Treatment Technology Evaluation for Aqueous Metal and Cyanide Bearing Hazardous Wastes (F007)." *J. Air Waste Mgmt. Assoc.* 41: 710–715 (1991).

HELFFERICH, F., *Ion Exchange.* Chapter 5. McGraw-Hill, New York (1962).

HELFFERICH, F. G. "Multicomponent Ion Exchange in Fixed Beds—Generalized Equilibrium Theory for Systems with Constant Separation Factors." *Ind. Eng. Chem. Fund.* 6, pp. 362–364 (1967).

HOIGNÉ, J. "Mechanisms, Rates, and Selectivities of Oxidations of Organic Compounds Initiated by Ozonation of Water." *Handbook of Ozone Technology and Applications.* Vol. 1 (R. G. Rice and A. Netzer, editors). Ann Arbor Sci. Publ., Ann Arbor, Mich. 1982).

HOIGNÉ, J., AND BADER, H. "The Role of Hydroxyl Radical Reactions in Ozonation Processes in Aqueous Solutions." *Water Res.* 10: 377 (1976).

HOIGNÉ, J., AND BADER, H. "Rate Constants of Reactions of Ozone With Organic and Inorganic Compounds in Water—I. Non-Dissociating Organic Compounds." *Water Res.* 17: 173–183 (1983-1).

HOIGNÉ, J., AND BADER, H. "Rate Constants of Reactions of Ozone With Organic and Inorganic Compounds in Water—II. Dissociating Organic Compounds." *Water Res.* 17: 185–194 (1983-2).

HOIGNÉ, J., BADER, H., HAAG, W. R., AND STAEHELIN, J. "Rate Constants of Reactions of Ozone with Organic and Inorganic Compounds in Water—III. *Water Res.* 19: 993–1004 (1985).

JONES, B. M., LANGLOIS, G. W., SAKAJI, R. H., AND DAUGHTON, C. G. "Effect of Ozonation and UV Irradiation on Biorefractory Organic Solutes in Oil Shale Retort Water." *Environmental Progress.* Vol. 4, No. 4, pp. 252–257 (Nov. 1985).

KLEIN, G., TONDEUR, D., AND VERMEULEN, T. "Multicomponent Ion Exchange in Fixed Beds—General Properties of Equilibrium Systems." *Ind. Eng. Chem. Fund.* 6, pp. 339–351 (1967).

KUNIN, R. *Ion Exchange Resins,* 2nd ed. John Wiley & Sons, New York, N.Y. (1958).

LATIMER, W. M. *Oxidation Potentials.* Prentice Hall, Englewood Cliffs, N.J. (1952).

LAUGHLIN, R. G. W., GALLO, T., AND ROBEY, H. "Wet Air Oxidation for Hazardous Waste Control." *J. Haz. Mat.* 8: 1–9 (1983).

LEWIS, N., TOPUDURTI, K., WELSHANS, G., AND FOSTER, R. "Control Technology: A Field Demonstration of the UV/Oxidation Technology to Treat Ground Water Contaminated With VOCs." *J. Air Waste Manage. Assoc.* 40, 4: 540–547 (1990).

LI, L., GLOYNA, E. F., AND SAWICKI, J. E. "Treatability of DNT Process Wastewater by Supercritical Water Oxidation." *Water Env. Res.* Vol. 65, No. 3, pp. 250–257 (1993).

MANEVAL, J. E., KLEIN, G., AND SINKOVIC, J. *Selenium Removal from Drinking Water by Ion Exchange.* U.S. EPA, Water Engineering Research Laboratory, Office of Research and Development, Cincinnati, Ohio (1985).

MANGELSON, K. A., AND LAUCH, R. P. "Removing and Disposing of Radium from Well Water." *J. Am. Water Works Assoc.* 82, 6: 72–76 (1990).

MARCUS, Y., AND HOWERY, D. G. *Ion Exchange Equilibrium Constants.* IUPAC, Butterworths, London (1975).

MARON, S. H., AND LANDO, J. B. *Fundamentals of Physical Chemistry.* Macmillan Publishing Co., New York, N.Y. (1974).

MATTER-MÜLLER, C., GUJER, W., AND GIGER, W. "Transfer of Volatile Substances from Water to the Atmosphere." *Water Research.* 15, pp. 1271–1279 (1981).

MEITES, L. *Handbook of Analytical Chemistry.* McGraw-Hill, New York, N.Y. (1963).

MONTGOMERY, J. M. *Water Treatment Principles and Design.* John Wiley & Sons, New York, N.Y. (1985).

MURPHY, J. S., AND ORR, J. R. *Ozone Chemistry and Technology.* The Franklin Institute Press, Philadelphia, Pa. (1975).

NORTH, J. C., RUSSEL, W. G., AND MATHIS, R. F. "Ion Exchange/Metal Precipitation Wastewater Treatment System: Case History." *Proceedings of the 39th Industrial Waste Conference, May 1986,* pp. 549–563. Lewis Publishers, Chelsea, Mich. (1985).

NYER, E. K. *Groundwater Treatment Technology.* Van Nostrand Reinhold Co., New York, N.Y. (1985).

ORION RESEARCH. *Handbook of Electrode Technology.* Orion Research, Inc., Cambridge, Mass. (1982).

PALMER, S. A. K., BRETON, M. A., NUNNO, T. J., SULLIVAN, D. M., AND SURPRENANT, N. F. *Technical Resource Document: Treatment Technologies for Metal/Cyanide-Containing Wastes, Volume III,* EPA/600/S2–87/106, U.S. EPA, Cincinnati, Ohio, February (1988).

PATTERSON, J. W. *Industrial Wastewater Treatment Technology,* 2nd ed. Butterworth, Stoneham, Mass. (1985).

PATTERSON, J. W., ALLEN, H. E., AND SCALA, J. J. *Carbonate Precipitation for Heavy Metals Pollutants.* J. Water Pollut. Control Fed. 49, pp. 2397–2410, December 1977.

PERRY, R. H., GREEN, D. W., AND MALONEY, J. O., eds. *Perry's Chemical Engineers' Handbook,* 6th edition. McGraw-Hill, New York, N.Y. (1984).

PITZER, K. S. "Thermodynamics of Electrolytes. I. Theoretical Basis and General Equations." *J. Phys. Chem.* 77: 268–277 (1973).

PITZER, K. S., AND MAYORGA, G. "Thermodynamics of Electrolytes. II. Activity and Osmotic Coefficients for Strong Electrolytes with One or Both Ions Univalent." *J. Phys. Chem.* 77, No. 19: 2300–2308 (1973).

REICHENBERG, D. "Ion Exchange Selectivity." In *Ion Exchange.* Vol. 1, Chapter 7. Jacob Marinsky, Marcel Dekker, New York, N.Y. (1966).

REID, R. C., PRAUSNITZ, J. M., AND SHERWOOD, T. K. *The Properties of Gases and Liquids,* 3rd ed. McGraw-Hill, New York, N.Y. (1977).

REYNOLDS, T. D. *Unit Operations and Processes in Environmental Engineering.* Brooks/Cole, Monterey, Calif. (1982).

RICE, R. G. "Ozone for the Treatment of Hazardous Materials. Water—1980." *Symposium Series: American Institute of Chemical Engineers.* 209 (77): 79–107 (1981).

RICE, R., AND BROWNING, M. *Ozone Treatment of Industrial Wastewater. Pollution Control Technology Review No. 84.* Noyes Data Corp., Park Ridge, N.J. (1981).

RICE, R. G., AND NETZER, A., eds. *Handbook of Ozone Technology and Applications.* Vol. 1. Ann Arbor Sci. Publ., Ann Arbor, Mich. (1982).

ROBERTS, P. V., HOPKINS, G. D., MUNZ, C., AND RIOJAS, A. H. "Evaluating Two Resistance Models for Air Stripping of Volatile Organic Contaminants in a Countercurrent, Packed Column." *Environmental Sci. and Tech.* 19: 2: 164–173 (1985).

ROSEN, H. M. "Use of Ozone and Oxygen in Advanced Wastewater Treatment." *J. Wat. Poll. Control Fed.* 45: 2521 (1973).

ROY, A., EATON, H. C., CARTLEDGE, F. K., AND TITTLEBAUM, M. E. "Solidification/Stabilization of Hazardous Waste: Evidence of Physical Encapsulation." *Env. Sci. and Tech.* 26, 7: 1349–1353 (1992).

SEDLAK, D. L., AND ANDREN, A. W. "Oxidation of Chlorobenzene with Fenton's Reagent." *Env. Sci. and Tech.* 25, 4: 777–782 (1991a).

SEDLAK, D. L., AND ANDREN, A. W. "Aqueous Phase Oxidation of Polychlorinated Biphenyls by Hydroxyl Radicals." *Env. Sci. and Tech.* 25, 8: 1419–1427 (1991b).

SIEGEL, S. K., AND CLIFFORD, D. A. *Removal of Chromium from Ion Exchange Regenerant Solution.* U.S. EPA, EPA/600/S2-88/007, Cincinnati, Ohio (1988).

SILLEN, L. G., AND MARTELL, A. E. *Stability Constants of Metal-Ion Complexes.* Special Publication, No. 17, Chemical Society, London (1964).

SINGER, P. C. "Assessing Ozonation Research Needs in Water Treatment." *J. Amer. Water Works Assoc.* 82: 9: 78–88 (1990).

SMITH, R. M., AND MARTELL, A. E. *Critical Stability Constants.* Plenum, New York, N.Y. (1976).

SNOEYINK, V. L., AND JENKINS, D. *Water Chemistry.* John Wiley & Sons, New York, N.Y. (1980).

SPIVEY, J. J., ALLEN, C. C., GREEN, D. A., WOOD, J. P., AND STALLINGS, R. L. *Preliminary Assessment of Hazardous Waste Pretreatment as an Air Pollution Control Technique.* U.S. EPA, EPA/600/2-86/028, Cincinnati, Ohio, March, 1986.

STAEHELIN, J., AND HOIGNÉ, J. "Decomposition of Ozone in the Presence of Organic Solutes Acting as Promoters and Inhibiters of Radical Chain Reactions." *Env. Sci. and Tech.* 19: 1206 (1985).

STUMM, W., AND MORGAN, J. J. *Aquatic Chemistry,* 2nd ed. John Wiley & Sons, New York, N.Y. (1981).

SUBRAMMONIAN, S., CLIFFORD, D., AND VIJJESWARAPU, W. "Evaluation Ion Exchange for Removing Radium from Groundwater." *J. Am. Water Works Assoc.* 82, 5: 61–70 (1990).

SURPRENANT, N., NUNNO, T., KRAVETT, M., AND BRETON, M. *Technical Resource Document: Treatment Technologies for Halogenated Organic Containing Wastes, Vol. 1.* U.S. EPA, EPA/600/S2-87/098, Cincinnati, Ohio, Jan. (1988).

THORNTON, T. D., LADUE, III, D. E., AND SAVAGE, P. E. "Communications: Phenol Oxidation in Supercritical Water: Formation of Dibenzofuran, Dibenzo-p-dioxin, and Related Compounds." *Env. Sci. Tech.* Vol. 25, No. 8, pp. 1507–1510, (1991).

TOMIYASU, H., FUKUTOMI, H., AND GORDON, G. "Kinetics and Mechanism of Ozone Decomposition in Basic Aqueous Solution." *Inorg. Chem.* 24: 2962 (1985).

TONDEUR, D., AND BAILLY, M. (RODRIGUES, A. E., ED.) "Design Methods for Ion-Exchange Processes Based on the Equilibrium Theory." In *Ion Exchange: Science and Technology,* pp. 147–197. Martinus Nijhoff Publishers, Boston, Mass. (1986).

TONDEUR, D., AND KLEIN, G. "Multicomponent Ion Exchange in Fixed Beds—Constant Separation Factor Equilibrium." *Ind. Eng. Chem. Fund.* 6, pp. 351–361 (1967).

TOPUDURTI, K. V., LEWIS, N. M., AND HIRSCH, S. R. "The Applicability of UV/Oxidation Technologies to Treat Contaminated Groundwater." *Env. Progress* 12: 1: 54–60 (1993).

UNTERBERG, W., WILLMS, R. S., BALINSKY, A. M., REIBLE, D. D., WETZEL, D. M., AND HARRISON, D. P. *Analysis of Modified Wet-Air Oxidation for Soil Detoxification.* U.S. EPA, EPA/600/S2-87/079, Cincinnati, Ohio, Jan. 1988.

U.S. EPA. *Development Document for Effluent Limitations Guidelines and New Source Per-*

formance Standards for the Major Inorganic Products Segment of the Inorganic Chemicals Manufacturing Point Source Category. EPA-440/1-74-007-a, March 1974.

U.S. EPA. *Treatment of Metal Finishing Wastes by Sulfide Precipitation.* EPA-600/2-77-049, February 1977.

U.S. EPA. *Summary Report: Control and Treatment Technology for the Metal Finishing Industry. Ion Exchange.* EPA/625/8-81-007, Industrial Environmental Research Laboratory, Cincinnati, Ohio, June 1981.

U.S. EPA. *Development Document for Effluent Limitations Guidelines and Standards for the Metal Finishing Point Source Category.* EPA440/1-83/091, June 1983.

U.S. EPA. *Environmental Regulations and Technology: The Electroplating Industry.* EPA/625/10-85/001, Cincinnati, Ohio, (1985).

U.S. EPA. *Systems to Accelerate In-Situ Stabilization of Waste Deposits.* EPA/540/2-86/002, Hazardous Waste Engineering Laboratory, Cincinnati, Ohio, Sept. 1986.

U.S. EPA. *Technical Resource Document: Treatment Technologies for Metal/Cyanide-Containing Wastes, Volume III.* EPA/600/S2-87/106, Feb. 1988.

U.S. EPA. *Stabilization/Solidification of CERCLA and RCRA Wastes, Physical Tests, Chemical Testing Procedures, Technology Screening, and Field Activities.* EPA/625/6-89/022, Office of Research and Development, Washington, D.C., May 1989.

U.S. EPA. *Ultrox International Ultraviolet Radiation/Oxidation Technology, San Jose, California.* EPA/540/S5-89-012, U.S. EPA Risk Reduction Engineering Laboratory, Cincinnati, Ohio, May 1990.

U.S. EPA. *Supercritical Water Oxidation Engineering Bulletin.* EPA/540/S-92/006, U.S. EPA Risk Reduction Engineering Laboratory, Cincinnati, Ohio, September 1992.

VASLOW, F., AND BOYD, G. E. *J. Amer. Chem. Soc.* 75: 4691 (1952).

WALKER, C. A., AND ZABBAN, W. "Disposal of Plating Room Wastes, VI. Treatment of Plating Room Waste Solutions with Ozone." *Plating* 40: 777 (1953).

WALTER, J. E. "Multiple Adsorption from Solutions." *J. Chem. Phys.* 13, 6: 229–234 (1945).

WALTON-DAY, K., MACALADY, D. L., BROOKS, M. H., AND TATE, V. T. "Field Measurements for Measurements of Ground Water Redox Chemical Parameters." *Groundwater Monitoring Review.* Fall (1990).

WEATHINGTON, B. C. *Destruction of Cyanide in Wastewaters: Review and Evaluation,* EPA/600/S2-88/031, U.S. EPA Water Engineering Research Laboratory, Cincinnati, Ohio, July (1988).

WEBER, W. J., JR. *Physicochemical Processes for Water Quality Control,* Wiley-Interscience, New York, N.Y. (1972).

WHITLOW, J. E., AND ROTH, J. A. "Heterogeneous Ozonation Kinetics of Pollutants in Wastewater." *Environmental Progress.* Vol. 7, No. 1, pp. 52–57 (1988).

WILK, L., PALMER, S., AND BRETON, M. *Technical Resource Document: Treatment Technologies for Corrosive-Containing Wastes, Volume II.* EPA/600/S-87-099, U.S. EPA, Cincinnati, Ohio, January 1988.

WPCF, *Pretreatment of Industrial Wastes, Manual of Practice No. FD-3.* Water Pollution Control Federation, Washington, D.C., 1981.

YAO, C. C. D., AND HAAG, W. R. "Rate Constants for Direct Reactions of Ozone with Several Drinking Water Contaminants." *Water Res.* 25: 7: 761–773 (1991).

ZEFF, J. D., LEITIS, E., AND HARRIS, J. A. "Chemistry and Application of Ozone and Ultraviolet Light for Water Reuse—Pilot Plant Demonstration." *Proceedings of the 38th Purdue Industrial Waste Conference.* Lewis Publishers, Chelsea, Mich., pp. 105–116 (1983).

ZIMMERMAN, F. J. *U.S. Patent No. 2,824,058.* (1958).

6

Biological Treatment Processes

Organic materials may be transformed, removed, and/or completely converted to inorganic byproducts by the use of microorganisms. While microbial processes have long been used for the reduction of gross organic content (BOD, COD), their application to the deliberate removal of specific organic compounds is relatively recent. While certain general principles can be outlined, and will be presented in this chapter, the engineer contemplating use of a biological process generally must perform substantial treatability studies on a specific waste to be managed. The objective of these tests is to ascertain the following:

- Will microorganisms grow under the contemplated conditions?
- What are the rate parameters for the process that should be used for scaleup?
- What is the degree of removal of various specific pollutants of concern, and how does this vary over the range of operating variables?
- What are the requirements for other nutrients (including oxygen)?

Generally any testing for biodegradation potential is conducted in a tiered fashion. Simple batch tests are used to establish whether microbial growth can occur on a given waste stream, and to suggest the limits to such growth. Further tests of a more sophisticated nature can establish the kinetic relationships for organic removal (Grady, 1986).

WHY AND HOW MICROORGANISMS GROW

Like any other process, the growth of microorganisms occurs due to both thermodynamically and kinetically favorable conditions. The potential for microbial growth, and to a limited degree the stoichiometric nature of such growth (should it occur), can be ascertained from thermodynamic considerations. However, the kinetics of growth are a function of the microbial populations themselves, and only limited *a priori* generalizations can be made.

For the success of any biological waste treatment process it is necessary that conditions favor continued microbial growth. This is necessary since microorganisms continually decay, and since (as will be shown below) some excess sludge production is always required from thermodynamic considerations. Absent such favorable conditions, the continual addition of microorganisms from an external source would be required. Such processes have yet to be competetive with ones in which the continual multiplication of a standing crop of microorganisms occurs.

Stoichiometry

Microorganisms require a variety of nutrients for their growth. These are generally subdivided into major nutrients, minor nutrients, and trace nutrients. The elements C, H, O, and N are considered major nutrients. P, K, S, and Mg are minor elements. Trace nutrients may include a variety of trace elements as well as vitamins. Roughly speaking, the major nutrients constitute more than 2% each of the microbial dry weight, while minor nutrients constitute between 0.1 and 2% (Pirt, 1975).

These nutrients are used in the synthesis of new cellular mass. In addition, microorganisms require a source of energy. This may be chemical, in which case an electron donor and electron acceptor are required. The nutrients required for synthesis of new cells may fill these roles. Alternatively, microorganisms may gain energy via sunlight (actually by the use of sunlight to conduct certain oxidation–reduction processes which are not otherwise energetically favorable). Energy sources and classifications are discussed below.

Regardless of the source of energy, if the composition of the resulting microbial biomass is known and if the nature and composition of all the reactants (nutrients) and products are known, a number of conclusions may be reached. In the following analysis, only major nutrients are considered. However, the approach can clearly be extended to include any of the minor or trace elements.

Let the empirical composition of the biomass which is produced be donated as $CH_aO_bN_c$. A commonly used choice is $CH_{1.4}O_{0.4}N_{0.2}$ (Hoover and Porges, 1952). However, this can also be determined experimentally. Suppose that an organic-laden waste with the empirical formula of $CH_dO_eN_f$ is to be biologically treated. Depending upon the class of microorganisms, either oxygen, nitrate, sulfate, or no other compound might be needed for respiration. Byproducts from these additional compounds might include water, nitrogen, or hydrogen sulfide, respectively. Ammonia may be needed as an additional nitrogen source (or may, in fact, be produced as a waste from excess nitrogen contained within the waste). Additional byproducts might also include carbon dioxide and water. From this information, one can write the following chemical equations (given as examples of the above types of processes):

Aerobic Degradation with Ammonia as a Nitrogen Source

$$CH_dO_eN_f + qO_2 + rNH_3 = sCH_aO_bN_c + tH_2O + uCO_2 \quad (6.1)$$

Degradation by Nitrifying Organisms

$$CH_dO_eN_f + qHNO_3 + rNH_3 = sCH_aO_bN_c + tH_2O + uCO_2 + vN_2 \quad (6.2)$$

Degradation by Sulfate Reducing Organisms

$$CH_dO_eN_f + qH_2SO_4 + rNH_3 = sCH_aO_bN_c + tH_2O + uCO_2 + vH_2S \quad (6.3)$$

Anaerobic Degradation by Methane-Forming Organisms

$$CH_dO_eN_f + rNH_3 = sCH_aO_bN_c + tH_2O + uCO_2 + vCH_4 \quad (6.4)$$

For any of the above possible processes, termed "growth equations," the unknown stochiometric coefficients (q, r, s, t, u, v) are related by a set of mass balances on each of the elements (Battley, 1987). For example, for the process in Equation 6.1, the following can be written:

Element	Mass balance equation		
C	1	$=$	$s + u$
H	$d + 3r$	$=$	$sa + 2t$
O	$e + 2q$	$=$	$sb + t + 2u$
N	$f + 3r$	$=$	sc

Therefore, if the empirical compositions of the organic feed and the microorganisms are known (a, b, c, d, e, f), the above results in a set of four equations in five unknowns, in the case of Equations 6.1, 6.3, and 6.4 and four equations in six unknowns in the case of Equation 6.2. In the case of Equation 6.2, involving nitrifying organisms, an additional constraint arises due to the fact that nitrate is used only as an energy source and not a nitrogen source (i.e., $q = 2v$). Hence, in all of the above processes there is one degree of freedom.

The one degree of freedom means that the stoichiometric ratios may vary with process conditions. In other words, it is possible to operate an aerobic process, for example, at various ratios of oxygen consumed per unit waste degraded or sludge produced per unit waste degraded. However, once one additional constraint is placed on the system, such as the above, all stoichiometric ratios are fixed.

For any given process, if the stoichiometric equation that results is compared with the concentration of nutrients present in a waste to be treated, a limiting nutrient (for growth yield) can be ascertained. This is defined (Brock, 1979) as that nutrient that is present in least amount relative to the needs of the organisms. In other words, this is the nutrient that would first be depleted if growth according to the stoichiometric equation is allowed to proceed indefinitely. In general, the limiting nutrient in treatment processes should be that for which the greatest degrees of removal are desired, and other nutrients should be present in excess.

Not considered in the above formulation are the necessity for minor and trace nutrients to be present in abundance. It is frequently the case that phosphorous is yield limiting. On average, 1.5% of dry weight microbial biomass consists of phosphorous, although this can vary substantially with metabolic conditions. Requirements for potassium are similar to phosphorous. Magnesium and sulfur requirements are about 0.3% of microbial biomass. Other elements which have frequently been found to be essential for growth (in lesser amounts) are calcium, manganese, iron, copper, and zinc. A number of other elements may rarely, depending upon type of microorganism, be required for growth (Pirt, 1975).

Example 6-1

An industrial waste has a degradable organic material present with the empirical formula CH_2ON. Your are asked to design an aerobic biological process for removal of this material with a sludge (excess microorganism) production of 0.1 kg/kg organic removed. Develop the growth equation under these conditions and compute: (a) kg oxygen required per kg organic removed and (b) additional ammonia nitrogen required, if any.

Solution

Rewriting Equation 6.1 for the problem conditions, assuming the cell composition of Hoover and Porges:

$$CH_2ON + qO_2 + rNH_3 = sCH_{1.4}O_{0.4}N_{0.2} + tH_2O + uCO_2$$

Therefore, the following balances are written:

$$\begin{aligned} &C:\ 1 = s + u \\ &H:\ 2 + 3r = 1.4s + 2t \\ &O:\ 1 + 2q = 0.4s + t + 2u \\ &N:\ 1 + r = 0.2s \end{aligned}$$

The stipulation of 0.1 kg sludge produced/kg organic yields a value of s if conversion to molar units is done. First, the molecular weight of CH_2ON is computed as $12 + 2 + 16 + 14 = 44$. Then the molecular weight of cells is computed as $12 + 1.4 + 0.4(16) + 0.2(14) = 22.6$

Therefore:

$$\begin{aligned} s &= (0.1 \text{ g sludge/g organic} \times 44 \text{ g organic/mole})/(22.6 \text{ g sludge/mole}) \\ &= 0.195 \end{aligned}$$

By successive elimination in the above equations, the following is found:

$$\begin{aligned} u &= 0.805 \\ r &= -0.961 \\ t &= -0.578 \\ q &= 0.555 \end{aligned}$$

Negative coefficients for ammonia and water mean that they appear on the opposite side of the equation (i.e., there is excess ammonia, which is a product, and water is required as a reactant). Therefore, the stoichiometric growth equation is:

$$CH_2ON + 0.555\ O_2 + 0.578\ H_2O = 0.961\ NH_3 + 0.195\ CH_{1.4}O_{0.4}N_{0.2} + 0.805\ CO_2$$

The oxygen requirement is therefore 0.555 m/m organic or

$$(0.555 \text{ m/m} \times 32 \text{ g oxygen/mole})/44 \text{ g organic/mole} = 0.403 \text{ g } O_2\text{/g organic}$$

Energetics of Microbial Growth

The growth of microorganisms represents a balancing of two processes—anabolism and catabolism. Anabolism, or biosynthesis, is the process by which new cellular materials (proteins, nucleic acids, etc.) are synthesized from simple materials. For this synthesis to occur, energy is required. Catabolism is the process of conversion of energy into a form in which it can be utilized for biosynthesis.

Microorganisms can be broadly classified according to the materials that they use for anabolic and catabolic processes. For biosynthesis, organisms that can subsist solely on inorganic carbon sources (carbon dioxide, carbonate, and bicarbonate ions) are termed lithotrophs. Organisms that require some external sources of organic carbon for biosynthesis are termed organotrophs. For catabolism, organisms that can obtain energy from sunlight are termed phototrophs. Organisms that require some external chemical materials for energy are termed chemotrophs. This leads to four categories of organisms classified by nutritional characteristics.

Chemoorganotrophs—require organic carbon for biosynthesis and obtain energy from externally supplied chemicals. This class includes most bacteria, all fungi and higher organisms, and is probably the most important for waste treatment.

Chemolithotrophs—do not require organic carbon for biosynthesis but require some external chemical energy source. This class includes the nitrifying bacteria.

Photolithotrophs—capture energy from sunlight and do not require organic carbon. Includes algae and higher plants.

Photoorganotrophs—capture energy from sunlight but require organic carbon for biosynthesis. This group includes a number of photosynthetic bacteria, none of which appear to be of importance for waste treatment at the current time.

For chemotrophs, energy is obtained from external chemicals by an overall redox reaction in which an electron-rich material ("electron donor") reacts with an electron-poor material ("electron acceptor"). If the electron acceptor is a distinct chemical compound obtained from the external environment, the process is said to be a "respiration"; if there is no distinct external electron acceptor then the process is said to be "fermentative." Since the process is a redox process, it may be described by a pair of half-reactions. Similarly, the process of biosynthesis may be viewed as a redox half-reaction. Based on calorimetric measurements, the free energy of each half reaction under "standard" conditions can be assessed. Table 6-1 summarizes a number of these reactions. For the last reaction in Table 6-1, the free energy change (designated as ΔG_s) must be computed by considering the intermediates from which synthesis of cellular material commences using the following relationship.

$$\Delta G_s = (\Delta G_p + \Delta G_n)/\text{k} + \Delta G_c \tag{6.5}$$

where ΔG_c is the effective free energy change of biosynthesis from pyruvate and a

TABLE 6-1 FREE ENERGIES FOR VARIOUS HALF-REACTIONS (adapted from McCarty, 1972).

Half-reaction			$\Delta G^0(W)$ (Kcal/mole electrons)
$1/3\ NO_2^- + 4/3\ H^+ + e^-$	=	$1/6\ N_2 + 2/3\ H_2O$	−22.263
$1/4\ O_2 + H^+ + e^-$	=	$1/2\ H_2O$	−18.675
$1/5\ NO_3^- + 6/5\ H^+ + e^-$	=	$1/10\ N_2 + 3/5\ H_2O$	−17.128
$1/8\ SO_4^{-2} + 19/16\ H^+ + e^-$	=	$1/16\ H_2S + 1/16\ HS^- + 1/2\ H_2O$	5.085
$1/8\ CO_2 + H^+ + e^-$	=	$1/8\ CH_4 + 1/4\ H_2O$	5.763
$1/8\ CO_2 + 1/8\ HCO_3^- + H^+ + e^-$	=	$1/8\ CH_3COO^- + 3/8\ H_2O$ (acetate)	6.609
$1/4\ CO_2 + H^+ + e^-$	=	$1/24\ C_6H_{12}O_6 + 1/4\ H_2O$ (glucose)	10.020
$1/5\ CO_2 + 1/20\ HCO_3^- + 1/20\ NH_4^+ + H^+ + e^-$	=	$1/20\ C_5H_7O_2N + 9/20\ H_2O$	See text

Note: Free energies are for all reactants and products at unit activity, except $[H^+] = 10^{-7}$

simple ammonia source; estimated at 7.5 kcal/mole electrons, ΔG_p is the free energy change for conversion of the carbon source to pyruvate (without any electron acceptor), which is obtained by adding 8.545 kcal/mole to the negative of the energy of the half reaction in Table 6.1; and ΔG_n is the free energy change associated with conversion of nitrogen source to ammonia, and is equal to 0, 4.17, 3.25, and 3.78 kcal/mole, respectively (for ammonia, nitrate, nitrite, and atmospheric nitrogen as nitrogen sources. The constant k is a measure of inefficiency of energy conversion, and has been assumed to equal 0.6 (McCarty, 1972).

From this information, a stoichiometry for a biological process can be computed, provided that the fraction of electrons from the electron donor that is used for biosynthesis is known. This corresponds to the one degree of freedom that exists when the elemental mass balance approach is used. Also, since the free energy changes of the reaction are known, or can be estimated, from the thermodynamic principle that a feasible process must have a zero or negative net free energy, feasible values of f can be computed. This process is illustrated by example.

Example 6-2

Consider the aerobic degradation of acetate using the method of half reactions, and estimate the maximum feasible yield of microorganisms for this process.

Solution

First, the three half reactions are written:

$$R_d\ \tfrac{1}{8}\ CH_3COO^- + \tfrac{3}{8}\ H_2O = \tfrac{1}{8}\ CO_2 + \tfrac{1}{8}\ HCO_3^- + H^+ + e^- \quad \text{kcal/mole}$$
$$R_a\ \tfrac{1}{4}\ O_2 + H^+ + e^- = \tfrac{1}{2}\ H_2O \quad -6.609$$
$$R_s\ \tfrac{1}{20}\ C_5H_7O_2N + \tfrac{9}{20}\ H_2O = \tfrac{1}{5}\ CO_2 + \tfrac{1}{20}\ HCO_3^- + \tfrac{1}{20}\ NH_4^+ + H^+ + e^- \quad -18.675$$

Now, if f represents the fraction of electrons from the donor devoted to synthesis, then the reactions may be added by summing $R_d\,(1-f)$ times R_a and f times R_s to yield the following:

$$\tfrac{1}{8}\ CH_3COO^- + \tfrac{1}{4}(1-f)O_2 + \tfrac{f}{20}\ NH_4^+ = (\tfrac{1}{8} - \tfrac{f}{5})CO_2 + (\tfrac{1}{8} - \tfrac{f}{20})HCO_3^- + \tfrac{f}{20}\ C_5H_7O_2N + (\tfrac{1}{8} - \tfrac{f}{20})H_2O$$

The free energy of synthesis ΔG_s is estimated from use of Equation 6.5, where $\Delta G_n = 0$, and ΔG_p is 8.545 − 6.609 kcal/mol. Therefore,

$$\Delta G_s = 7.5 + (1/0.6)(1.936) = 10.726 \text{ kcal/mole}$$

This can be used to determine the net free energy of the overall process by addition, weighted by the fraction contributed to the overall process.

$$\Delta G_{net} = -6.609 + (1 - f)(-18.675) + f(10.726)$$
$$= -25.284 + 29.401f$$

A process is thermodynamically feasible if and only if its net free-energy change is zero or negative. Applying this constraint to the above, the process is thermodynamically feasible only if $f \leq 0.86$ (0.86 is designated as the "thermodynamic limit"). Note also from the balanced equation that, if f is greater than $\frac{5}{8}$ (0.625), inorganic carbon (carbon dioxide and bicarbonate) would be required as reactants. Since, in general, aerobic heterotrophic microorganisms cannot fix inorganic carbon, the maximum yield would be set by this constraint (designated as the "practical stoichiometric limit")

From the balanced equation, the molar yield of microorganisms (moles organisms/mole acetate) equals $2f/5$. Hence, for the two limits above, the estimated maximum yields are 0.344 and 0.25 moles/mole, respectively.

Experimentally, the concentration of microorganisms in a system is measured using either suspended solids or volatile suspended solids as an analytical technique. Suspended solids are those materials that are filterable under standard conditions and dried to a constant weight. In addition to carbon, hydrogen, oxygen, and nitrogen, this also includes elements that would be included as ash, such as metals, alkali cations, and halides. By determining the amount of material remaining after ignition, the ash content can then be subtracted and the resulting mass is referred to as volatile suspended solids. In general, for biological solids, volatile suspended solids are 85% of the suspended solids measurement (Grady and Lim, 1980). However, for wastewaters that contain large amounts of inert solids, the measurement of VSS is subject to less interference. VSS is generally assumed to be equivalent to the measurement of the carbon + hydrogen + oxygen + nitrogen (and perhaps + phosphorous) of cellular materials.

Kinetics

The rates at which microorganisms grow correspond to the rates at which nutrients are removed from a material undergoing decomposition. The magnitude of the rates will determine the length of time required for treatment, or, in a continuous system, the required volume of the unit in which treatment will be conducted.

If an inoculum of microorganisms is placed into a new batch vessel containing nutrient medium and the subsequent population is measured as a function of time, a curve such as Figure 6-1 would be obtained. There are several distinct phases to this curve.

Initially, there is a very low rate of increase of microorganisms. This phase is termed "lag phase," during which time microbial populations are adapting to their new environment and preparing for rapid growth. Then there commences a logarithmic (or exponential) growth period, referred to as the "log growth phase." If the

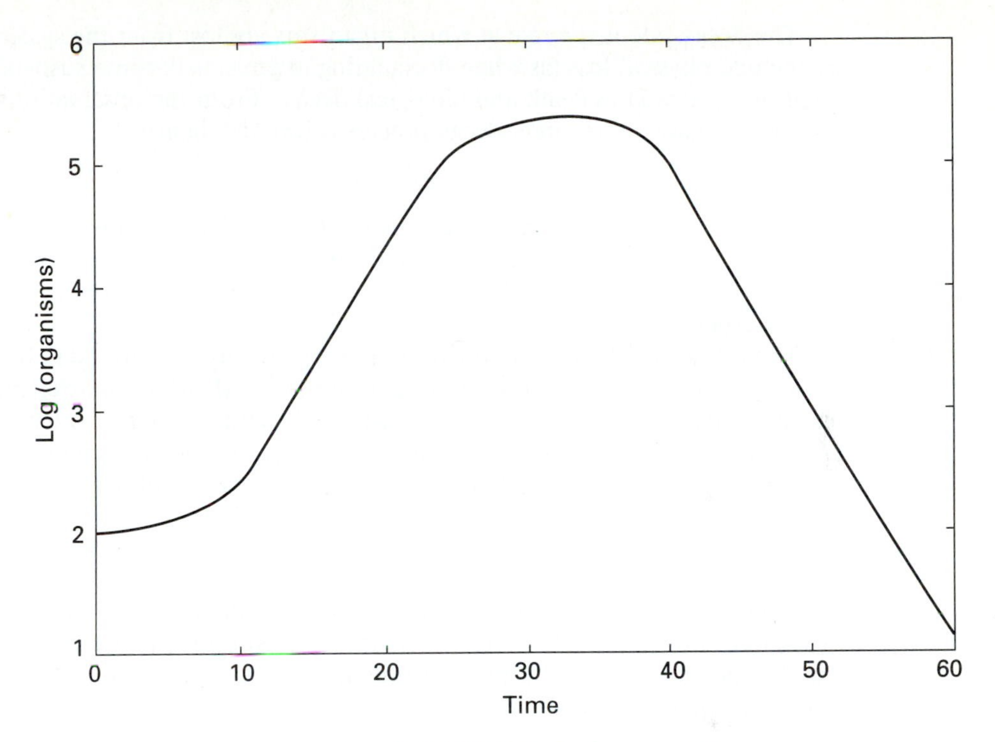

Figure 6-1 Schematic of batch microbial growth.

initial populations of microorganisms is low relative to the initial substrate present, the increase of microorganisms will not deplete substrate for some period of time, allowing this phase to be prolonged. As nutrients become exhausted, and perhaps as toxic byproducts accumulate in the culture medium, growth rate decreases. At a certain point, the rate of multiplication to form new cells will balance out the rate at which old cells are declining, resulting in a steady state population called the "stationary phase." The duration of this phase may be long or short, but eventually the overall number of cells will start to decline at a logarithmic rate during "endogenous" or "decay" phase.

This process can be described quantitatively by three rate processes involving microorganisms and the nutrient that limits the growth rate. Note here that the growth rate limiting nutrient need not be the same as the growth yield limiting nutrient (O'Brien, 1972), but it most often is. The symbol S is the concentration of the growth rate limiting nutrient and X is the concentration of microorganisms. The three rate processes are a growth rate, r_g, a decay rate r_d, and a substrate utilization rate, r_s.

The growth rate is the rate at which new organisms are produced by biosynthesis (units of mass/volume-time). From the batch growth curve it is apparent that this process is first order in microorganisms concentration. Hence:

$$r_g = \mu X \tag{6.6}$$

where μ is termed the specific growth rate and has units of time^{-1}.

The decay rate is the rate at which organisms are lost from the system. This may include physical loss (as when flocculating organisms become suspended and "swept out") as well as death and biological decay. From the observations during batch decay phase, a first order decay process is implied, hence:

$$r_d = bX \quad (6.7)$$

where b is the specific decay rate, with units of time^{-1}. This parameter is generally thought to be independent of nutrient concentration but dependent upon temperature and stressful environmental conditions such as extremes in pH or the presence of toxic materials.

The rate at which nutrients are utilized has generally been found to be proportional to the rate at which new biomass is synthesized, which is reasonable if a constant composition for new biomass exists. If Y is this proportionality—termed true yield, formally dimensionless, but representing the mass of microorganisms synthesized per mass of nutrient utilized—then the substrate utilization rate can be given by:

$$r_s = r_g / Y \quad (6.8)$$

It has generally been found that the specific growth rate, and therefore the rate of substrate utilization, is a function of the concentration of nutrient(s) that is present. The most common kinetic formulation used to model this dependency is the Monod equation:

$$\mu = q_{max} S/(S + K_s) \quad (6.9)$$

where S is the concentration of limiting nutrient and q_{max} and K_s are constants called, respectively, the "maximum specific growth rate" and the "half-saturation constant." The Monod relationship may be approximated by a linear relation at low substrate concentrations ($S << K_s$) and by a constant (i.e., zero order relationship) at high substrate concentrations ($S >> K_s$).

The lag phase is not well depicted by Monod kinetics, nor is a prolonged phase of stationary growth. While some authors have attempted to modify the Monod model with a postulated first-order lag in response to growth rate, these attempts have not been regarded as being successful (Daigger and Grady, 1983). Alternatively, attempts have been made to develop structured models of microbial growth in which detailed physiological assumptions are used to formulate a mechanistic model of metabolism (Daigger and Grady, 1982; Domach et al., 1984). However, such models have yet to find utility in the analysis of biological waste treatment.

The designer should nonetheless recognize that, at best, the Monod model, as well as similar simple kinetic equations, are only very approximate descriptors of reality. While they may be useful for characterizing average conditions, they are particularly suspect during periods of fluctuating nutrient conditions. Some authors have suggested that, under some circumstances, the Monod parameters themselves may be functions of external conditions (Gaudy et al., 1971; Grady, 1986). Table 6-2 summarizes Monod parameters for a variety of situations. Table 6-3 summarizes data taken on several additional wastes for which the first-order simplification of Monod kinetics was applied. In this simplification, $S << K_s$ and Equation 6.9 are

TABLE 6-2 MONOD CONSTANTS FOR VARIOUS CONDITIONS

Waste	Process	q_{max} (day^{-1})	K_s (mg/L)	Y	b (day^{-1})	Reference
Domestic wastewater	Activated sludge	2.1–3.6	25–100	0.35–0.45	0.05–0.1	Benefield and Randall[1]
Phenol waste	Anaerobic dig.	0.665	0.03	0.16	0.019	Suidan et al. (1988)[2]
Primary sludge	Anaerobic dig.	0.26	1800	0.04	0.03	Parkin and Owen[1]
Packing house wastes	Anaerobic dig.	0.24	5.5	0.76	0.17	Lawrence and McCarty[1]

Notes: For aerobic processes, the basis temperature is 20°C; for anaerobic processes, the basis temperature is 35°C.

[1] Organic material COD basis, sludge VSS basis.

[2] Organic material phenol basis, sludge VSS basis—actually the data fit the Haldane model ($K_1 = 363$ mg/L).

replaced by:

$$\mu = KYSX \tag{6.10}$$

This model is often erroneously called the "first-order" kinetic model.

Some materials are capable of reversibly inhibiting microbial activity or growth rate. If their concentrations increase, growth decreases, and vice versa. These should be distinguished from actual inactivating agents (such as disinfectants) which irreversibly kill microorganisms. When inhibitors of this type are present they can be characterized using one of several modifications on the Monod relationship (Equation 6.9)(Cornish-Bowden, 1979). In competitive inhibition, the presence of an inhibitor (at a concentration I) results in a diminution of growth rate, but addition of sufficient nutrient will overcome this effect. Mathematically, this can be written as:

$$\mu = q_{max}S/[S + K_s(1 + I/K_i)] \tag{6.11}$$

In other cases, no amount of additional substrate will allow the microbial culture to achieve a growth rate of q_{max} in the presence of inhibitor. This situation is termed "mixed inhibition" (since there is a multiplicative term for both K_s and q_{max}. Mathematically, the relevant equation involves two additional constants as follows:

$$\mu = q_{max}S/[(1 + I/K_i)S + K_s(1 + I/K_i)] \tag{6.12}$$

If K_i equals K_s, the form of the inhibition is termed "noncompetitive." These models have been found useful in assessing the inhibitory actions of trace metals on activated sludge microorganisms (Patterson, 1971).

TABLE 6-3 KINETIC CONSTANTS USING THE "FIRST-ORDER" MODEL FOR VARIOUS CONDITIONS

Waste	Process	K (L/mg-d)	Y	b (day^{-1})	Reference
Petrochemical wastes	Activated sludge	0.35–0.40			Benefield and Randall[1]
Refinery waste	Activated sludge	1.44	0.53	0.24	Benefield and Randall[1]

Notes: For aerobic processes, the basis temperature is 20°C; for anaerobic processes, the basis temperature is 35°C.

[1] Organic material BOD_5 basis, sludge VSS basis.

A few organic substances are capable of being degraded at low concentrations, but functioning as inhibitors at higher concentrations. A major example of these compounds is phenols (Rozich et al., 1985), and possibly certain substituted phenols. For these materials, the use of Haldane kinetics, also called substrate-inhibition kinetics, is useful. This model substitutes the following equation for (6.9):

$$\mu = q_{max} S/(S + K_s + S^2/K_i) \tag{6.13}$$

where the final term in the denominator describes the increasing inhibition at high concentrations of the nutrient/inhibitor. Comparison of Equations 6.11 and 6.13 shows that substrate inhibition of this nature is a special form of competitive inhibition in which the substrate is also the inhibitor. This kinetic formulation predicts that a maximum attainable specific growth rate (μ^*) exists at a substrate concentration S^*. This condition is known as the critical point. It is not possible to treat a waste which obeys this kinetic formulation unless the growth rate and substrate concentrations are maintained below the critical point (Rozich and Gaudy, 1984). The equations for the critical point in terms of kinetic parameters are

$$S^* = (K_s K_i)^{1/2} \tag{6.14}$$

$$\mu^* = \mu_{max}/[1 + 2\,(K_s/K_i)^{1/2}] \tag{6.15}$$

PHYSICAL CLASSIFICATION OF MICROBIAL PROCESSES

Biological treatment processes can be conducted in a variety of physical systems, to some extent regardless of the nutritional type of process itself. The physical nature of the system may be classified by the presence or absence of a supporting medium onto which microorganisms adhere and grow (suspended versus fixed-film processes) and flow conditions (continuous versus cyclinical). The nature of the physical conditions will determine how the inherent kinetics of microbial growth are used to determine the performance of the process.

Suspended Growth Process

In the suspended growth processes, conditions are maintained to allow a flocculent suspension of microorganisms (often agglomerating along with suspended solids that are in the waste feed). This flocculent suspension facilitates physical separation of the biological solids from the treated liquid generally using sedimentation.

The simplest suspended-growth process, although of relatively recent origin, is the sequencing batch reactor (SBR) (Irvine and Busch, 1979). This process uses a single tank or basin. To a small amount of reserved sludge, waste is added at a determined rate, and thus volume is allowed to increase. This is called the "fill phase." When the total desired volume is reached, the fill is turned off, but mixing is continued (the "react phase"). Mixing is then turned off briefly to allow sedimentation, and in a subsequent decant phase, the treated supernate liquor is decanted. During the react phase, an aliquot of the suspension may be withdrawn to control the microbial population. The process may be kept totally aerobic, totally anaerobic, or cycled through various degrees of oxygenation by supplying air during the fill and react periods at various rates.

The SBR process was studied for application to treatment of leachate from the Hyde Park (N.Y.) landfill site (Ying et al., 1984). The leachate was neutralized, pre-aerated for two hours, and supplemented with ammonia and phosphorous to provide a nutrient ratio of TOC : N : P of 150 : 10 : 2 (weight basis). An aerobic fill and react period was used with 20% feed per day (i.e., one fifth of the total system volume was removed and replaced with feed daily), and a microorganism (MLSS) concentration of 10,000 mg/L. The cycle times were: fill—six hours, react—ten hours, settle—two hours, decant—five hours and idle—one hour (for a 24 hour total cycle time). The influent and effluent analyses were as indicated in Table 6-4.

In other suspended growth processes, a continuous feed of influent is received. By maintaining a constant system volume, a continuous treated stream results. There are three processes that operate in this mode: lagoons (or ponds), activated sludge, and anaerobic treatment. The use of lagoons for the treatment of hazardous waste is strongly discouraged in the current regulatory framework. Lagoon systems (a type of surface impoundment under the RCRA classification of treatment processes) currently require double liners and leachate collection systems, similar to landfill requirements. In addition, lagoons must be stringently monitored. Hence, solely from a treatment point of view, the economic and operational advantages that lagoons might otherwise have (and do, in fact, have when domestic wastewater is treated) are minimized in the context of hazardous waste management when compared to contained treatment processes (such as activated sludge).

The activated sludge process consists of two functional units, although they may frequently be present in the same physical structure. Waste flows into a tank (called the "aeration tank") where it is intimately mixed with microorganisms in a highly flocculent state. Vigorous transfer of oxygen is achieved either by the introduction of compressed air into the "mixed liquor" or by mechanical mixing using turbines. The suspension has an average residence time in the tank of 6-24 hours during which the organic materials in the influent have been oxidized (to carbon dioxide, water, and other inorganic products) or incorporated into new microorganisms (sludge). The overflow from this tank passes to a sedimentation tank. In this tank the flocculent microorganisms are separated by gravity and collected in a concentrated underflow, with the majority of them being returned to the aeration tank,

TABLE 6-4 RESULTS OF SBR TREATMENT OF HYDE PARK (NY) LANDFILL LEACHATE (Ying et al., 1984)

Parameter	Influent	Effluent
TOC (mg/L)	1784	219
Total organic halogen (mg/L)	210	98
Chlorendic acid (mg/L)	135	115
Phenol (mg/L)	390	6
Benzoic acid (mg/L)	590	11
Ortho-chlorobenzoic acid (mg/L)	220	2
Meta-chlorobenzoic acid (mg/L)	60	9
Para-chlorobenzoic acid (mg/L)	70	14

the remainder "wasted" and taken for further treatment and ultimate disposal. The supernate from the sedimentation tank, now containing a lower concentration of organic materials, can be sent either to further treatment or to disposal.

The three key design variables of an activated sludge process are the mean hydraulic residence time in the aeration tank (Θ)—defined as the tank volume divided by the influent waste flow, the mean residence time of microorganisms in the system (Θ_c)—defined as the mass of microorganisms in the system divided by the mass/day wasted, and the volumetric concentrations of microorganisms in the aeration tank. These three variables are interrelated by kinetic expressions, as will be discussed below. In particular, the value of the sludge residence time strongly influences the degree of organic removal, and at low sludge residence times flocculent microorganisms do not form well. Table 6-5 presents a range for these three variables in common activated sludge systems.

Anaerobic suspended growth processes are similar to activated sludge or lagoon processes except that a sealed tank is used from which oxygen is scrupulously excluded. Under such conditions a consortium of microorganisms predominates whose net effect is to degrade organic matter to carbon dioxide, methane, and other inorganic byproducts. The degradation proceeds via several types of bacteria, with organic acids (acetic and higher molecular weight) and hydrogen being important intermediates. In the degradation of wastes containing simpler, low molecular weight soluble compounds (sugars, starches), the rate of degradation is limited by the growth of the organisms which ultimately form methane (methanogens). With higher molecular weight polymers, such as greases and cellulosic materials, the rate of degradation may be limited by the rate of external hydrolysis or by the rate of formation of organic acids (Speece, 1983).

Anaerobic processes, either suspended growth or fixed film, often benefit from maintenance of elevated temperatures, particularly if methanogenesis is the limiting process. There are two optimum ranges generally employed. In mesophilic operation, temperature is maintained at approximately 35°C. In thermophilic operation, temperature is maintained at approximately 50–55°C. The excess energy required to achieve the elevated temperatures may be compensated for by a reduction in tankage, or an increase in gas production (Benefield and Randall, 1980).

The methane produced during anaerobic treatment can be recovered for energy or heating. For higher strength wastes, this potential for a credit strongly offsets the greater complexity of anaerobic processes and renders them more economic than aerobic processes. However, in general, the degree of treatment possible in anaerobic systems is not sufficiently great to permit direct discharge of the treated material to the environment.

TABLE 6-5 RANGES OF DESIGN VARIABLES IN ACTIVATED SLUDGE PROCESSES

Parameter	
Hydraulic residence time	2–24 hours
Sludge residence time	3–60 days
Microorganism concentration	1,000–10,000 mg/L MLVSS

In simple anaerobic digestion processes, waste is fed into and withdrawn from a closed tank on a continuous basis (or perhaps with "slugs" withdrawn and fed periodically). This process has been widely used for the treatment of municipal wastewater sludges. In these applications, typical hydraulic residence times are 10–60 days, and typical loadings are 1.6–8 kg/m^3-d, although there is a tradeoff between these parameters and the influent concentration and removal efficiency (Parkin and Owen, 1986).

If an external sedimentation tank is used to separate and recycle the anaerobic sludge (as in the activated sludge process), the anaerobic contact process results. The higher concentration of microorganisms thus permits a higher volumetric loading (i.e., a reduced hydraulic residence time). In this process, typical loadings are 1–6 kg COD/m^3-d (Speece, 1983). The resultant sludge age is about equal to the hydraulic residence time in conventional anaerobic digestion (i.e., 10–60 days, as would be anticipated from Monod kinetics), but the higher loading rates, which are possible even with relatively difficult-to-treat industrial wastes, arise from the ability to use a shorter hydraulic residence time.

Fixed-film processes rely upon the growth of a biological slime layer upon a fixed supporting medium. Nutrients from the waste diffuse into the film and products diffuse out from the film. Loss of microorganisms from the film occurs by shearing forces, and thus the biological population is maintained at a relatively constant level.

In fixed-film filter operations, the waste passes through a stationary bed of supporting medium. In the trickling filter process, this supporting medium is of relatively large size (2.5 — 10 cm), and the hydraulic flow is kept below 40 million gallons/acre-day (6 m/d) for high rate filters and under 4 million gallons/acre-day (0.6 m/d) for standard rate filters. Generally, the flow is introduced via a rotary nozzle system to allow periodic (on the order of once per minute) wetting and draining of the film. Under such conditions, both liquid and air can flow through the pores. Often the air flow is sufficiently large due to natural convection; however, fans can be used to provide greater degrees of air flow, particularly in treatment of high strength materials. The supporting medium for trickling filters can be either plastic, in the form of saddles, cylinders, rings, or various proprietary shapes or corrugated sheets; gravel or stone; or redwood slats or chunks. The use of plastic permits the construction of trickling filters to heights in excess of 20 feet. With other media, the height of the filter is generally limited to under 10 feet, due to structural considerations.

Anaerobic fixed films can be used for treatment. In these cases, the filters are usually allowed to become submerged. The flow may be upflow or downflow and, in some cases, the liquid velocity may be maintained at a sufficiently high value (via recirculation) to allow expansion of the medium (and the adherent biomass). The use of granular activated carbon as a supporting medium in anaerobic expanded bed filtration may offer improved efficiency in the treatment of highly refractory wastes, or wastes containing toxic materials (Suidan et al., 1981). Table 6-6 summarizes design parameters that have been used in various anaerobic fixed film systems. In expanded bed systems, as biomass thickness on a supporting particle increases, the density of the particle decreases. Hence, biomass-rich particles tend to accumulate at

TABLE 6-6 SUMMARY OF ANAEROBIC FIXED-FILM DESIGN CONDITIONS

Parameter	Value	Reference
Hydraulic residence time	1–5 hours (e)	Switzenbaum and Jewell (1980)
	1–3 days (f)	Young and McCarty (1969)
	9 hours (ec)	Suidan et al. (1981)
Volumetric loading	Up to 8 kg COD/m^{3}d (e)	Switzenbaum and Jewell (1980)
	0.8–3.5 kg COD/m^{3}d (f)	Young and McCarty (1969)
	1–5.6 kg COD/m^{3}d (ec)	Suidan et al. (1981)
Biomass concentration	Up to 30 g VSS/L (e)	Switzenbaum and Jewell (1980)
	2–6 g VSS/L (f)	Young and McCarty (1969)
Solids residence time	85–660 days (f)	Young and McCarty (1969)

Note:
(e)—expanded bed process
(f)—fixed (unexpanded) filter
(ec)—expanded bed process with activated carbon

the top of an expanded bed. They can be removed at a controlled rate at this point, excess biomass sheared off and the support returned to the reactor.

Rotating biological contractors (RBC's) consist of packing with an overall cylindrical cross-section arranged about a central shaft. The packing may consist of parallel disks, or it may be an integral mass of corrugated material. The shaft rotates at a slow rate such that the shafts dip into a tray through which waste to be treated is passing. In aerobic RBC's, for a portion of each rotation, the packing is allowed to travel above the tray, thus permitting exposure to the atmosphere and oxygenation. In anaerobic RBC's, the biological packing remains submerged throughout the entire cycle. Figure 6-2 provides a schematic of an aerobic RBC.

The rotational action provides improved mass transport (for both air and organic material) as compared to a trickling filter. Furthermore, use of RBC's permits a higher biomass to be exposed to a given volume of waste.

The principal design characteristics of RBC's are the areal hydraulic loading, the areal organic loading, and the rotational speed. There is some disagreement about the relative importance of areal and hydraulic loading in RBC design and operation (Friedman et al., 1979; Lin et al., 1986; Opatken, 1986). Friedman et al.

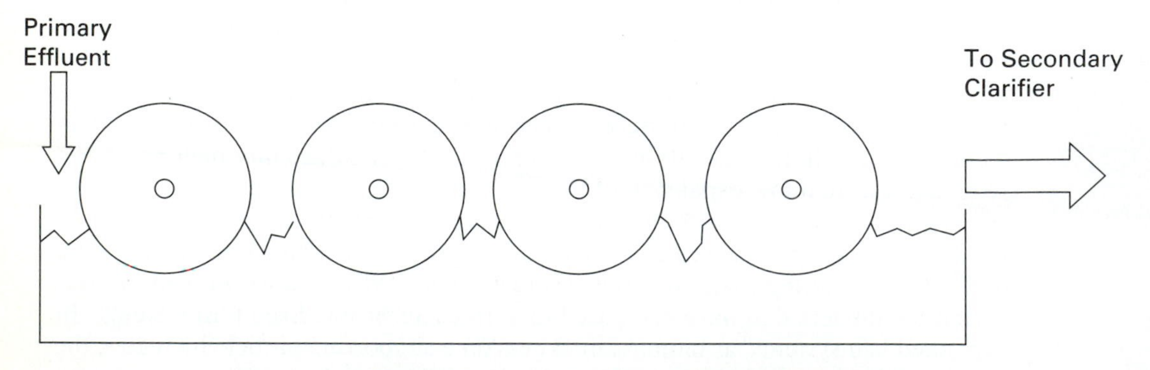

Figure 6-2 Schematic of a rotating biological contactor.

(1979) showed that the rotational speed (in the range of 5 – 30 RPM) could influence the performance characteristics of aerobic RBC's, perhaps by an effect on oxygen transfer. However, in practice it is likely that the rotational speed of an RBC will be constrained by mechanical considerations and not subject to precise specification by the designer. Typical values of the design parameters for aerobic RBC's would be:

Organic loading: 2 – 50 g BOC/m^2-d
Hydraulic loading: 100 L/m^2-d
Hydraulic residence time: 0.5 – 2 hours

Some degree of "staging" (use of multiple units in series) may be used to improve efficiency (Opatken, 1986). The design bases for anaerobic RBC's are less well established.

DESIGN EQUATIONS FOR PROCESSES

It is desirable to have quantitative relationships between various design variables and the degree of removal of contaminants in biological treatment. While the numerical values of parameters used in these relationships may vary as a function of the waste being treated, the functional forms of individual relationships are generally suited to describe the processes being modeled. Hence, from laboratory and/or pilot studies, these numerical values can be obtained, and on this basis the full-scale system can be specified.

The basis for development of process-design equations is a mass balance. In verbal terms, conservation of mass requires that the rate of change of mass of a constituent within a system must equal the net rates of production/consumption due to reaction within the system plus the net rate at which the material is transported across system boundaries. The rate of change of mass is generally a differential (if spatial gradients are absent) or a partial differential (if spatial gradients are present). Note also that the mass in a system is the product of concentration and volume, and hence, if system volume is not constant then its pattern of change must be known or specified.

Batch Suspended

Batch suspended growth systems can be considered as closed systems with respect to both microorganisms and substrate. Therefore, the mass balance equations for this system can be written as differential equations. This procedure will be illustrated using a system which obeys Monod kinetics (Equations 6.5 through 6.8).

The general mass balances on a batch system can be written as:

$$d(VX)/dt = (r_g - r_d)V \quad (6.16)$$

$$d(VS)/dt = r_s V \quad (6.17)$$

However, in a closed batch system at constant volume, these can be written as

$$dX/dt = (r_g - r_d) \quad (6.18)$$

$$dS/dt = r_s \quad (6.19)$$

Substitution of the relevant rate expressions yields:

$$dX/dt = [q_{max}S/S + K_s) - b]X \quad (6.20)$$

$$ds/dt = -q_{max}SX/Y(S + K_s) \quad (6.21)$$

The above general system is not amenable to an analytical solution. However, it may be integrated numerically using, for example, a Runge–Kutta technique. If endogenous decay phase can be neglected, however, an analytical solution can be found in which the values of S and X are implicit functions of time (Simkins and Alexander, 1981):

$$Y = (X - X_0)/S_0 - S) \quad (6.22)$$

$$K_sX \ln(S/S_0) = (S_0 + X_0/Y + K_s) \ln(X/X_0) - (S_0 + X_0/Y)q_{max}t \quad (6.23)$$

where S_0 and X_0 are the initial concentrations of nutrient and microorganisms, respectively. Thus, either by numerical integration of Equations 6.20 and 6.21, or by solution of Equations 6.22 and 6.23, the time course of microbial growth (and decline) in a closed batch system can be computed.

These equations can be used to estimate kinetic parameters from the results of batch experiments. Due to the above mentioned difficulties with Monod kinetics describing growth during transients, however, the performance of the model during lag and stationary phase, in particular, may be poor. Furthermore, there are two ways in which batch experiments can be used to obtain kinetic parameters and these may lead to differing results.

In the first method, a series of experiments at various initial substrate concentrations is conducted. Ideally, the concentration of microorganisms initially present should be low relative to the substrate concentration (so that the substrate concentration does not diminish significantly during the initial phases of the experiment (Gaudy et al., 1971). The specific growth rate is determined from the slope during exponential phase growth (e.g., by plotting microbial concentrations against time on semilogarithmic paper). Then the observed values of μ as a function of S_0 can be used to obtain K_s and q_{max} either by nonlinear regression of Equation 6.9 or by various linearizations. If measurements of the change in nutrient concentrations are made simultaneously, the yield coefficient can be determined. By measurements of the change in microorganism concentration with time during the latter stages of such experiments, the decay coefficient can be obtained.

In the second method for assessing kinetics, Equations 6.22 and 6.23 are used along with batch data of the substrate depletion curve obtained from a single-batch experiment. If the initial concentration of microorganisms is known, and if the initial substrate concentration (S_0) is greater than K_s (but not extremely greater), then a nonlinear least squares approach can be used to determine K_s, Y, and q_{max}. If the substrate concentration is on the order of K_s, or less, then only the ration q_{max}/Y will be determined precisely by the data (Robinson and Tiedje, 1983). In principle, this method can be extended to the use of any kinetic model coupled with numerical integration of the differential equations.

For a batch reactor which is receiving feed, such as a sequencing batch reactor (SBR), the mass balances developed above must be modified. Equations 6.16 and 6.17 must be changed to reflect the input of material across the system boundaries and volume now becomes a function of time (due to variable influent flows). If it is assumed that the influent flow is Q (which may be time dependent) and contains a nutrient concentration S_i, and no microorganisms, then the revised mass balance equations are:

$$d(VX)/dt = (r_g - r_d)V \tag{6.24}$$

$$d(VS)/dt = r_s V + QS_i \tag{6.25}$$

Generally, the density of liquid in a biological process may be regarded as constant, hence dV/dt equals Q. Therefore, the above equations can be rearranged:

$$dX/dt = (r_g - r_d) - XQ/V \tag{6.26}$$

$$dS/dt = r_s + (S_i - S)Q/V \tag{6.27}$$

with $dV/dt = Q$.

The above system of differential equations constitute an initial value problem. If the rate expressions (with their parameters) are specified, along with the initial concentrations of X and S, the initial volume, V, and the time course of the flow rate, numerical integration can be used to compute the time dependency of substrate and microorganism concentration. Use of this approach with Monod kinetics is not recommended due to the difficulties with modeling transient growth. However, approximate process performance may be assessed from batch kinetic data.

Batch film

The passage of a nutrient-rich liquid over a supporting surface results in a biological growth adherent to the surface. The situation is depicted, schematically, in Figure 6-3. There may be a roughness associated with both the biofilm itself as well as the liquid layer. However, to a first approximation, both layers may be considered to consist of a certain uniform (average) depth. In aerobic systems, four distinct phases may be discerned: the gas phase, the liquid phase, the biomass, and the solid support. In anaerobic systems, only the latter three phases exist.

In order for nutrients to be biologically decomposed in such a system, a number of processes must occur. Sequentially these may be considered as:

1. [in aerobic systems] transport of oxygen into liquid
2. transport of dissolved nutrients (including electron acceptors) within the liquid to the surface of the biofilm (liquid transport).
3. transport of nutrients (including electron acceptors) within the biofilm.
4. metabolic reactions
5. back diffusion (biofilm and liquid) of products

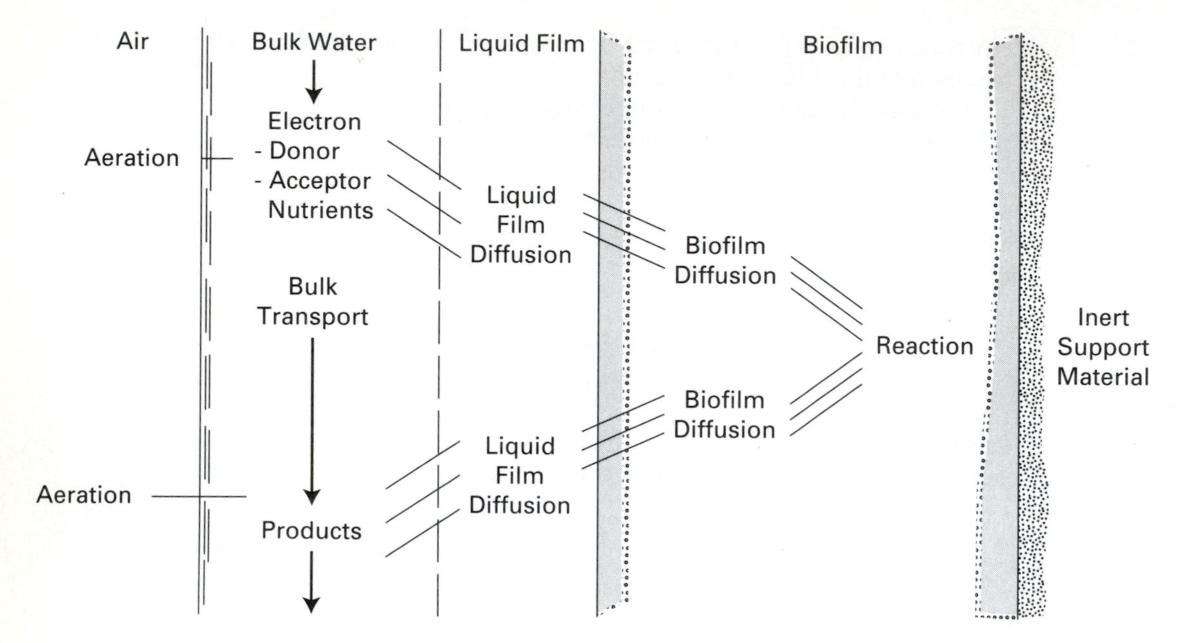

Figure 6-3 Schematic of biofilm processes. (From Harremoes, 1984, by permission of John Wiley & Sons, Inc., Copyright 1984.)

In general, processes two and three are the major ones likely to be quantitatively controlling process performance. The ability (in aerobic processes) for oxygen to diffuse at a sufficiently high rate through the gas into the liquid will be a perquisite to process success. However, once certain minimum thresholds are met, this will not control performance. Hence, biofilm performance will be considered in situations where processes two through four need to be considered.

The model describing these processes uses Monod kinetics, and neglects the process of endogenous decay by assuming that the biofilm (with respect to both its thickness and the concentration profiles contained within) is at steady state (Williamson and McCarty, 1976). It is assumed that diffusion within the biofilm is with depth only (the z coordinate—where $z = 0$ is the biofilm/liquid interface), and characterized by Fick's law with an apparent diffusivity D_b. The quantity D_b has generally been assumed to be identical to the corresponding diffusivity in water, but limited measurements suggest a range of 10–100% of this value (Williamson and McCarty, 1976). If a mass balance is performed over a differential element of biofilm (depth dz, and horizontal dimensions dx and dy), then under the conditions of the problem, a mass balance on biofilm concentration of substrate (S_b) can be written as:

$$\underbrace{(\partial S_b/\partial)dx\,dy\,dz}_{\text{change in mass}} = \underbrace{(-D_b\,\partial S_b/\partial z)dx\,dy}_{\text{diffusion through top plane}} + \underbrace{[D_b\,\partial S_b/\partial z + D_b(\partial^2 S_b/\partial z^2)dz]dx\,dy}_{\text{diffusion through bottom plane}} - \underbrace{r_s dx\,dy\,dz}_{\text{substrate utilization}} \tag{6.28}$$

The term for diffusion through the bottom plane of the biofilm arises from a Taylor series expansion of the flux due to Fick's law. After cancellation of the terms, the following arises:

$$(\partial S_b/\partial) = D_b(\partial^2 S_b/\partial z^2) - r_s \tag{6.29}$$

$$dS_b/dz = -q_{max}SX/YD_b(S + K_s) \tag{6.30}$$

where X is the local concentration of biomass (i.e., at a given depth). This equation is subject to the following boundary conditions:

$$S_b = S_s \text{ at } z = 0$$

$$dS_b/dz = -J/D_b \text{ at } z = 0$$

where $J = -K_L\,(S_s - S_L)$.

J is the flux (mass/area-time) of nutrient to the biofilm surface. S_s is the bulk solution concentration of nutrient, and K_L is the liquid phase mass transport coefficient (which may also be written as an eddy diffusivity divided by the liquid layer depth, D_w/L_w). For the purpose of solving this equation, the biofilm is considered to be of depth L_b and the liquid layer is considered to be of depth L.

The general solution to the nutrient concentration profile is not analytically obtainable. However, Suidan and Huang (1985) have presented a useful approximation to the analytical solution. The following dimensionless variables are defined (where S_{bo} is the biofilm concentration at the liquid interface and X is the average concentration of biomass in the biofilm):

$$S_f^* = S_{bo}/K_s$$

$$L_b^* = L_b\,(q_{max}X/YD_bK_s)^{1/2}$$

$$J^* = J/(K_s q_{max} X D_b/Y)^{1/2}$$

Then the approximate general solution to the biofilm model can be expressed by:

$$L_b^* = J^* + \tanh^{-1}\left\{\frac{0.5\,J^{*2} + J^*\,[1 + (J^*/3.4)^{1.19}]^{-0.61}}{S_f^*}\right\} \tag{6.31}$$

Furthermore, if an overall decay rate is assumed (b—which is due to both endogenous decay as well as physical loss of microorganisms from the surface, and which has units of 1/time), an additional relationship may be developed from the microbial mass balance:

$$L_b^* = J^*(q_{max}/b) \tag{6.32}$$

From these relationships, the dependency of substrate flux into the biofilm on concentration of substrate at the interface may be computed. Then, given the mass transfer coefficient (from turbulent correlations), the corresponding bulk substrate concentration may be determined. Alternatively, by suitable rearrangement, the following can be obtained directly for a dimensionless bulk solution substrate concentration:

$$S_l^* = J^*[L^*D_b/D_w) + 0.5\,J^*] + J^*\,[1 + (J^*/3.4)^{1.19}]^{-0.61} \tag{6.33}$$

where $S_l^* = S_l/K_s$.

In a batch process, providing that the rate at which nutrients are being removed from solution is slow compared to the mass transport and diffusion times, the above relationship can be used in an overall mass balance. If a total interfacial area of A_s (on which there is biomass) is contacted with a system volume V, then the substrate mass balance can be written as:

$$dS_l/dt = -JA/V \tag{6.34}$$

The system consisting of Equations 6.34 and 6.34 can be numerically integrated to produce time dependent estimates of solution nutrient concentration.

If the thickness of biofilm is large, so that the concentration of nutrient at the solid-biofilm interface can be considered negligible, then the biofilm is said to be "deep." In this case, an alternative to Equation 6.34 can be written as (Suidan, 1986):

$$J^{*2}/2 = S_l^* - J^*L^* - \ln(1 + S_l^* - J^*L^*) \tag{6.35}$$

As noted above, particularly when the solution concentration of nutrients are quite low, the overall process of nutrient removal by biofilms may be regarded as a first order process. In this circumstance Equation 6.34 can be simplified to the following:

$$dS_l/dt = -K''S \tag{6.36}$$

where K'' is regarded as an empirical treatability constant. In this case, the decline in substrate with time should be logarithmic—that is, it will be a straight line on a plot of natural log (substrate) versus time, with the slope equaling $-K''$.

Continuous Suspended

A continuous suspended microbial system consists of a volume, with a continuous flow of material entering and leaving the tank. The influent contains nutrient material to be degraded, and possibly microorganisms. The effluent contains residual nutrients, byproducts of biological activity, and microorganisms. In the simplest case, the liquid in the tank can be regarded as completely mixed (in other words, the concentration of any and all materials is identical at any point in the tank). By definition, resistance due to biofilms must be negligible since there are no concentration gradients.

A schematic of this completely mixed case appears in Figure 6-4. There is a constant input of feed (and effluent) with a flow rate Q. The feed contains a concentration S_0 of substrate and X_0 of microorganisms. The effluent, which also discharges at a rate Q (and thus the system volume is constant) contains concentrations S and X. By definition, since there are no concentration gradients in the system, the effluent concentrations must equal their concentrations in the system volume. Mass balances over substrate and microorganisms can be written as:

$$d(VX)/dt = (r_g - r_d)V + Q(X_0 - X) \tag{6.37}$$

$$d(VS)/dt = r_sV + Q(S_0 - S) \tag{6.38}$$

However, the system volume is constant. Also, the ratio V/O has dimensions of time,

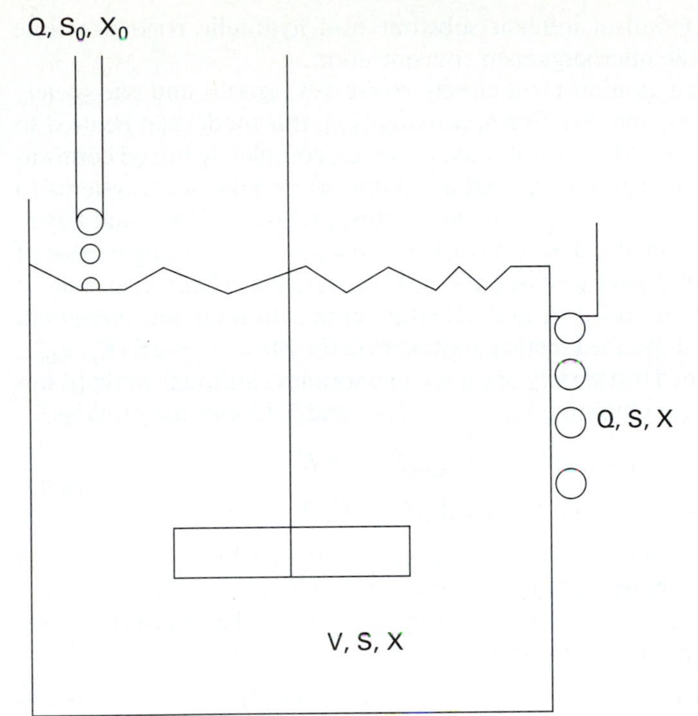

Figure 6-4 Schematic of CSTR system with no recycle.

and is physically equal to the mean residence time of fluid in the system. If this is denoted as Θ, the above equations can be written as:

$$dX/dt = (r_g - r_d) + (X_0 - X)/\Theta \qquad (6.39)$$

$$dS/dt = r_s + (S_0 - S)/\Theta \qquad (6.40)$$

At steady state, the time derivatives can be set equal to zero. If the Monod relationships are substituted into the resulting equations, the following solution results:

$$\mu = (1/\Theta) + b - X_0/X \qquad (6.41)$$

$$X = Y(S_0 - S)/\Theta\mu \qquad (6.42)$$

$$\mu = q_{max}S/(S + K_s) \qquad (6.43)$$

Most commonly, the influent microorganism concentration is negligible ($X_0 = 0$). In this case, Equation 6.41 becomes:

$$\mu = (1/\Theta) + b \qquad (6.44)$$

This results in the following explicit relationships for the composition of the reactor:

$$X = Y(S_0 - S)/(1 + b\Theta) \qquad (6.45)$$

$$S = K_s(1 + b\Theta)/[(q_{max} - b)\Theta - 1] \qquad (6.46)$$

Hence, the hydraulic residence completely determines the residual substrate concentration and the combination of influent substrate and hydraulic residence time completely determines the microorganism concentration.

This hydraulic configuration most closely resembles lagoons and completely mixed anaerobic digestors, and, as a first approximation, this model can be used to estimate their performance. Most commonly, however, completely mixed continuous flow systems (without recycle) are used as laboratory or pilot scale systems to determine kinetic parameters from experimental testing programs. The usual way in which such tests are conducted is to feed a single wastewater (i.e., at a single value of S_0) to a series of systems at varying hydraulic residence times. After steady state is reached, the effluent microorganism and substrate concentrations are measured. From the above relationships, the kinetic parameters in the Monod model (K_s, q_{max}, b and Y) can be determined in a variety of ways. The simplest, although perhaps not the most precise, involves combining Equations 6.42 and 6.43 and rearranging:

$$\mu = Y(S_0 - S)/\Theta X = q_{max} S/(S + K_s)$$
$$(S_0 - S)/\Theta X = (q_{max}/Y)/(1 + K_s/S) \tag{6.47}$$

The quantity on the left contains only experimental variables, while the quantity on the right contains the parameters to be determined. By either nonlinear regression or linearization of Equation 6.47, K_s and the quantity (q_{max}/Y) can be determined. For example, taking reciprocals of both sides (Metcalf and Eddy, 1979):

$$\Theta X/(S_0 - S) = (Y/q_{max}) + K_s(Y/q_{max})\,(1/S) \tag{6.48}$$

Therefore a plot of $\Theta X/(S_0 - S)$ (its reciprocal is often called the specific removal rate, U) versus $1/S$ will yield a straight line; from its slope and intercept, the kinetic parameters can be obtained.

The remaining kinetic parameters are determined by combining Equations 6.42 and 6.43 as follows:

$$\mu = Y(S_0 - S)/\Theta X = (1/\Theta) + b$$
$$(S_0 - S)/X = (1/Y) + (b/Y)\Theta \tag{6.49}$$

From this, a plot of the quantity $(S_0 - S)/X$ versus the hydraulic residence time can be used to estimate the yield (Y) and decay coefficient (b) by linear regression. Therefore, since the ratio of q_{max}/Y is already known, all four parameters can be determined.

In a CSTR, Equation 6.46 shows that, as Θ increases, S decreases. However, if the hydraulic residence time is allowed to increase without limit, S *does not* decrease indefinitely. Computation of the limit of Equation 6.46 results in the following relationship for the minimum substrate concentration achievable in a CSTR:

$$S_{min} = K_s b/(q_{max} - b) \tag{6.50}$$

Hence, it is impossible to reduce the soluble substrate concentration below this value in a CSTR. As hydraulic residence time decreases, a point will be reached where the loss of microorganisms due to dilution from the system will be greater than the maximum possible growth rate at the influent reactor concentration. This is obtained by substituting $S = S_0$ in Equation 6.46 and solving. The resulting minimum hydraulic residence time is given by:

$$\theta^{min} = \frac{1}{\left(\frac{q_{max}S_0}{S_0 + K_s}\right) - b} \quad (6.51)$$

Conditions at which the hydraulic residence time falls below this value are referred to as "washout," since any microbial population will be washed out of the system and, at steady state, $S = S_0$ and $X = 0$.

Example 6-3

A treatability study of an aqueous waste is performed using a suspended growth system designed as a CSTR without recycle. The system is fed wastewater containing 200 mg/L of biodegradable COD, and is operated at a series of hydraulic residence times. After steady state occurs, the effluent is filtered (so as to remove biomass) and COD and reactor VSS concentrations are measured. Given the results recorded below, estimate the kinetic parameters in the Monod model.

Residence time		Effluent soluble COD	VSS
hours	days	(mg/L)	(mg/L)
9	0.375	42.7	60.9
12	0.5	18.3	75.0
18	0.75	10.2	78.3
24	1	7.4	72.2
36	1.5	4.8	75.6
48	2	2.8	72.0

Solution

By use of Equation 6.48, the reciprocal of the specific removal rate is plotted versus the reciprocal of the substrate concentration. The following table illustrates the computation and Figure 6-5 shows that a relatively good straight line fit is obtained.

U^{-1} (days)	$1/S$ (L/mg)
0.145	0.0234
0.206	0.0546
0.309	0.0976
0.375	0.135
0.581	0.209
0.731	0.353

By a least square analysis, the slope of the line is 1.838 and the intercept is 0.124. The slope divided by the intercept yields an immediate estimate of K_s—14.8 mg/L.

The remaining kinetic parameters may be computed by the use of Equation 6.49 in which the ratio of COD removed to microorganisms is plotted against residence

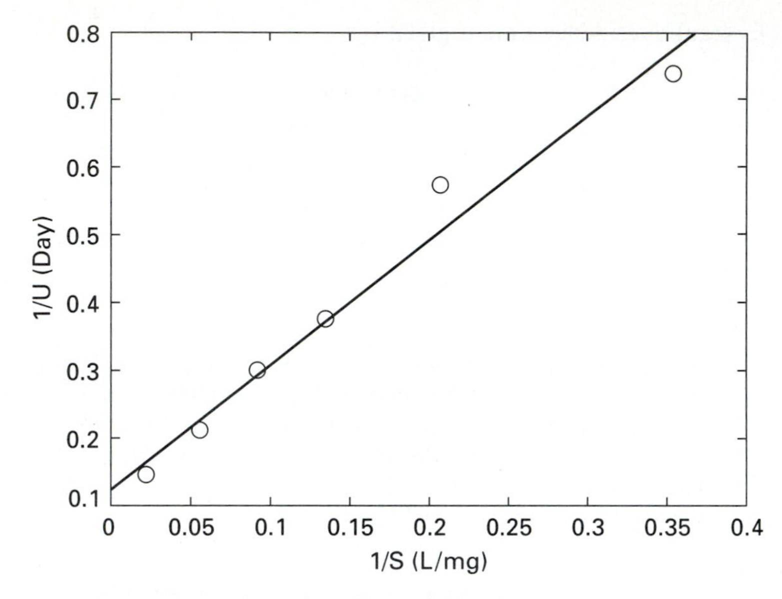

Figure 6-5 Reciprocal specific loading versus reciprocal substrate.

time. The table below provides the coordinates of this plot and Figure 6-6 shows the corresponding results. Note that the plot in Figure 6-6 shows rather more scatter than the plot in Figure 6-5. This is typical for treatability studies.

$(S_0 - S)/X$	Residence time (days)
2.582	0.375
2.421	0.5
2.424	0.75
2.670	1
2.583	1.5
2.737	2

By a least squares analysis of this plot, the slope and y-intercept are, respectively, 0.142 and 2.425. The yield is the reciprocal of the intercept, or 0.412. The maximum specific growth rate (q_{max}) is the yield divided by the intercept from the first regression, that is, 0.412/0.124, or 3.324 d^{-1}. The decay coefficient, b, is the slope from this equation multiplied by the yield, that is, 0.142×0.412, or 0.0585 d^{-1}.

For use in process operations, it is often desirable to use a complete mix system with separation and recycle of biological mass. This enables retention of a larger population of active microorganisms, and thus a more effective use of system volume. A schematic of this type of system is given in Figure 6-7. This system is similar to the CSTR with no recycle, except that the effluent from the reaction vessel is collected into a separator in which the microorganism (generally desired to be in a flocculent condition to facilitate separation) are separated from supernate. The su-

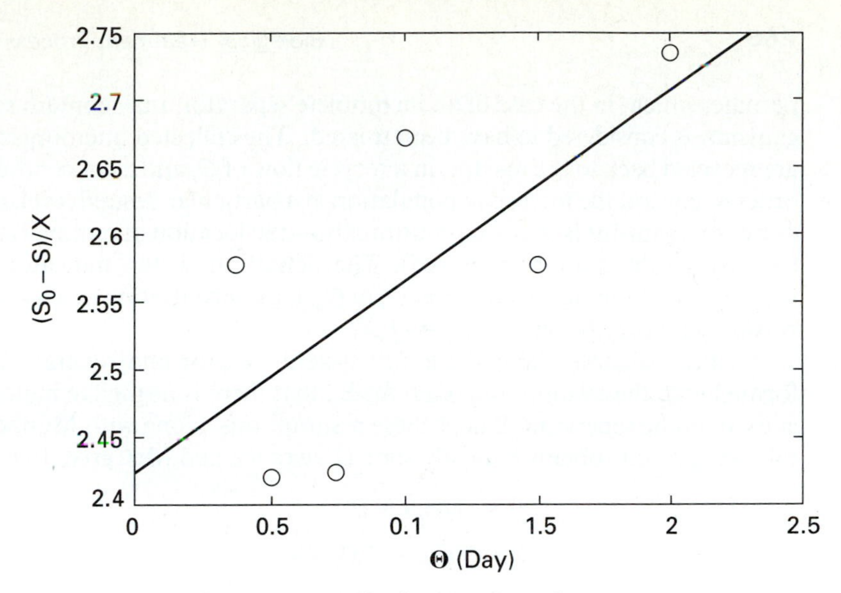

Figure 6-6 Second kinetic plot to determine parameters.

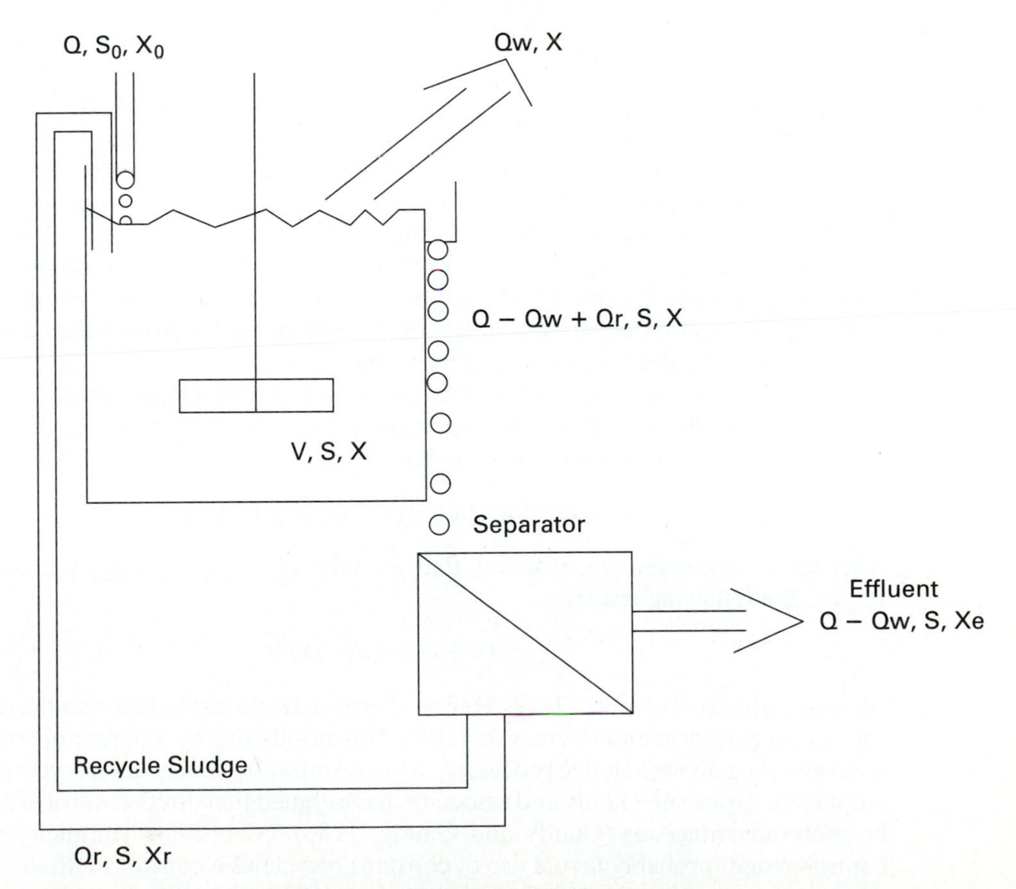

Figure 6-7 Schematic of CSTR with recycle.

pernate, which, in the case of an incomplete separator, may contain some microorganisms, is considered to have been treated. The collected microorganisms (sludge) are recycled back to the reactor, in a recycle flow of Q_r and a concentration of X_r. In order to control the microbial population at a particular desired level, a certain mass of microorganisms is removed continually—the location of removal is unimportant (relative to the amount removed). The definition sketch indicates a volumetric wasting rate (from the reaction vessel) of Q_w at a concentration of X—hence the mass wasting rate may be written $P_x = Q_w X$.

Mass balances about the entire system (reactor and separator) may be performed with the assumption, often made, that there is negligible biological removal activity in the separator. Under these assumptions, along with Monod kinetics, the following results obtain at steady state (Lawrence and McCarty, 1970):

$$\mu = (1/\Theta_c) + b \tag{6.52}$$

$$X = Y(S_0 - S)\Theta_c/\Theta\mu \tag{6.53}$$

$$\mu = q_{\max}S/(S + K_s) \tag{6.54}$$

$$X = Y(S_0 - S)\Theta_c/\Theta(1 + b\Theta_c) \tag{6.55}$$

$$S = K_s(1 + b\Theta_c)/[(q_{\max} - b)\Theta_c - 1] \tag{6.56}$$

where Θ_c, termed the sludge age (or mean sludge residence time) is defined as the mass of solids in the system (which, if there is assumed to be minimal activity in the separator is equal simply to VX) divided by the wasting rate of solids by deliberate and inadvertent action ($P_x + QX_e$). Note that in this system, the residual substrate is a function solely of the *sludge* residence time, and not of either the influent substrate concentration, nor the *hydraulic* residence time. Equation 6.50 remains valid for the CSTR with recycle. Equation 6.51 must be modified by replacing $\Theta^{\min}$ with $\Theta_c^{\min}$. In other words, washout occurs when the *solids* residence time drops below a critical value, rather than the hydraulic residence time.

An additional equation can be obtained by performing a mass balance on solids around the separator itself. From inspection of Figure 6-7, by equating the flux of solids into and from the separator, one has:

$$(Q - Q_w + Q_r)X = (Q - Q_w)X_e + Q_rX_r \tag{6.57}$$

After some rearrangement, realizing that the term $(Q - Q_w)X_e + Q_wX$ is equal to VX/Θ_c, the following results:

$$X_r = (1 + \alpha - \Theta/\Theta_c)X/\alpha \tag{6.58}$$

where α is the recycle ratio, Q_r/Q. Hence, there is a relationship between the recycle solids concentration and the recycle ratio. This results in two different operational strategies for activated sludge processes. Most commonly, a constant recycle ratio is employed. However, Gaudy and associates have argued that direct control of X_r may be more advantageous (Gaudy and Gaudy, 1980). At present, simplicity of the control system probably favors use of constant recycle as a control method.

Example 6-4

You are asked to design an activated sludge process (CSTR with recycle) for the treatment of the waste described in Example 6-3. The influent biodegradable COD is 200 mg/L. This is to be reduced to an effluent COD (consisting of both microorganisms that wash over the clarifier weir and residual soluble organic matter) of 5 mg/L or less. In addition, the recycle ratio is to be 0.25, and a maximum concentration of 7,500 mg/L of recycle solids (VSS) is to be carried. Previous testing has shown that the effluent VSS is a function of the VSS fed to the clarifier according to the following specific equation for this waste:

$$X_e = 0.0001\ X^{1.5}\ (X \text{ and } X_e \text{ in mg/L})$$

Based on this information, compute the feasible range of SRT values for the process.

Solution

If the effluent suspended solids were zero, then the effluent quality would be determined solely by application of Equation 6.56, which, with substitutions based on the problem specification, becomes:

$$10 = 14.8(1 + 0.0585\ \Theta_c)/[(3.324 - 0.0585)\Theta_c - 1]$$

$$\Theta_c = 0.78\ d$$

Therefore, the solids residence time must be at least $0.78d$ in order to meet the design conditions. Since the clarifier is not ideal, however, the actual effluent quality must be computed at a series of assumed sludge ages. At each sludge age, Equations 6.56 and 6.55 are used to compute S and X, respectively:

$$S = 14.8(1 + 0.0585\ \Theta_c)/(3.266\ \Theta_c - 1)$$

$$X = 0.412\ \Theta_c(200 - S)/[0.5(1 + 0.585\ \Theta_c)]$$

Given X, from the problem specification, X_e and its COD equivalent can be computed, leading to an estimate for the total effluent COD:

$$S_{tot} = S + 1.42\ X_e$$

The recycle sludge concentration is computed by Equation 6.58:

$$X_r = (1 + 0.25 - 0.5/\Theta_c)/0.25$$

Figure 6-8 shows these values as a function of sludge residence time. In order to maintain a value of S_{tot} at or below 5 mg/L, the sludge age must be between 1.5 and 9 days. In order to maintain the recycle solids at or below 5,000 mg/L, the sludge age must be below 10 days. It is therefore recommended to hold the sludge age just below 9 days to achieve the design constraints. While a somewhat shorter age can be used (down perhaps to 7 days), there is an increased risk of poor flocculation of solids at low values of Θ_c.

The final check that should be made on this system is a comparison of the sludge age with the wasting due to effluent solids. The sludge age should be no greater than that which could be achieved by no additional deliberate sludge wasting, that is,

$$\Theta_c^{max} = \Theta X/X_e$$

Since Θ is 0.5 day, and X/X_e is in the range of several hundred, the design sludge age is far less than this maximum value.

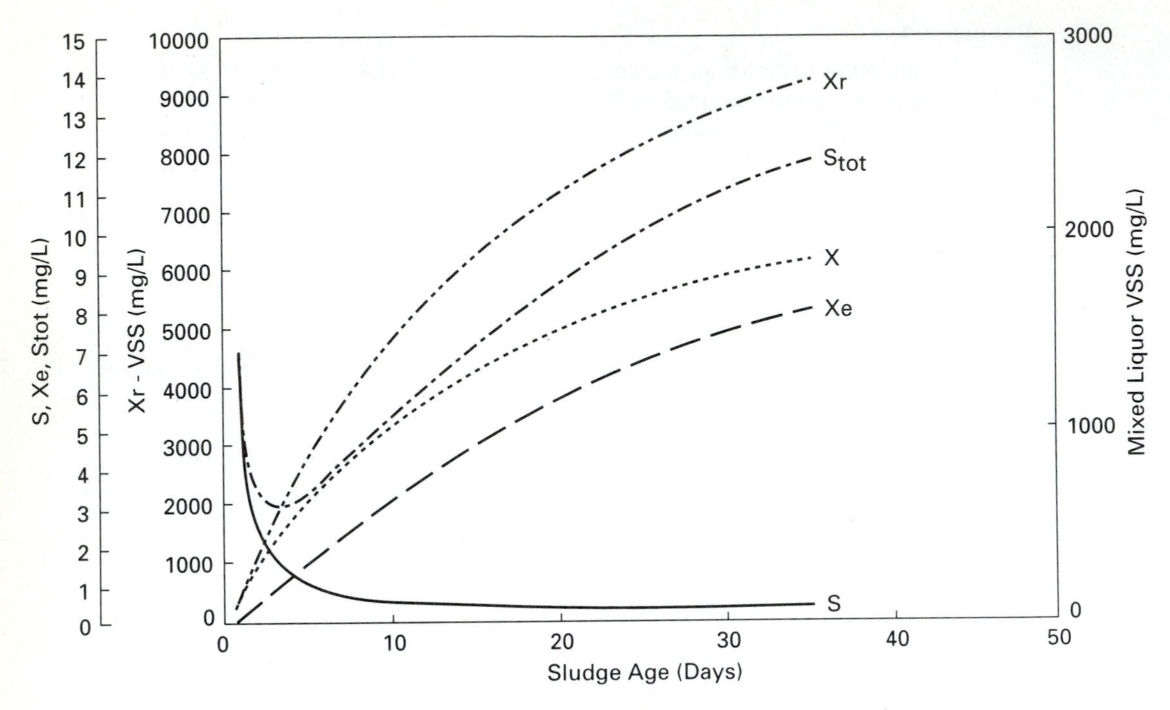

Figure 6-8 Effect of sludge age on CSTR behavior.

The other extreme in continuous processes from the complete mix system (either with or without recycle) is the plug flow system. In such a process, there is a total absence of mixing of influent with any material that entered the reactor either earlier or later. By definition, the mean hydraulic residence time in such a system (Θ) assumes a value equal to the residence time of each and every element of material that enters the system. Such a process can be conceptualized as a series of batch reactors moving along a conveyor, each obeying batch kinetics and being discharged at an identical time—that is, after holding for time Θ. At steady state it is therefore necessary only to consider what happens to a single hypothetical bottle as it travels through the system.

In the case of a biological system, the only relevant scenario is when recycling exists. Otherwise, in a true plug flow system, there would be sterility due to the absence of an inoculum. Figure 6-9 provides a definition for this system. In this particular case, sludge wasting is shown as coming from the separator (i.e., the return sludge line). At the inlet, after mixing of the recycle and the incoming flows, the combined conditions (denoted by the subscript i) are:

$$Q_i = Q + Q_r \qquad = Q(1 + \alpha) \tag{6.59}$$

$$S_i = (QS_0 + Q_rS)/Q_i \quad = (S_0 + \alpha S)/(1 + \alpha) \tag{6.60}$$

$$X_i = (QX_0 + Q_rX_r)/Q_i \quad = (X_0 + \alpha X_r)/(1 + \alpha) \tag{6.61}$$

From the above argument, the profile (with position in the reactor vessel) of substrate and microorganisms are described by a modified version of Equations 6.18 and 6.19,

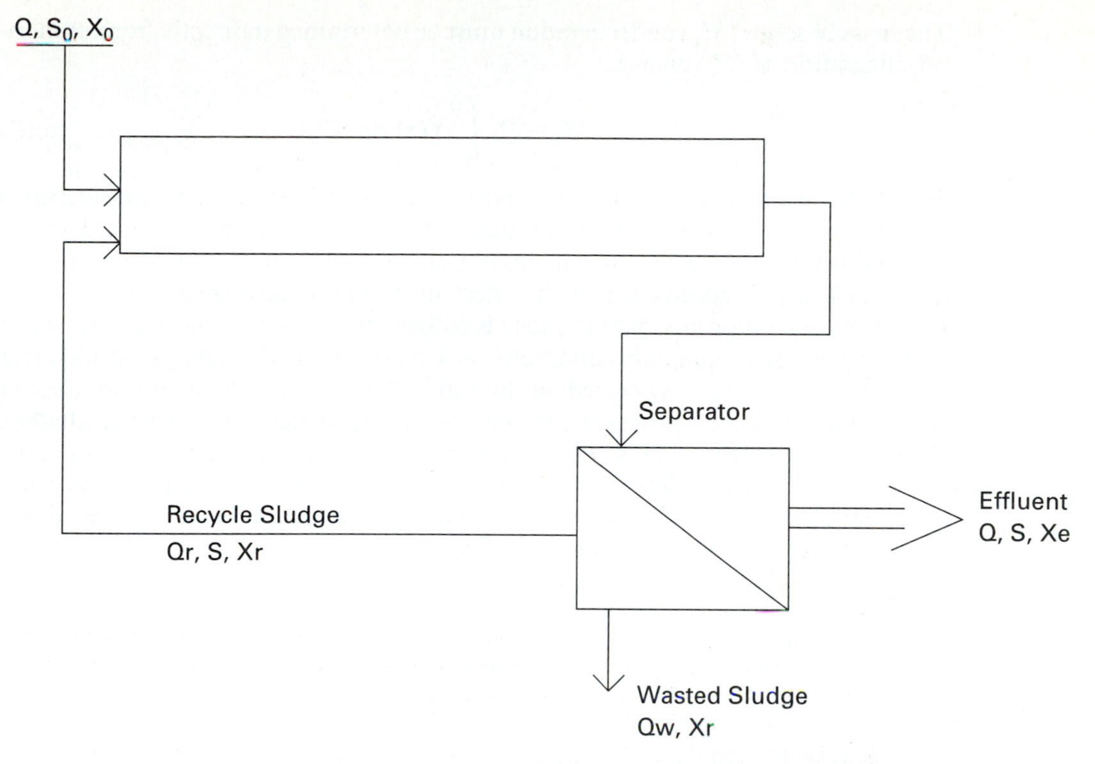

Figure 6-9 Definition sketch for plug flow reactor with recycle.

in which the position in the reactor is described by the cumulative residence time (τ—which ranges between 0 and Θ). The variables S^* and X^* are used to denote these position-dependent values. If the Monod kinetic assumptions are employed, the relationships for these values are:

$$dX^*/d\tau = [q_{\mathbf{max}}S/(S + K_s) - b]X \tag{6.62}$$

$$dS^*/d\tau = -q_{\mathbf{max}}SX/Y(S + K_s) \tag{6.63}$$

However these equations must be solved subject to a set of initial and boundary conditions as follows:

$$\tau = 0,\ S^* = (QS_0 + Q_rS)/Q_i,\ X^* = (QX_0 + Q_rX_r)/Q_i$$

or

$$S^* = (S_o + \alpha S)/(1 + \alpha),\ X^* = (X_0 + \alpha X_r)/(1 + \alpha)$$

$$\tau = \Theta,\ S^* = S$$

Also, Equation 6.58, the separator mass balance, remains exactly the same for this case as well. Therefore, an alternative form of the first initial condition for microorganisms can be written as:

$$\tau = 0, \qquad X^* = [X_0/(1 + \alpha)] + X_{\mathbf{exit}}[1 - (\Theta/\Theta_c)\alpha/(1 + \alpha)]$$

where $X_{\mathbf{exit}}$ is the concentration of biomass which exits the reactor at $\tau = \Theta$.

The mass of solids (M_s) under aeration must be determined indirectly from numerical integration of X^* versus τ.

$$M_s = Q_i \int_0^{\Theta} X(\tau)\, d\tau \tag{6.64}$$

The sludge wasting rate (P_w) is $Q_i(X(\Theta) - X(0))$, which arises from considering a mass balance over the reactor. From this, the sludge age may be computed.

Direct use of the plug flow equations is difficult, since, in the typical case (where S_0 is known, S is specified, and the effect of hydraulic residence time or sludge residence time on process performance is sought) the equations have unknowns in either the initial or boundary conditions, which must be evaluated by trial and error integration. However, as pointed out by Schroeder (1977) in the common range at which many biological processes are operated (most particularly, activated sludge), there is little difference in the predicted performance using a plug flow *model* and a complete mix *model* to described kinetics. Hence, for computational ease, the complete mix models are generally used, even though an actual system may have mixing patterns more closely resembling plug flow.

Example 6-5

You are asked to design a plug flow activated sludge process to treat a waste that has identical biokinetic coefficients to those discussed in Examples 6-3 and 6-4. However, in this case, the waste has an initial biodegradable COD of 500 mg/L and you are asked to reduce the soluble COD remaining to 0.5 mg/L. Compare the result to the required design parameters for a CSTR activated sludge process. Assume that the sedimentation process is ideal, and that the recycle ratio is 0.2 and the hydraulic residence time is 4 hours.

Solution

First, the inlet conditions are examined. If zero microorganisms are in the influent, then at $\tau = 0$, $S = (500 + 0.2 \times 0.5)/(1 + 0.2)$. The inlet condition for microorganisms is not known, since there is an unknown value for both X_r and Θ_c. The governing differential equations will be integrated assuming a variety of values for $X(0)$ until the value for S at $\tau = 0.1666$ day is equal to the desired (and assumed) value of 0.5 mg/L. The forward finite difference approach is used for the integration in which the derivatives are used to compute the values for the dependent variables at a small successive time step. By using a small enough time step, truncation errors can be controlled. A time step of 0.001 day is suitable for this purpose, and the computations can be performed using a spreadsheet. A small portion of the computation is illustrated below.

Time (days)	X (mg/L)	S (mg/L)	μ (d^{-1})	dx/dt (mg/L-d)	ds/dt (mg/L-d)	Integral (mg-d/L)
0	294.00	416.75	3.210	926.542	−2290.634	
0.001	294.93	414.46	3.209	929.283	−2297.417	0.294
0.002	295.86	412.16	3.209	932.028	−2304.213	0.295
0.003	296.79	409.86	3.208	934.779	−2311.022	0.296
0.004	297.72	407.55	3.208	937.534	−2317.843	0.297
0.005	298.66	405.23	3.207	940.295	−2324.675	0.298

At $\tau = 0$, it is assumed that X is 294 mg/L and S is computed to be 416.75 mg/L by the problem conditions. The derivatives are computed from Equations 6.62 and 6.63, given the values of S and X (note that μ is defined by Equation 6.54). At all subsequent time steps, say $\tau + \Delta\tau$, X is computed as $X' + (dX/dt)'\Delta\tau$ and S is computed as $S' + (dS/dt)'\Delta\tau$, where the primes indicate the values from the immediate prior step are used in the evaluation. The term "INTEGRAL" represents the contribution of that row to the integral in Equation 6.64, evaluated by the trapezoidal rule. After trial and error at various assumed values for $X(0)$, it was found that 294 mg/L resulted in a substrate concentration of 0.5 mg/L at the assumed hydraulic residence time of 0.1666 day. The microorganism concentration exiting the reactor under these conditions is computed as 461 mg/L. The summation of all terms in the "INTEGRAL" column and divided by the difference between the final and initial sludge concentrations (461 – 294 mg/L) yields a sludge age of 0.379 day.

For a CSTR, the use of Equations 6.55 and 6.56 results in a computed sludge age of 19.9 days and a solids concentration of 11,400 mg/L. Note that the plug flow configuration is predicted to be more efficient (i.e., to yield the desired reduction while using a lower concentration of microorganisms). This is an intrinsic property of the difference between a plug flow and a CSTR configuration. In practice, performance of a physical system designed as a PFR would be intermediate between the two computed numbers due to practical factors which limit the degree to which backmixing can be prevented (such as the mixing required for solids suspension and aeration).

It should also be emphasized that the above analyses focus most particularly on the degradation of soluble (or most probably, readily soluble) material. The presence of nondegradable organic matter or degradable but suspended organic matter (including nonsettleable solids originating from growth in the reactor) may contribute to the *total* organic matter in the effluent—and, in fact, in some cases may be the dominant portion of that load. Therefore, the adequate design and operation of the process to encourage efficient separation may be more important than high level removal of soluble organics.

Continuous Film

By definition, continuous film processes involve diffusional resistance in the biofilm layer. This must be considered in the analysis of such processes. A rational analysis of biofilm processes must start with relationships between flux into a biofilm and liquid phase concentrations, as embodied in Equations 6.33 or 6.35, for the more general case or for deep biofilms. The use of these models will be discussed below. However, due to their complexity, a variety of more empirical relationships have arisen for biofilm processes.

For trickling filters, a very frequent design equation is the modified Velz relationship—also called the Eckenfelder equation (Parker and Merrill, 1984):

$$\ln(S_e/S_i) = -k_{20}\Theta^{(T-20)}A_sD/Q^n \tag{6.65}$$

where S_e is the effluent substrate concentration (generally, in this equation, considered the sum of soluble and nonsoluble BOD_5), S_i is the *applied* influent concentration (corrected, if necessary, for the dilution due to recycle). A_s is the media specific surface area, D is the media depth, and Q'' is the total (waste + recycle) hydraulic flow rate. The parameters k_{20} and n are media and waste specific treatability constants (at a reference temperature, T, of 20°C). The parameter Θ is a temperature correction coefficient to reflect the effect of temperature on degradation rate. In general, pilot tests must be performed to obtain these parameters (or, possibly, they may be available from the manufacturer of the trickling filter medium). Table 6-7 provides sample kinetic data for various media and wastewaters.

Pescod et al. (1984) reviewed a number of studies on the treatment of high organic strength wastes by RBC's, and developed an empirical correlation between removal rate and organic areal loading (providing that the wastes had not received prior treatment and that the hydraulic residence time was at least 45 minutes). Their correlations, based on either BOD_5 (which gives better fit) or COD are:

$$\text{g/m}^2\text{d BOD}_5 \text{ removed} = 0.986\ (\text{g/m}^2\text{-d BOD}_5 \text{ loaded})^{0.965} \tag{6.66}$$

$$\text{g/m}^2\text{d COD removed} = 0.573\ (\text{g/m}^2\text{-d COD loaded}) + 5.597 \tag{6.67}$$

The generalizability of these equations to a variety of wastewaters and rotating contactors is unclear. From the concentrations in the applied wastewater, the loading rate to a given RBC can be computed, and hence the above model gives an estimate for removal rate—from which the effluent concentration can be obtained.

A rational analysis of fixed film processes requires specification of the flow pattern in the system. In certain cases, such as fluidized beds, and, possible for RBC's (at least for a single "module"), the system may be considered completely mixed. In other words, there is a uniform concentration of degradable nutrient in the liquid. If the influent substrate concentration is S_{10}, and the effluent is S_{1e}, the flow rate through the system is Q, and the interfacial area of biofilm is A, then conservation of mass requires:

$$Q(S_{10} - S_{1e}) - JA = 0 \tag{6.68}$$

TABLE 6-7 KINETIC PARAMETERS FOR THE MODIFIED VELZ EQUATION (20°C) (Benefield and Randall)

Medium	Waste	k_{20}	n	A_s (ft^{-1})
rock, 2.5–4 in.	domestic	0.036	0.49	20–40
1-in. Raschig rings	domestic	0.031	0.63	58
2 1/4-in. Raschig rings	domestic	0.08	0.274	25
corrugated plastic (SURFPAC™)	domestic	0.05–0.079	0.45–0.5	28
corrugated plastic (SURFPAC™)	coke plant	0.0211	0.5	28
corrugated plastic (SURFPAC™)	textile waste	0.016–0.04	0.5	28
corrugated plastic (SURFPAC™)	pharmaceutical	0.029	0.5	28

Notes: These constants are valid with Q^H in gpm, A_s in ft^{-1}, and D in ft. Organic concentrations are on a BOD_5 basis. Specific surface for Raschig rings from Rich (1961).

or, rearranging

$$J = (Q/A)(S_{10} - S_{le}) \tag{6.69}$$

This can also be written as:

$$J = (V/A)(S_{10} - S_{le})/\Theta \tag{6.70}$$

where Θ is the mean hydraulic residence time in the system. This equation must be solved simultaneously with Equation 6.33, or, for deep biofilms, Equation 6.35 to describe the system.

In the case of a plug-flow biofilm, the constraining mass balance relationship may be derived as follows. By arguments similar to those used to develop Equation 6.62, one has the following equation for S_1 as a function of residence time (τ) in the

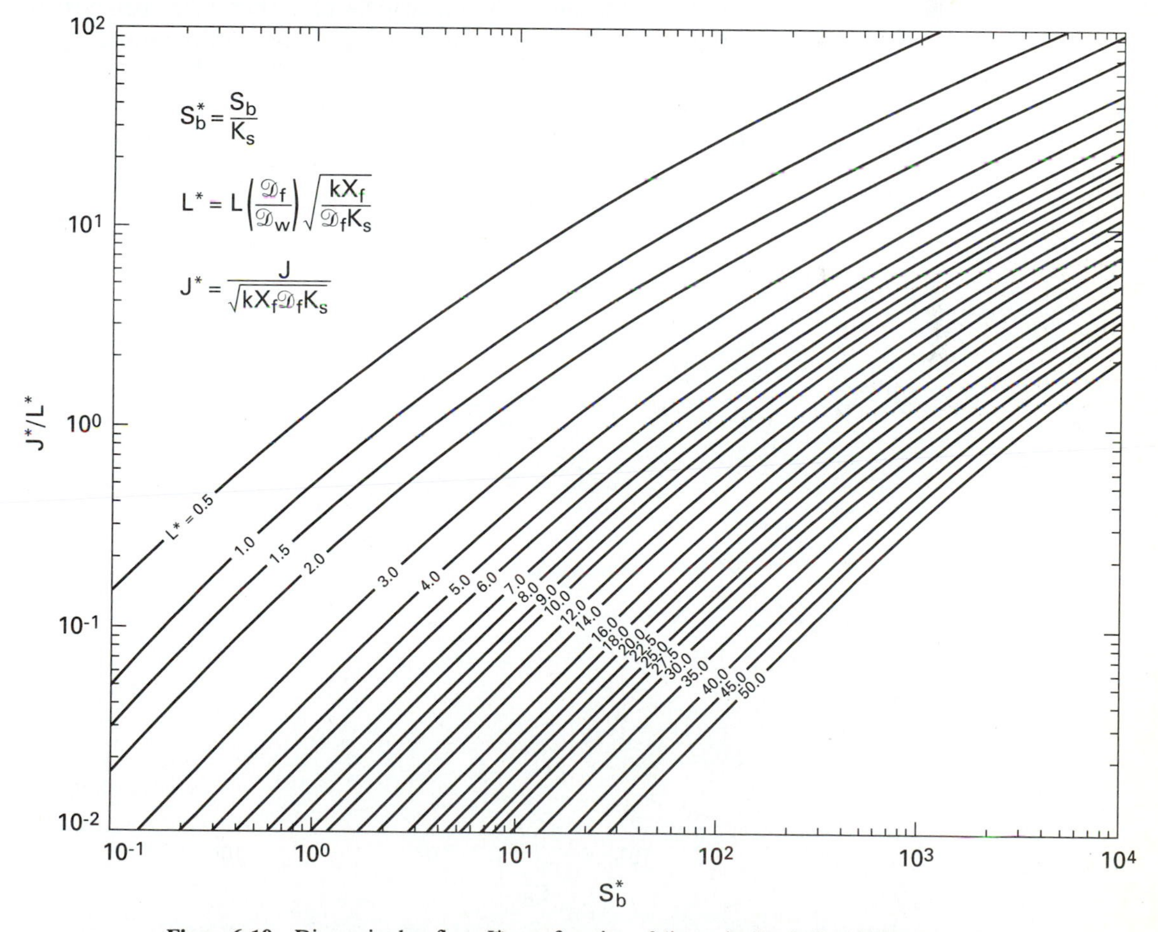

Figure 6-10 Dimensionless flux, J^*, as a function of dimensionless bulk substrate concentration and stagnant layer thickness for a deep biofilm. (Adopted from Suidan, 1986, by permission of ASCE.)

reactor:

$$dS_1/d\tau = -JA/V$$

By integration,

$$\Theta = -(V/A)\int_{S_{10}}^{S_{1e}} dS_1/J \tag{6.71}$$

If the kinetic parameters are known, for either the plug flow or the complete mix biofilm system, the unknown design parameters are S_{10}, S_{1e}, Θ, J, A/V, and L. There are three equations that interrelate these parameters; either Equation 6.70 or 6.71 depending on the hydraulic configuration; either Equation 6.33 or Equation 6.35, depending upon the biofilm thickness; and a relationship for L (i.e., a correlating expression for liquid mass transfer coefficient). If the biofilm is not deep, then additional parameters must be specified. For a deep biofilm, therefore, specification of any three of the design parameters fixes the remaining ones. In the most typical case, the influent and effluent substrate concentrations are specified, and L is specified by hydraulic conditions—the remaining parameters are computed. Suidan (1986) prepared a useful set of nomographs in which the design parameters are

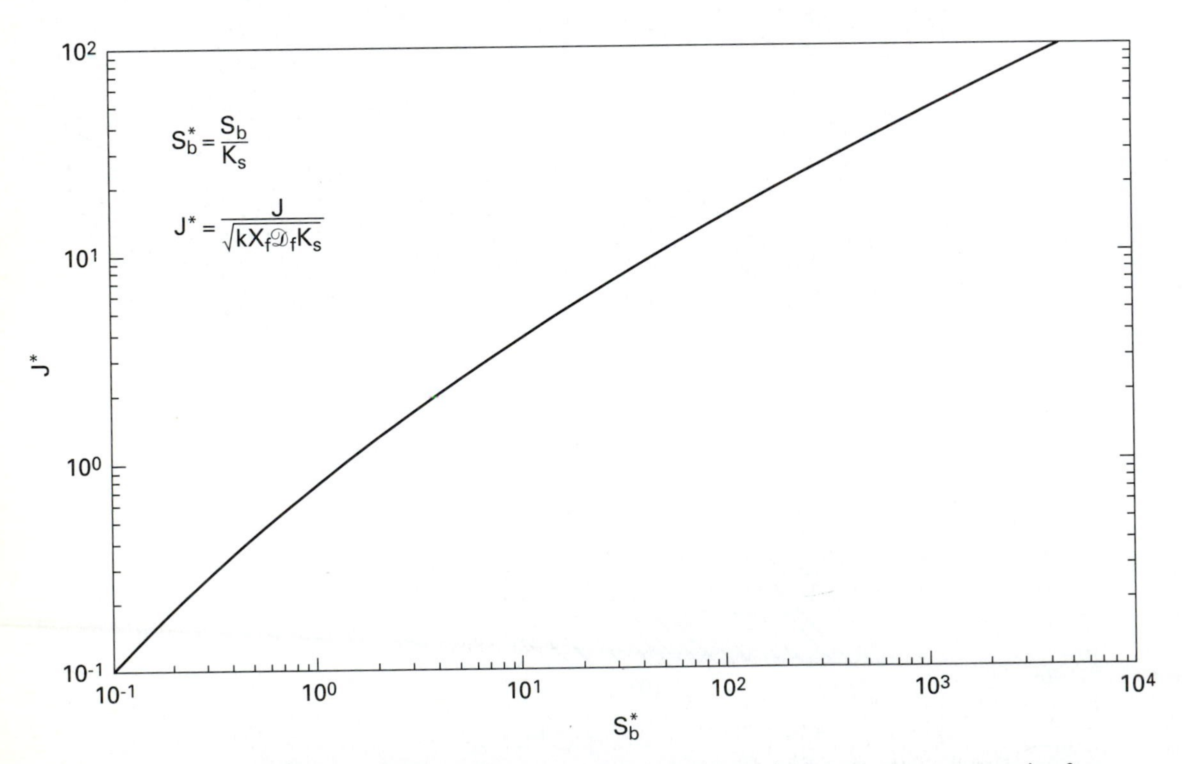

Figure 6-11 Dimensionless flux as a function of dimensionless substrate concentration for a deep biofilm where the dimensionless stagnant liquid-layer thickness is equal to zero. (After Suidan, 1986, by permission of ASCE.)

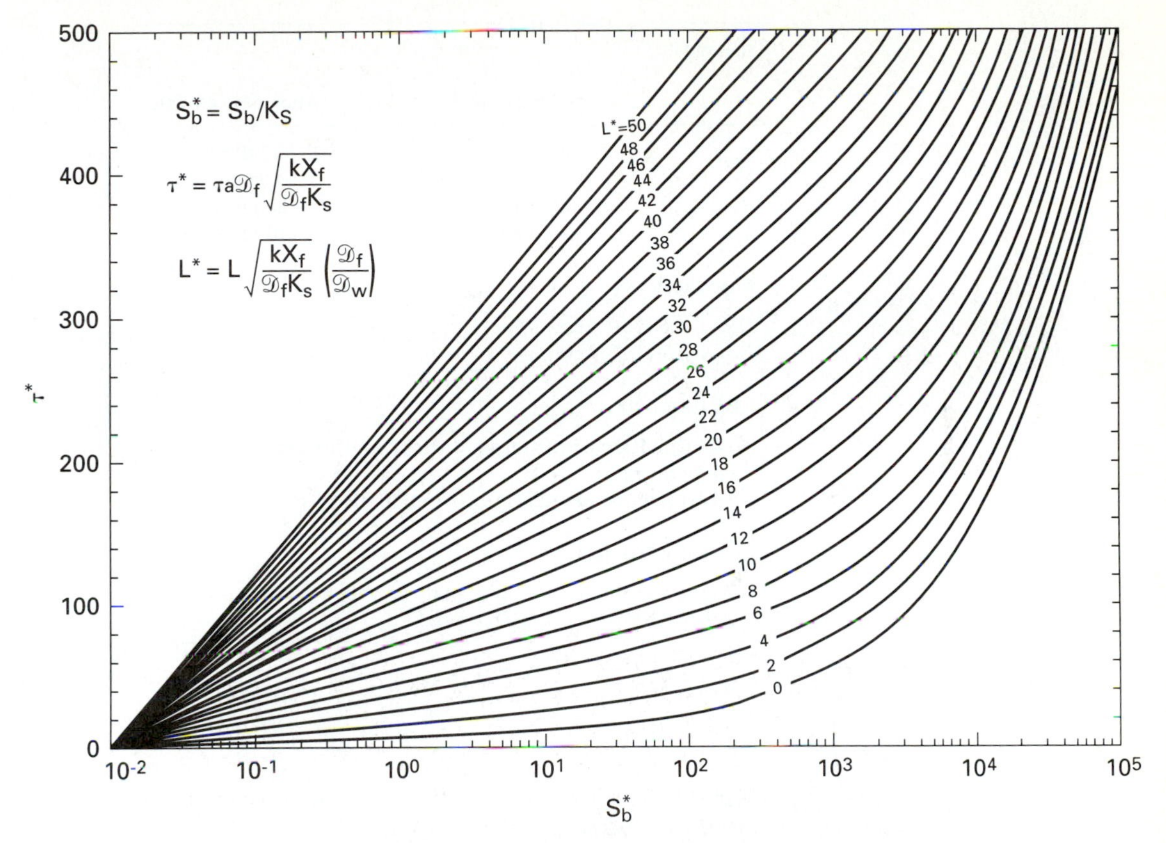

Figure 6-12 Dimensionless empty-bed detention time needed to achieve a dimensionless effluent substrate concentration of 0.01 in a deep biofilm plug flow reactor. (After Suidan, 1986, by permission of ASCE.)

related, in dimensionless form. Figure 6-10 pertains to a CSTR process, while Figures 6-11 through 6-14 pertain to a plug flow process. In these figures, the parameter "k" is equal to q_{max}/Y. Use of these figures will be illustrated in the following example problem.

Example 6-6

A waste leachate contains 300 mg/L of biodegradable COD. You are asked to estimate the required size of an RBC to treat this waste aerobically, and an anaerobic-packed bed reactor to treat this waste anaerobically. The RBC can be assumed to be hydraulically a complete mix system, while the anaerobic packed bed can be assumed to be a plug flow system. The desired effluent concentration of COD from either process is to be 10 mg/L, which is to be discharged to the municipal sewer. Given the following kinetic and physical parameters, estimate suitable design conditions to achieve the desired performance. The waste flow to be treated is 100 m^3/d. Assume, in both cases, that the biofilm is to be regarded as "deep."

Kinetic data:

	Aerobic growth	Anaerobic growth
q_{max}	4 d^{-1}	0.3 d^{-1}
Y	0.6 g/g	0.05 g/g
K_s	25 mg/L	200 mg/L
X_f	10 mg/cm^3	10 mg/cm^3

Miscellaneous:

D_w 1.6×10^{-5} cm^2/s
D_f 1.2×10^{-5} cm^2/s

RBC data:

Disks are 100 cm in diameter, with a central, unwetted zone of 20 cm in diameter
Disk rotational speed is to be 1 rpm (0.01667 s^{-1})

Anaerobic bed data:

Support medium	2 mm particles with a porosity of 0.4
Approach velocity	0.01 cm/s

Solution (RBC)

Note that the typical hydraulic residence time for RBC's given above is 30–120 minutes. Assume the low value of $\Theta = 30$ minutes.

Grady and Lim (1980) suggest the following correlation for the liquid film layer in RBC's at ambient temperature:

$$L = 2.144 \times 10^{-3}\,[\omega(r_0 + r_1)/2]^{2/3}$$

where L is in cm, ω is the rotational speed (in s^{-1}), and r_0 and r_1 are, respectively, the inner and outer radii of the wetted (biofilm-covered) areas of the rotating disk in cm. Therefore, the liquid film layer is computed as:

$$L = 2.144 \times 10^{-3}\,[0.01667\ \text{s}^{-1}(10\ \text{cm} + 50\ \text{cm})/2]^{2/3}$$

$$L = 1.35 \times 10^{-3}\ \text{cm}$$

The dimensionless liquid film length is computed by:

$$\begin{aligned} L^* &= L\,(D_f/D_w)\,(q_{max}X/YD_fK_s)^{1/2} \\ &= 1.35 \times 10^{-3}\text{cm}\,(1.2 \times 10^{-5}\ \text{cm}^2/\text{s}/1.6 \times 10^{-5}\ \text{cm}^2/\text{s})X \\ &\quad \left[\frac{4d^{-1}\,(1/24\ d/\text{hr})(1/3600\ \text{hr/s})\,(10\ \text{mg/cm}^3)}{0.6 \times 1.2 \times 10^{-5}\ \text{cm}^2/\text{s} \times 0.025\ \text{mg/cm}^3}\right]^{1/2} \\ &= 0.05 \end{aligned}$$

This is sufficiently small as to be assumed equal to zero—in other words, the liquid film resistance is not controlling. Hence, Figure 6-12 should be used in this computation.

The dimensionless bulk liquid substrate concentration is computed by:

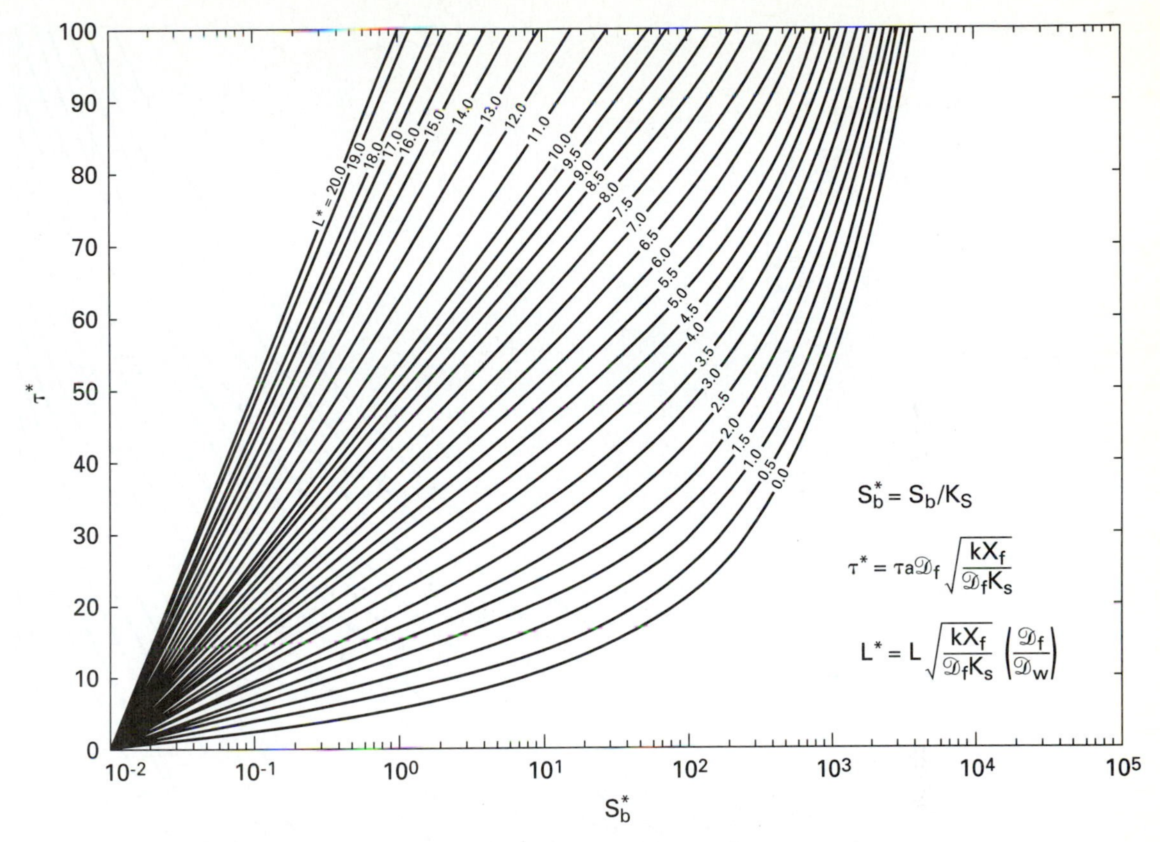

Figure 6-13 Dimensionless empty-bed detention time needed to achieve a dimensionless effluent substrate concentration of 0.01 in a deep biofilm plug flow reactor. (After Suidan, 1986, by permission of ASCE.)

$$S_b^* = 10 \text{ mg/L}/25 \text{ mg/L} = 0.4$$

From the figure, J^* equals 0.3. The flux into the biofilm is computed by:

$$J = J^* \, (q_{\max} X D_f K_s / Y)^{1/2}$$

$$J = 0.3[4 \, d^{-1}(1/24 \, d/\text{hr})(1/3600 \text{ hr/s}) \\ (1.2 \times 10^{-5} \text{ cm}^2/\text{s})(10 \text{ mg/cm}^3)(0.025 \text{ mg/cm}^3)/0.6]^{1/2}$$

$$J = 4.56 \times 10^{-6} \text{ mg/cm}^2\text{s} = 3940 \text{ mg/m}^2\text{d}$$

This may be coupled with Equation 6.70 to yield:

$$3940 = (V/A)(300 - 10) \times 1{,}000 \; L/\text{m}^3/(30 \text{ min}/24 \times 60)$$

$$3940 = (V/A)(13.92 \times 10^6)$$

$$V/A = 2.83 \times 10^{-4} \text{ m}$$

For the given conditions, the volume of the system is given by the product of the flow

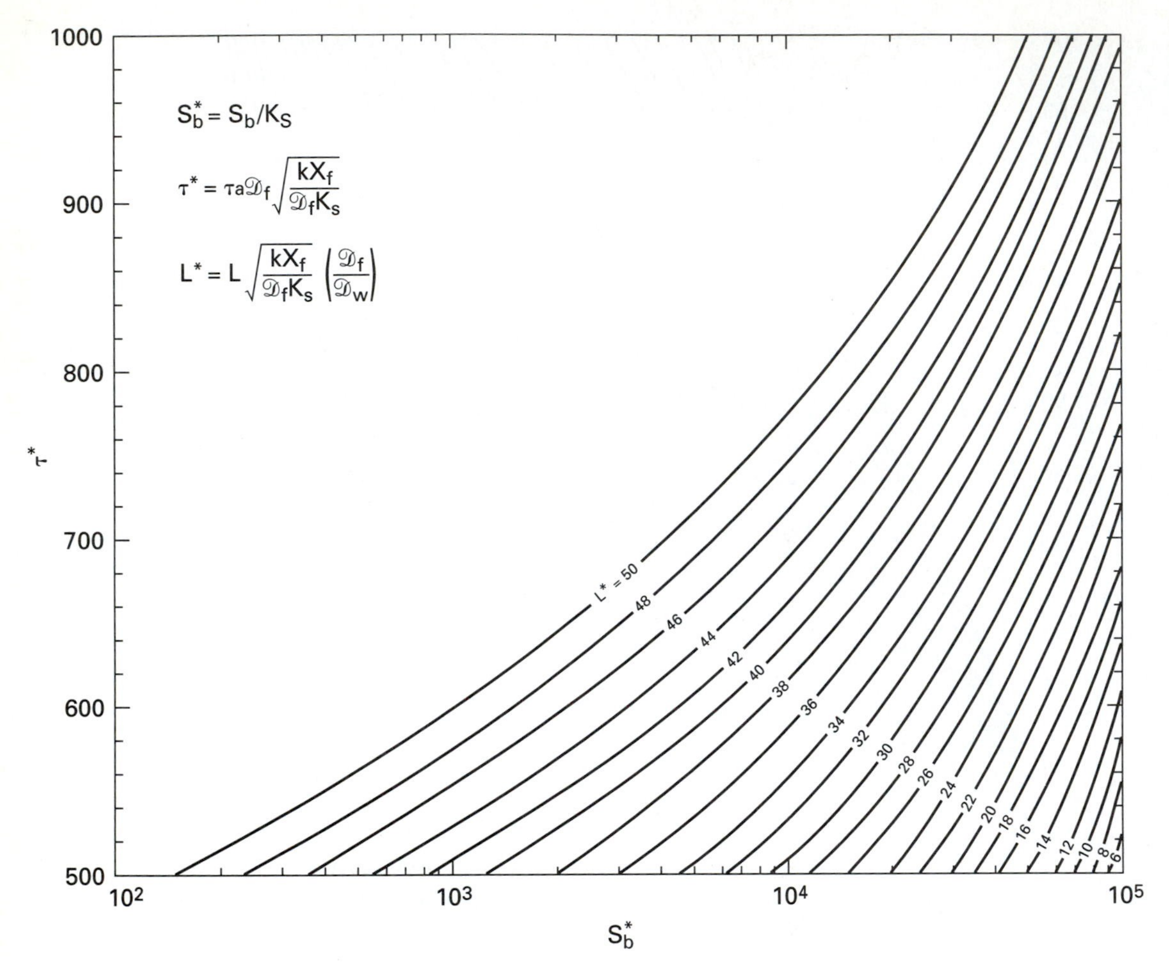

Figure 6-14 Dimensionless empty-bed detention time needed to achieve a dimensionless effluent substrate concentration of 0.01 in a deep biofilm plug flow reactor. (After Suidan, 1986, by permission of ASCE.)

rate and the hydraulic residence time, and appropriate conversions—as 2.08 m^3. Therefore the required surface area is equal to 7,349 m^2.

The available area per disk (considering both sides) is given as:

$$2\pi(r_1^2 - r_0^2) = 1.51 \text{ m}^2$$

Thus nearly 5,000 disks would be needed—far too impractical. The use of a greater hydraulic residence time does not change this problem. Since the liquid film resistance is already negligible, it is concluded that there is little that can be done with this design. However, several alternatives are available—these include use of a series of completely mixed RBC's or the use of an RBC as a roughing process preliminary to further treatment. These alternatives can be explored by use of the model presented.

Solution — Anaerobic Packed Bed

In this case, note that the specific surface area (area per unit volume) is fixed. If the medium is assumed spherical, then the following must hold:

$$A/V = 6(1-\epsilon)/d_p = 6(1-0.4)/0.2\ \text{cm}$$
$$= 18\ \text{cm}^2/\text{cm}^3$$

The thickness of the liquid film can be estimated using correlations presented for a packed bed by Suidan (1986):

$$d_p/L = (1.09/\epsilon)\,(\text{Re}^{1/3} + \text{Sc}^{1/3}\text{Re}^{-2/3})$$

where

Re (Reynolds number) = $\rho d_p V/\mu$
Sc (Schmidt number) = $\mu/\rho D_w$.

From the problem statement, the Reynolds and Schmidt numbers can be computed:

$$\text{Re} = (1\ \text{g/cm}^3)(0.2\ \text{cm})(0.01\ \text{cm/s})/0.01\ \text{g/cm-s}$$
$$= 0.2$$

$$\text{Sc} = (0.01\ \text{g/cm-}s)/(1\ \text{g/cm}^3 \times 1.6 \times 10^{-5}\ \text{cm}^2/\text{s})$$
$$= 625$$

Hence,

$$\frac{0.2\ \text{cm}}{L} = (1.09/0.4)(0.2^{1/3} + 625^{1/3}0.2^{-2/3})$$

$$L = 0.00286\ \text{cm}$$

The dimensionless length is now computed as in the RBC example:

$$L^* = 0.00286\ \text{cm}\ (1.2 \times 10^{-5}\ \text{cm}^2/\text{s}/1.6 \times 10^{-5}\ \text{cm}^2/\text{s})X$$
$$\left[\frac{0.3\ d^{-1}\ (1/24\ d/\text{hr})(1/3600\ \text{hr/s})\ (10\ \text{mg/cm}^3)}{0.05 \times 1.2 \times 10^{-5}\ \text{cm}^2/\text{s} \times 0.2\ \text{mg/cm}^3}\right]^{1/2}$$
$$= 0.036$$

which also indicates minimal liquid film resistance.

The dimensionless substrate concentrations are 10/200 and 300/200—that is, 0.05 and 1.5. From Figure 6-10, the dimensionless residence times corresponding to these values (τ^*) are, respectively, 2 and 8. The dimensionless version of the integral in Equation 6.71 is evaluated as the difference between these two values; hence τ^* is 6.0. Substitution in the definition yields the value of the empty bed contact time (τ) as follows:

$$\tau^* = \tau(A/V)D_f(q_{\max}X/YD_fK_s)^{1/2}$$

$$6 = \tau(18\ \text{cm}^{-1})(1.2 \times 10^{-5}\ \text{cm}^2/\text{s})x$$
$$[0.3\ d^{-1} \times 10\ \text{mg/cm}^3/(0.05 \times 24\ \text{hr/d}$$
$$\times 3{,}600\ \text{s/hr} \times 1.2 \times 10^{-5}\ \text{cm}^2/\text{s} \times 0.2]^{1/2}$$

$$\tau = 1{,}632\ \text{sec}$$
$$= 45\ \text{min}$$

Therefore, the required height is 0.01 cm/s × 1,632 sec = 16.3 cm.

OXYGEN TRANSFER CONSIDERATIONS

In aerobic systems, it is necessary to supply sufficient oxygen to the system to satisfy the requirements for microbial respiration, and to maintain a concentration of oxygen in the system such that dissolved O_2 is not rate limiting. Fortunately, the Monod half saturation constant for oxygen, in the case of aerobic microorganisms, is considered less than 0.1 mg/L (Pirt, 1975)—or less than 1% of the solubility at ambient temperatures. Hence, oxygen is very rarely rate limiting.

Nevertheless, oxygen must be supplied to a system at a proportionate rate to the amount of organic material degraded—as indicated by Equations 6.1 and 6.2. A more direct computation of oxygen requirements can be made by considering a mass balance over oxygen demand—which is an alternative way of describing reducing power. If microbial biomass of composition $CH_aO_bN_c$ were to be completely degraded, aerobically, by non-nitrifying microorganism, it would have an equivalent oxygen demand as described by the following equation:

$$CH_aO_bN_c + \left[\frac{a}{4} - \frac{b}{2} - \frac{3c}{4} + 1\right]O_2 = CO_2 + (1/2)(a - 3c)\, H_2O + cNH_3 \qquad (6.72)$$

Thus, for the empirical composition $C_5H_7O_2N$ (or, equivalently, $CH_{1.4}O_{0.4}N_{0.2}$), each gram of cell mass (in the form of volatile suspended solids) is "equivalent" to 1.42 grams of oxygen. Hence, if a waste flow Q (m^3/d) which contains S_i (mg/L or g/m^3) of biodegradable oxygen demand (ultimate BOD, or biodegradable COD) is treated to a level of S (mg/L or g/m^3) of oxygen demand, and in the process a quantity P_x of biomass (mg/L or g/m^3 VSS) is produced (as waste sludge), then the oxygen, which must be supplied to the biomass to satisfy respiration, m_R (g/d) can be given by the following mass balance:

$$m_R = Q(S_i - S) - 1.42\, P_x \qquad (6.73)$$

The actual oxygen which must be supplied to a biological system will consist of this contribution, due to respiration, as well as a contribution due to the difference between the oxygen in the influent (DO_i — in g/m^3) and the effluent (DO_e — also in g/m^3). Thus, the total oxygen which must be transferred (m_r — in g/d) is:

$$m_T = m_R + Q(DO_e - DO_i) \qquad (6.74)$$

In many suspended growth systems, such as the activated sludge process, the oxygen requirement is satisfied by either mechanical mixing to vigorously contact the slurry of biomass with air, or by the use of compressed air (or oxygen) introduced through spargers—or possibly by a combination of these two devices. The specification of aeration equipment for such systems is beyond the scope of this text; however, the power required to transfer oxygen can be estimated from several rules of thumb. For mechanical aerators, the efficiency, in lbs of oxygen transferred per horsepower-hour (shaft), can be as high as 3.5–5, and is dependent upon the specific aerator and the power per unit volume used. For diffused aeration, a slightly lower efficiency, 1.5–2.5 lbs/hp-hr (6.9–11.5 kg/MJ) is typical (Benefield and Randall, 1980).

For fixed-film systems, estimation of aeration capacity is more difficult. From simple mass transfer assumptions, Schroeder and Tchobanoglous (1976) estimated a maximum expected oxygen flux into a biofilm on trickling filters to be 3.3×10^{-5} mg/sec-cm^2, where the area is the wetted interfacial area in the trickling filter. This can be compared to the organic removal rate to estimate whether supplemental forced aeration (by a fan above the filter) is required. In addition, the air flow rate required to supply this oxygen (typically at least an order of magnitude above the amount which is transferred) may be compared to the velocity of air that flows through the trickling filter due to natural convection (Schroeder and Tchobanoglous, 1976) to determine if forced ventilation is required. As an alternative to forced aeration, a high degree of recirculation can be used to reduce the organic loading—at the expense of building a larger trickling filter.

In the case of anoxic or anaerobic processes, and particularly when nitrate respiration is important, there may be a need to supply additional nitrate. This can be computed directly from the stoichiometric equations. However, in some circumstances, the concentration of nitrate can be rate limiting. Grady and Lim (1980) have suggested a Monod product relationship for the simultaneous limitation by a carbonaceous substrate (BOD) and nitrate as follows (to replace Equation 6.9):

$$\mu = q_{\max} [S/(S + K_s)][N/(N + K_N)] \qquad (6.75)$$

where N is the concentration of nitrate and K_N is the half saturation constant. K_N has a value of 0.1–0.2 mg/L nitrate as nitrogen. Hence, little excess nitrate is required (from a stoichiometric point of view), and to prevent rate limitations a residual of about 0.5 mg/L would be sufficient.

OPERATIONAL STABILITY OF ANAEROBIC PROCESSES

Anaerobic processes have the potential advantage, *vis à vis* aerobic processes, of generating methane gas which can be recovered for energy value. In addition, these processes tend to produce less sludge than aerobic processes. However, the operation of anaerobic processes is somewhat more difficult than aerobic processes, particularly when the potential for toxic inputs or slug shock loads exists.

A methane-generating anaerobic process can, at an elementary level, be considered as consisting of three types of microorganisms—hydrolyzing, acid-forming, and methane forming, with each class of organism dependent upon the previous class for supply of substrate. However, it is generally believed that the rate of growth of the methane forming organisms tends to be slower than the other classes, and that the methane forming organisms are more sensitive to toxic materials than other classes of organisms. In particular, Table 6-8 summarizes inhibiting concentrations of inorganic materials for methane forming organisms. While certain organic materials may also be inhibiting, these concentrations are not well characterized. In addition, anaerobic processes will operate only within a relatively narrow pH range, as indicated in Figure 6-15.

In the presence of a high slug nutrient concentration, the growth of acid-forming organisms will accelerate and result in an accumulation of acids at a rate in excess of the capacity of methane-forming organisms to assimilate these compounds. This will result in a pH reduction that may serve to further depress the rate of

TABLE 6-8 INHIBITORY CONCENTRATIONS OF INORGANIC MATERIALS TO ANAEROBIC PROCESSES. (Adapted from U.S. EPA, 1979.)

Material	Inhibitory concentration in solution (mg/L)
Ammonia nitrogen	1500–3000
Calcium	2500–4500
Magnesium	1000–1500
Potassium	2500–4500
Sodium	3500–5500
Copper	0.5
Zinc	1.5
Chromium(VI)	3.0
Nickel	2.0

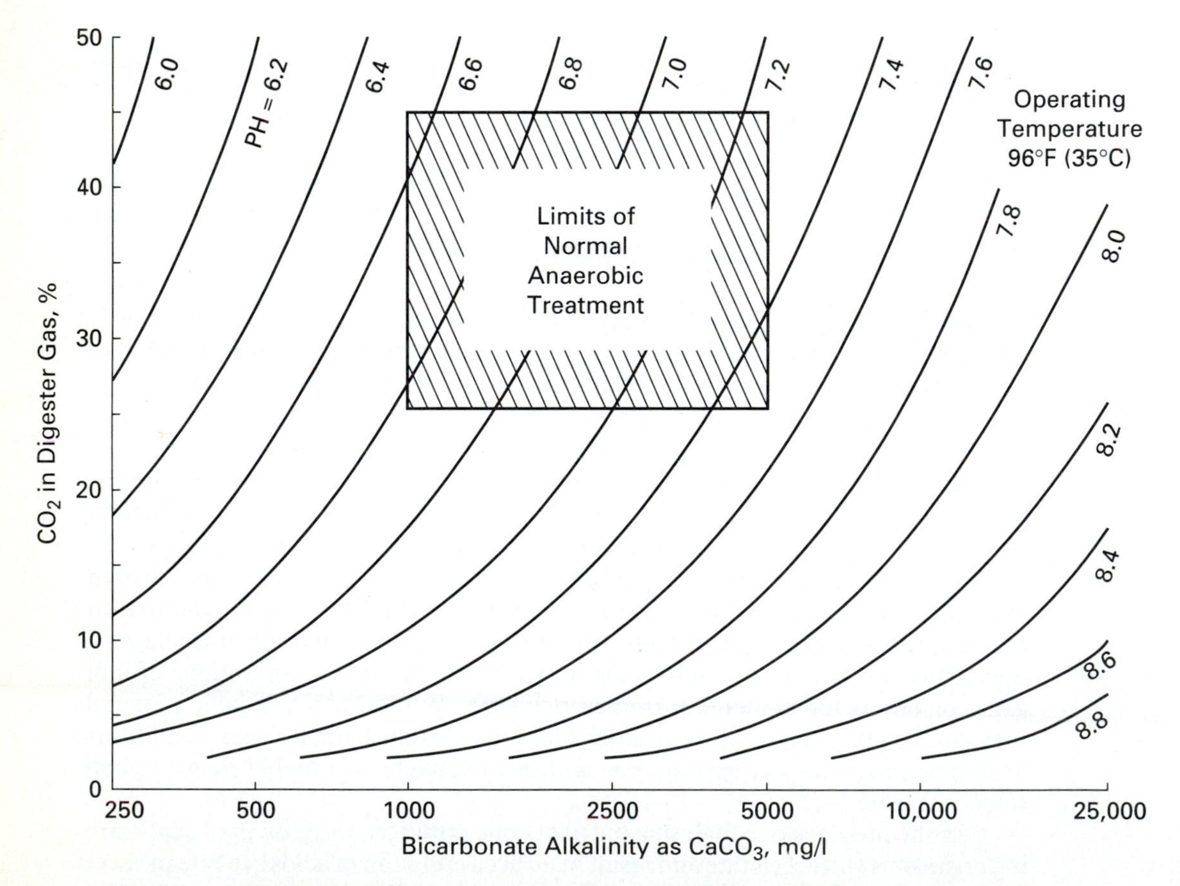

Figure 6-15 Relationship between pH and bicarbonate concentration near 95°F. (From U.S. EPA, 1979.)

methanogenesis—if unchecked, this may lead to a catastrophic collapse of the anaerobic system. Similarly, if toxic materials are present, the rate of methane formation will be selectively depressed, resulting in an acid accumulation and a pH drop. In order to prevent such catastrophic failure, continuous monitoring of anaerobic gas production is advisable to detect the onset of inhibition. At this point, countermeasures can be taken. These include adding buffering materials to maintain pH in the desirable region, adding sulfide compounds, particularly ferrous sulfide, to precipitate toxic heavy metal cations, and diluting high ionic-strength feed material (Parkin and Owen, 1986).

BIOLOGICAL DEGRADATION OF XENOBIOTICS

The above analysis focuses on the biological removal of gross organic components, such as BOD or COD. Of particular concern, however, in the context of hazardous waste treatment is the ability of biological processes to reduce the concentration of particular organic materials. Since these are generally of man-made origin, these are referred to as "xenobiotics."

The removal mechanisms for these materials may be complex. Biological degradation may occur using the material as a primary substrate (i.e., deriving energy from the material as a primary source), or as a secondary substrate (deriving energy primarily from other compounds while simultaneously degrading the xenobiotic). The xenobiotic may merely be adsorbed to biomass and thus "removed" from any supernate when physical separation occurs. Highly volatile compounds may be removed by stripping into the gas phase above a biological reactor. In this section, biological degradation will be discussed. In a later section the physical removal mechanisms (adsorption, stripping) that occur in biological systems will be considered.

Numerous studies of individual compound degradation processes have been conducted. These have generally been of two types. In batch tests, cultures of microorganisms (which may or may not have been acclimated) are exposed to a waste containing xenobiotic, and the loss of material from the system is monitored over time. This may or may not include the processes of adsorption (depending upon the analytical methodology) or volatilization. The study reported by Tabak et al. (1981) is typical of this approach. In this work, the degradability of 96 "priority pollutants" was assessed. In a somewhat more sophisticated version of this technique, the rate of pollutant disappearance in batch systems may be used to compute an apparent second-order biodegradation constant, k_b, based on the following model (Paris and Wolfe, 1987):

$$r_s = k_b X S_x \tag{6.76}$$

where S_x is the concentration of xenobiotic.

In the second type of study, of a more complex, and thus less frequently conducted nature, a continuous system is fed waste and the resulting removal is measured. This type of study can be performed either in laboratory, pilot, or even full-scale installations. Typical of this approach is the study of Petrasek et al. (1983) who also investigated the removal of priority pollutants in activated sludge systems.

Based upon these types of studies, the following qualitative rules of thumb for biodegradation have been formulated (Lyman et al., 1982):

1. Highly branched compounds are more resistant to biodegradation.
2. Short chain substituents are less rapidly degraded than long chain substituents.
3. Highly oxidized materials, such as halogenated or oxygen-rich materials, may resist further oxidation under aerobic conditions (in the presence of oxygen) but may be more rapidly degraded under anaerobic conditions.
4. More highly polar compounds tend to be more biodegradable than less polar compounds.
5. Unsaturated aliphatic compounds are more readily biodegraded than saturated aliphatic compounds.
6. Alcohols, aldehydes, acids, esters, amides, and amino acids are more susceptible to biological decomposition than their analogous alkanes, olefins, ketones, dicarboxylic acids, nitriles, and chloroalkanes.
7. Increased substitution on aliphatic compounds impedes biodegradation.
8. Increased halogenation impedes biodegradation.
9. Methyl, chloro, nitro, and amino substituents on aromatic rings inhibit biodegradation of the ring.

This approach has been made somewhat more formalistic in recent work in which semiquantitative estimates of aerobic degradation rate have been correlated with molecular properties. In particular, use of molecular connectivity indices coupled with degree of chlorination, presence of hetrocyclic nitrogen, and certain other functional groups produced a model which explained over 90% in the variability of degradation rate (Boethling and Sabijic, 1989).

While few quantitative predictive models of biodegradation kinetics exist, work in this field is rapidly developing and promises to assist in the *a priori* design process. As an example, Paris and Wolfe (1987) found good quantitative correlation between the apparent second-order degradation (aerobic) rate of a series of anilines and the Van Der Waals' radius of these compounds. However, although there have been proposals for the establishment of a comprehensive, uniform database on biodegradation rates (Grady, 1986), the absence of such a resource hinders development of such models. Similarly, there is only limited work on the appropriate kinetic model to be used for assessing the degradation of xenobiotics which are not the primary substrate for microbial growth (Schmidt et al., 1985) and thus no general conclusions can be drawn.

GENETIC ALTERATIONS, ACCLIMATION, AND SELECTION

It is widely accepted that, by proper adaptation of microbial populations, it may be possible to improve the degree of degradation of specific types of xenobiotic compounds. However, there is some suggestion of a tradeoff to acclimation. Grady (1986), based on his review of the literature, postulates that microbial populations that become insensitive to the effects of high concentrations of a toxic, but biodegradable, xenobiotic may also have an increased value for the K_s of that compound—thereby leading to higher residual concentrations for a given cell residence time.

The mechanisms leading to population adaptation to degradation of xenobiotics are complex. For some compounds, acclimation of a microbial population may involve synthesis of either a greater copy number of the relevant enzyme systems, or induction of such systems. In some cases, when multiple populations are present, acclimation may involve a selection for competent strains which may only exist at low concentrations. Such acclimation can be accomplished in the field for hazardous waste applications (e.g., Ying et al., 1984). For acclimation to occur, a high concentration of microorganisms and target compound are desirable, but the compound concentrations should be maintained below an inhibitory level (Kim and Maier, 1986). In continuous systems, however, if the xenobiotic is allowed to fall to low concentrations for a period of time, competency of a biological population to degrade the material can be lost until another acclimation phase occurs; this process can be analyzed using simple kinetic models for the loss and growth of the acclimated population (Nieuwstad and van t'Hof, 1986).

The process of acclimation, in many cases, appears to be determined by the inheritance of a competent plasmid. Plasmids are mobile units of DNA which can be transferred among bacteria, often across species lines. A number of such degradative plasmids are known (Hardman, 1987). It has been proposed that, by the use of recombinant DNA technology, the degradative ability of microbial populations can be enhanced (Olson and Goldstein, 1988). However, four principal factors appear to limit this approach (Hardman, 1987):

1. The multistep nature of degradative pathways.
2. The novel nature of the degradative organisms.
3. The limited knowledge of individual catabolic pathways.
4. Considerations about the release of engineered organisms into the environment.

Thus, while the potential for the use of these organisms in treatment of xenobiotics certainly exists, much basic scientific investigation needs to be performed to realize this potential (Johnston and Robinson, 1984; Hardman, 1987).

VOLATILIZATION AND PHYSICAL REMOVAL MECHANISMS

During biological treatment, and particularly during activated sludge treatment, many compounds can be removed from the supernate waste by processes other than biodegradation. In particular, the processes of adsorption (to biological solids) and volatilization can be of major importance.

For metals, whose volatility can generally be neglected, the primary removal mechanism available in biological treatment is adsorption to solids. Removal of metals is quite dependent upon the metal itself, and likely the chemical constituents of the waste. However, from 20–90% removals are generally noted (Petrasek and Kugelman, 1983).

In the case of organic materials, the processes of biodegradation, adsorption, and volatilization of specific compounds can be viewed as competitors. It has been found, at least in laboratory systems for a limited series of compounds, that the overall removal rate for organics can be described as the summation of the biodegradation rate, per se (i.e., r_s), the stripping rate, and the adsorption rate. The stripping rate can be developed from standard gas–liquid mass transfer considerations, and the adsorption rate can be computed by assuming equilirium adsorption (perhaps with a partition coefficient estimated from the octanol–water partition coefficient) (Blackburn et al., 1984). In general, compounds with Henry's law constants in excess of 0.001 m^3-atm/mole readily volatilize in activated sludge treatment (Bishop and Jaworski, 1986), to an extent such that 90% of the influent may exit in the off-gas.

With increasing concerns for toxic air pollutants, it may be necessary to consider either enclosure of a biological treatment system with containment and treatment of exhaust gases. Alternatively, when anaerobic processes are suited for degradation, their inherent enclosed nature may make them more advantageous.

Similarly, if toxic contaminants are removed primarily via an adsorption mechanism, the subsequent options for disposal of the sludge may be limited. It may be required, for example, in extreme situations, to incinerate this material. In such circumstances, biological treatment might be regarded as pretreatment for incineration.

COUPLED PHYSICAL AND BIOLOGICAL PROCESSES

It is possible to enhance the efficiency of a biological process by the addition of adsorbent material. The most successful type of coupled biological-physical process is the addition of powdered activated carbon to activated sludge. This is patented under the name of the PACT process.

The addition of carbon to an activated sludge reactor enhances the removal of both biodegradable and nonbiodegradable organic material, and also apparently increases the settleability of the sludge (DeWalle et al., 1977). Similar improvements can also be found when carbon is added to a SBR process. In a pilot test, using hazardous landfill leachate, Ying et al. (1987) found that a PAC-SBR process was more cost-effective than use of an SBR followed by granular carbon adsorption—both for the removal of total organic carbon and for the removal of specific organic compounds, including chlorinated organic hydrocarbons.

There is limited information on the rational design of PACT processes. In particular, the available data suggests that at least for gross organic material, it may not be possible to predict the rate of adsorption from other data taken in non-biological systems (Garcia-Orozco et al., 1986). Thus, for design of this process, laboratory or pilot scale data are required.

PROBLEMS

6.1. A solution containing 500 mg/L butanol ($C_4H_{10}O$), 40 mg/L ammonium nitrogen, and no inorganic carbon is oxidized biologically. At the end of the experiment, 25 mg/L of soluble COD and 5 mg/L of ammonium nitrogen remain. During the process, 400 mg/L of oxygen was consumed and 75 mg/L of inorganic carbon evolved as gaseous CO_2. From these data, deduce a reasonable stoichiometric equation and propose an empirical cell composition consistent with the results.

6.2. In the absence of oxygen, acetate can be anaerobically degraded by either methanogens (organisms using carbon dioxide as an electron acceptor and producing methane) or by sulfate reducers (organisms producing hydrogen sulfide from sulfate). Estimate the maximum cell yield from acetate using both of these organisms—which has a higher cell yield?

6.3. The following data were taken on a CSTR with recycle treating a wastewater with an influent biodegradable COD of 500 mg/L.

Θ d	Θ_c d	X mg/L	S mg/L
0.2	3	1549.9	17.9
0.2	5	1442.4	13.9
0.2	7	2451.7	7.4
0.2	14	5684.2	6.5
0.3	20	4182.1	3.0
0.3	30	8790.9	3.2

Determine the parameters for the Monod model from these data. If it is desired to design a CSTR with recycle to treat this wastewater to a final effluent COD (soluble) concentration of 10 mg/L, determine the Θ_c which should be used and estimate the uncertainty in this estimate.

6.4. The Monod constants for a particular waste stream have been determined to be the following:

$$q_{max} = 2.5\ d^{-1} \qquad K_s = 75 \text{ mg/L COD}$$

$$Y = 0.37 \text{ mg VSS/mg COD} \qquad b = 0.03\ d^{-1}$$

Compute the following:

a) the minimum cell residence time in a CSTR with recycle

b) the minimum cell residence time to reduce the effluent soluble (residual) COD to 15 mg/L—you may assume that the COD is completely degradable

c) if each mg of MLVSS is equivalent to 1.42 mg/L COD, and if the clarifier is 99% efficient (i.e., one percent of the applied solids pass through to the effluent), what is the value of the cell residence time which minimizes the total effluent COD (residual plus solids)?

6.5. Consider the waste stream with kinetic coefficients given in problem 6.4. If a plug flow process with recycle is to be used, compute the required sludge age under the following assumptions:

- ideal sedimentation

- recycle ratio = 0.2
- hydraulic residence time = 4 hours

(Note: Use of a spreadsheet will be required for this problem)

6.6. Consider the aerobic RBC process from Example 6.6. It was found that a single stage required an extremely large number of disks. Compute the required number of disks in each stage, if a first stage is to reduce the COD from 300 mg/L to 100 mg/L and a second stage is to reduce the 100 mg/L to 10 mg/L. Assume a liquid residence time of 30 minutes in each stage, with all other conditions as described in Example 6-6.

6.7. Consider the anaerobic filter process from Example 6.6. If the influent COD is increased to 1,000 mg/L, how does this affect the depth of the filter required to achieve the same effluent concentration (10 mg/L). How does your finding compare with the effect that a similar change in influent would have on the performance of a CSTR with recycle?

REFERENCES

BABCOCK, R. W., RO, K. S., HSIEH, C. C., AND STENSTROM, M. K. "Development of an Off-Line Enricher-Reactor Process for Activated Sludge Degradation of Hazardous Wastes." *Water Environment Research* 64(6): 782–791 (1992).

BATTLEY, E. H. *Energetics of Microbial Growth.* John Wiley and Sons, New York (1987).

BENEFIELD, L. D., AND RANDALL, C. W. *Biological Process Design for Wastewater Treatment.* Prentice Hall, Englewood Cliffs, N.J. (1980).

BISHOP, D. F., AND JAWORSKI, N. A. "Biological Treatment of Toxics in Wastewater: The Problems and Opportunities." Presented at the International Conference on Innovative Biological Treatment of Toxic Wastewaters, Arlington, Va. (1986).

BLACKBURN, J. W., TROXLER, W. L., AND SAYLER, G. S. "Prediction of the Fates of Organic Chemicals in a Biological Treatment Process—An Overview." *Environmental Progress* 3, 163–175 (1984).

BOETHLING, R. S., AND SABIJIC, A. "Screening Level Model for Aerobic Biodegradability Based on a Survey of Expert Knowledge," *Environmental Science and Technology* 23, 672–679 (1989).

BROCK, T. D. *Biology of Microorganisms,* 3rd edition. Prentice Hall, Englewood Cliffs, N.J. (1979).

CORNISH-BOWDEN, ATHEL. *Fundamentals of Enzyme Kinetics.* Butterworths, London (1979).

DAIGGER, G. T., AND GRADY, C. P. L., JR. "The Dynamics of Microbial Growth on Soluble Substrate. A Unifying Theory." *Water Research* 16, 4, 365–382 (1982).

———."An Evaluation of Transfer Function Models for the Transient Growth Response of Microbial Cultures." *Water Research* 17, 11, 1661–1667 (1983).

DEWALLE, F. B., CHIAN, E. S. K., AND SMALL, E. M. "Organic Matter Removal by Powdered Activated Carbon Added to Activated Sludge." *Journal of the Water Pollution Control Federation* 49, 593–599 (1977).

DOMACH, M. M., LEUNG, S. K., CAHN, R. E., COCKS, G. G., AND SHULER, M. L. "Computer Model for Glucose-Limited Growth of a Single Cell of *Escherichia coli* B/r-A." *Biotechnology and Bioengineering* 26, 203–216 (1984).

FRIEDMAN, A. A., ET AL "Effect of Disk Rotational Speed on Biological Contactor Efficiency." *Journal of the Water Pollution Control Federation* 51, 11, 2678–2690 (1979).

GARCIA-OROZCO, J. H., FUENTES, H. R., AND ECKENFELDER, W. W., JR. "Modeling and Performance of the Activated Sludge-Powdered Activated Carbon Process in the Presence of 4,6-dinitro-o-cresol." *Journal of the Water Pollution Control Federation* 58, 4, 320–325 (1987).

GAUDY, A. F., JR. ET AL. "Control of Growth Rate by Initial Substrate Concentration at Values Below Maximum Rate." *Applied Microbiology* 22, 6, 1041–1047 (1971).

GAUDY, A. F., JR. AND GAUDY, E. T. *Microbiology for Environmental Engineers and Scientists.* McGraw-Hill, New York (1980).

GRADY, C. P. L., JR. "Biodegradation of Hazardous Wastes by Conventional Biological Treatment." *Hazardous Waste and Hazardous Materials* 3(4):333–365 (1986).

GRADY, C. P. L., JR., AND LIM, H. C. *Biological Wastewater Treatment.* Marcel–Dekker, New York (1980).

HAAS, C. N. "Biological Process Diffusional Limitations." *Journal of the Environmental Engineering Division.* ASCE, 107, 269–273 (1981).

HARDMAN, D. J. "Microbial Control of Environmental Pollution: The Use of Genetic Techniques to Engineer Organisms with Novel Catalytic Capabilities." Chapter 7 in *Environmental Biotechnology.* C. F. Foster and D. A. J. Wase [eds.]. Ellis Horwood, Chichester, UK (1987).

HOOVER, S. R. AND PORGES, N. "Assimilation of Dairy Wastes by Activated Sludge." *Sewage and Industrial Wastes* 24, 306–312 (1952).

IRVINE, R. L., AND BUSCH, A. W. "Sequencing Batch Biological Reactors—An Overview." *Journal of the Water Pollution Control Federation* 51, 2, 235–243 (1979)

JOHNSTON, J. B., AND ROBINSON, S. G. *Genetic Engineering and New Pollution Control Technologies.* Noyes Publications, Park Ridge, N. J. (1984).

KIM, C. J., AND MAIER, W. J. "Acclimation and Biodegradation of Chlorinated Organic Compounds in the Presence of Alternate Substrates." *Journal of the Water Pollution Control Federation* 58, 2, 157–164 (1986).

LAUCH, R. P., HERRMANN, J. G., MAHAFFEY, W. R., JONES, A. B., DOSANI, M., AND HESSLING, J. "Removal of Creosote from Soil by Bioslurry Reactors." *Environmental Progress* 11(4): 265–271 (1992).

LAWRENCE, A. W., AND P. L. MCCARTY. "Unified Basis for Biological Treatment Design and Operation." *Journal of the Sanitary Engineering Division, ASCE* 96, 757–778 (1970).

LIN, S. D. "A Close Look at Changes of BOD_5 in an RBC System." *Journal of the Water Pollution Control Federation* 58, 7, 757–763 (1986).

LYMAN, W. J., REEHL, W. F., AND ROSENBLATT, D. H. *Handbook of Chemical Property Estimation Methods: Environmental Behavior of Organic Compounds.* McGraw-Hill, New York (1982).

MCCARTY, P. L. "Energetics of Organic Matter Degradation." Chapter 5 in *Water Pollution Microbiology.* R. Mitchell [ed.]. Wiley-Interscience, New York (1972).

METCALF AND EDDY. *Wastewater Engineering.* McGraw-Hill, New York (1979).

NIEUWSTAD, T. J., AND VAN'T HOF, O. "Modeling Activated Sludge Nitrilotriacetic Acid Accumulation." *Journal of the Water Pollution Control Federation* 58, 10, 1000–1004 (1986).

O'BRIEN, W. J. "Limiting Factors in Phytoplankton Algae: Their Meaning and Measurement." *Science* 178, 616–617 (1972).

OLSON, B. H., AND GOLDSTEIN, R.A. "Applying Genetic Ecology to Environmental Management." *Environmental Science and Technology* 22, 4, 370–372 (1988).

OPATKEN, E. L. "An Alternative RBC Design—Second Order Kinetics." *Environmental Progress* 5, 1, 51–56 (1986).

PARIS, D. F., AND WOLFE, N. L. "Relationship Between Properties of a Series of Anilines and Their Transformation by Bacteria." *Applied and Environmental Microbiology* 53, 5, 911–916 (1987).

PARKER, D. S., AND MERRILL, D. T. "Effect of Plastic Media Configuration on Trickling Filter Performance." *Journal of the Water Pollution Control Federation* 56, 8, 955–961 (1984).

PARKIN, G. F., AND OWEN, W. F. "Fundamentals of Anaerobic Digestion of Wastewater Sludges." *Journal of Environmental Engineering* 112, 5, 867–920 (1986).

PATTERSON, J. W. "Modes of Toxicity Analysis in Activated Sludge." Preprints of the Division of Water, Air and Waste Chemistry of the American Chemical Society (1971).

PESCOD, M. B. ET AL. "Treatment of Strong Organic Wastewater Using Aerobic and Anaerobic Packed-Cage RBC's." *Proceedings of the Second Annual Conference on Fixed Film Biological Processes,* vol. 2, pp. 1315–1335 (1984).

PETRASEK, A. C., AND KUGELMAN, I. J. "Metals Removals and Partitioning in Conventional Wastewater Treatment Plants." *Journal of the Water Pollution Control Federation* 55, 9, 1183–1190 (1983).

PETRASEK, A. C., KUGELMAN, I. J., AUSTERN, B. M., PRESSLEY, T. A., WINSLOW, L. A., AND WISE, R. H. "Fate of Toxic Organic Compounds in Wastewater Treatment Plants." *Journal of the Water Pollution Control Federation* 55, 10, 1286–1296 (1983).

PIRT, S. J. *Principles of Microbe and Cell Cultivation.* John Wiley and Sons, New York (1975).

RICH, L. G. *Unit Operations of Sanitary Engineering.* John Wiley, New York (1961).

ROBINSON, J. A., AND TIEDJE, J. M. "Nonlinear Estimation of Monod Growth Kinetic Parameters from a Single Substrate Depletion Curve," *Applied and Environmental Microbiology* 45, 5, 1453–1458 (1983).

ROZICH, A. F., AND GAUDY, A. F., JR. "Critical Point Analysis for Toxic Waste Treatment." *Journal of Environmental Engineering* 110, 3, 562–572 (1984).

ROZICH, A. F., ET AL. "Selection of Growth Rate Model for Activated Sludges Treating Phenol." *Water Research* 19, 481–490 (1985).

SCHMIDT, S. K., SIMKINS, S., AND ALEXANDER, M. "Models for the Kinetics of Biodegradation of Organic Compounds not Supporting Growth." *Applied and Environmental Microbiology* 50, 2, 323–331 (1985).

SCHROEDER, E. D. *Water and Wastewater Treatment.* McGraw-Hill, New York (1977).

SCHROEDER, E. D., AND TCHOBANOGLOUS, G. "Mass Transfer Limitations on Trickling Filter Design." *Journal of the Water Pollution Control Federation* 48, 4, 771–775 (1976).

SPEECE, R. E. "Anaerobic Biotechnology for Industrial Wastewater Treatment." *Environmental Science and Technology* 17, 416A–427A (1983).

SUIDAN, M. T. "Performance of Deep Biofilm Reactors." *Journal of Environmental Engineering* 112, 1, 78–93 (1986).

SUIDAN, M. T., ET AL. "Anaerobic Carbon Filter for Degradation of Phenols." *Journal of the Environmental Engineering Division, ASCE* 107, 563–579 (1981).

SUIDAN, M. T., AND HUANG, Y. "Unified Analysis of Biofilm Kinetics." *Journal of Environmental Engineering* 111, 5, 634–636 (1985).

Suidan, M. T., Najm, I. M., Pfeffer, J. T., and Wang, Y. T. "Anaerobic Biodegradation of Phenol: Inhibition Kinetics and System Stability." *Journal of Environmental Engineering* 114, 6, 1359–1376 (1988).

Switzenbaum, M. S., and Jewell, W. J. "Anaerobic Attached-Film Expanded Bed Reactor Treatment." *Journal of the Water Pollution Control Federation* 52, 7, 1953–1965 (1980).

Tabak, H. H., Quave, S. A., Mashni, C. I., and Barth, E. F. "Biodegradability Studies With Organic Priority Pollutant Compounds." *Journal of the Water Pollution Control Federation* 53, 10, 1503–1518 (1981).

U.S. EPA *Process Design Manual: Sludge Treatment and Disposal.* U.S. EPA 625/1–79–011 (1979).

Williamson, K., and McCarty, P. L. "A Model of Substrate Utilization by Bacterial Films." *Journal of the Water Pollution Control Federation* 48, 1, 9–24 (1976).

Ying, W. et al. "Biological Treatment of a Landfill Leachate in Sequencing Batch Reactors." Presented at the 57th Annual Meeting of the Water Pollution Control Federation (1984).

Ying, W., Bonk, R. R., and Sojka, S. A. *Environmental Progress* 6, 1, 1–8 (1987).

Young, J. C., and McCarty, P. L. "The Anaerobic Filter for Waste Treatment." *Journal of the Water Pollution Control Federation* 41, 5, R160–R173 (1969).

7

Thermal Processes

Organic materials may be converted to inorganic matter, or changed in form, by high-temperature processes in the presence of an oxidizing agent. Such thermal processes can result in the partial or complete reduction in the degree of hazard of a material. The general class of such thermal processes is termed incineration. It has been estimated that 62 million metric tons, or approximately 25%, of the total hazardous waste generated per year in the United States is incinerable; about half of this is liquid and half is sludge or solid material (Oppelt, 1987).

Regulations for the operation of hazardous waste incinerators specify the control of certain pollutant emissions in the stack gas. These are as follows (40 C.F.R. Part 261 Subpart O):

1. For each principal organic hazardous constituent (POHC) in the waste stream, there must be at least a 99.99% ("four nines") destruction and removal efficiency (DRE) in the incinerator and the associated air pollution control devices.
2. At least 99% of the hydrogen chloride must be removed if the emissions are more than 1.8 kg/hr.
3. Particulate emissions must not exceed 180 mg/standard m^3 corrected to 7% oxygen in the stack gas.
4. Wastes containing chlorinated dioxins, chlorinated dibenzofurans and chlorinated phenols (RCRA codes F020–F028) require a 99.9999% DRE of these compounds.

In addition, for wastes containing polychlorinated biphenyls (PCB's), further regulations have been promulgated under the Toxic Substances Control Act (TSCA) (40 C.F.R. Part 761).

In incineration processes, the destruction of organic compounds is a function of the temperature attained during combustion, the duration (time) over which such high temperature is attained, and the turbulence (mixing) that occurs during combustion. Since these properties are determined, in large part, by the gross conditions of combustion, a central issue is estimation of the overall energy and mass balances that occur as a result of incineration. More specific computations of organic destruction require knowledge of compound specific kinetics.

ENERGETICS AND STOICHIOMETRY OF COMBUSTION PROCESSES

In an incineration process, the waste feed is co-combusted (either by premixing or by separate introduction) with fuel and air. Excess air may be used to ensure sufficient oxidation of trace compounds. During the course of combustion, there is a production of heated gases. Figure 7-1 provides a process schematic of incineration. Based on consideration of this schematic, overall mass and energy balances may be constructed.

Development of a Combustion Equation

As a result of combustion, based on elements present in the fuel, the waste, or the air, the following transformations in Table 7-1 will result if combustion is complete. The product distribution of hydrogen and the halogens (Cl, F) is ambiguous and dependent upon the exact composition of the feed : fuel mixture. If, in the combined feed there is an excess of hydrogen over halides, then the hydrogen halides (HCl, HF) will be produced, but if there is insufficient hydrogen then the elemental forms (Cl_2 or F_2) will be produced (Senkan et al., 1981). Note that even if no nitrogen is in the fuel : waste mixture, it will nevertheless be present in the air that is used as an oxidizer.

This information can be summarized in the form of a balanced stoichiometric equation based only on the ash-free component of the fuel : waste mixture, as well as

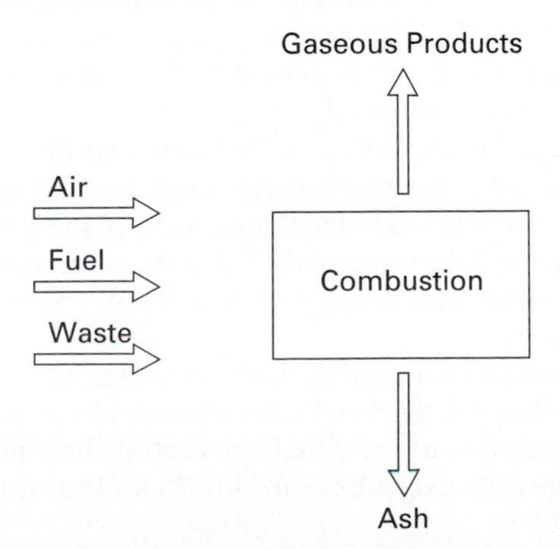

Figure 7-1 Schematic diagram of incineration processes.

TABLE 7-1 TRANSFORMATIONS RESULTING FROM COMPLETE COMBUSTION. (Modified from Lee and Huffman, 1989.)

Feed elements	Products
H	HC1, HF, H_2O
C	C
C1	HC1 or $C1_2$
F	HF or F_2
S	SO_2
N	N_2
Alkali metals (Na, K)	hydroxides
Non-alkali metals (e.g., Cu, Fe)	oxides

the following assumptions:

- Cl_2, SO_3, and NO_x formation is neglected
- oxygen in the waste-fuel mixture is chemically available as a oxidizer (and not, for example, bound up as an inorganic oxide)
- air composition is 79 mole % nitrogen and 21 mole % oxygen
- combustion is complete

With these assumptions, the following can be written (Theodore and Reynolds, 1987):

$$\begin{aligned} C_zH_yO_xCl_wS_vN_u &+ r(z+\phi+v-x/2)O_2 + r(79/21) \\ &(z+\phi+v-x/2)N_2 \longrightarrow z\,CO_2 + 2\phi\,H_2O + w\,HCl + v\,SO_2 + [u/2 \\ &\quad + r(79/21)(z+\phi+v-x/2)]N_2 + (r-1)(z+\phi+v-x/2)O_2 \\ &\text{with } \phi = (y-w)/4 \quad \text{if } y > w \\ &\qquad = 0 \quad \text{otherwise} \end{aligned} \tag{7.1}$$

In Equation 7.1, r must be greater than or equal to 1.0. If $r = 1$, the combustion is termed stoichiometric and exactly the amount of oxygen is supplied that is required for combustion. The term $(r - 1)$ represents the excess air that is used—usually expressed as a percentage. Excess air is generally used in incineration to ensure complete combustion of organic materials.

Knowing the elemental composition of the waste and fuel, the amount of gas produced per unit of waste fed can be computed as a function of excess air. From this information, and from a specification of the temperature of the system, the volumetric flow of gaseous products can be estimated. However, to do this, the temperature must first be estimated—thus, the overall computation of volumetric flows will be postponed to a later section.

To utilize this approach to computing stoichiometries, the elemental composition of the waste must be known, along with the elemental composition of possible fuels. Three fuels are in common use: residual (#6) fuel oil, distillate (#2) fuel oil, and natural gas. Table 7-2 presents average compositions and heating values for these fuels.

TABLE 7-2 AVERAGE PROPERTIES OF SUPPLEMENTAL FUELS. (Modified from Theodore and Reynolds, 1987 and Perry and Chilton, 1973.)

	Mole fraction on ash-free basis		
Element	Residual (#6) fuel oil	Distillate (#2) fuel oil	Natural gas
C	0.412	0.371	0.199
H	0.583	0.628	0.782
N	0.000	0.000	0.020
S	0.005	0.001	0.000
Net Heating Value (kJ/kg)	45,200	43,100	49,900

In the computation of combustion stoichiometries, and particularly for regulatory purposes, it is often necessary to correct gas-phase composition either for the presence of moisture, or for the presence of excess oxygen (Lee and Huffman, 1989a). The correction proceeds in two parts. Let x_x be the mole fraction (or volume fraction) of a particular gaseous product in a mixed gas stream, and let x and x_{H_2O} be the mole fractions of oxygen and water vapor in the gas, respectively. Then the corrected mole fraction composition of the component in the dry gas (x_x') is given by:

$$x_x' = x_x/(1 - x_{H_2O}) \tag{7.2}$$

If it is desired to correct the composition of either the dry gas or the original mixture to one in which the oxygen concentration had been set at a different value, a multiplication by a correction factor is necessary. If χ is the oxygen mole fraction to which the correction is to be made then the corrected composition (x_x''—wet gas) is given as:

$$x_x'' = x_x(0.21 - \chi)/(0.21 - x_{O_2}) \tag{7.3}$$

If it is desired to correct for both moisture and oxygen composition, then first the dry gas compositions should be found via Equation 7.2, including that of oxygen (x'_{O_2}). Then the values of x_x' and x'_{O_2} should be used in Equation 7.3 instead of x_x and x_{O_2}, respectively.

Example 7-1

A given waste is analyzed and found to have the following composition (ash-free mole fraction basis): C—0.5, H—0.2, N—0.05, S—0.01, O—0.15, and Cl—0.09. If this is to be incinerated using #6 fuel oil at a 1:1 weight ratio, and a 50% excess oxygen mixture, develop the stoichiometric combustion equation and compute the composition of the gaseous phase products.

Solution

From Table 7-2, using the composition of #6 fuel oil, one mole of the species $C_{0.412}H_{0.583}S_{0.005}$ would have a mass of 5.687 grams (note atomic weights are C = 12, H = 1, N = 14, S = 32, O = 16, Cl = 35.45). Similarly, one mole of the waste ($C_{0.5}H_{0.2}N_{0.05}S_{0.01}O_{0.15}Cl_{0.09}$) has a mass of 12.8105 grams. Therefore, if one mole of waste with a mass of 12.8105 grams was combusted with 12.8105 grams of fuel, it would be combusted with 2.2526 moles of fuel. First, the total number of moles in a mixture of

12.8105 grams of fuel and 12.8105 grams of waste is computed, then the mole fraction is obtained. Computations are illustrated in the following table. Also in this table, the coefficients are identified according to the scheme in Equation 7-1.

	Mixture	
	Moles	Mole fraction
C	1.428068	$0.439055 = z$
H	1.513262	$0.465247 = y$
N	0.05	$0.015372 = x$
S	0.021262	$0.006537 = w$
O	0.15	$0.046117 = v$
Cl	0.09	$0.027670 = u$

Since 50% excess air is specified, $r = 1.5$. Also, by definition, $\phi = 0.114677$ (i.e., $(y - w)/4$). Now, by substituting of the stoichiometric coefficients into Equation 7.1, the moles of each of the reactants and products can be determined. The results are given in the following table:

	Moles	
Reactants		
Waste + fuel	1	
O_2	0.888245	
N_2	3.341495	
Products		Mole percent
CO_2	0.439055	10.06
H_2O	0.229355	5.25
HCl	0.006537	0.0015
SO_2	0.046117	1.05
N_2	3.349182	76.71
O_2	0.296081	6.78

Finally, the mole percents of the products can be calculated by dividing the moles of a product produced (for one mole of the waste + fuel mixture) by the total moles of products produced.

Incineration Enthalpy Balances

During the combustion reaction, enthalpy is released by the actual chemical process itself—that is, combustion is an exothermic process. Incoming air and fuel are heated and discharged at an elevated temperature. In addition, there may be some loss of heat from the combustion box to the surroundings via losses through the walls. The temperature that is sustained during a combustion reaction in which a continuous feed of both fuel and air is kept results from balancing the heat release from the reaction against the heat losses due to transport of heated gases from the system and losses through the walls. This may be written in verbal and mathematical terms as follows (Bonner et al., 1981) which results from use of the first law of thermodynamics for continuous systems at constant pressure (Zemansky, Abbott, and Van Ness, 1975):

(rate of heat lost through refractory)
= (rate of enthalpy introduced in incoming feed streams)
+ (rate of enthalpy released by combustion)
− (rate of enthalpy exiting by combustion gases)

$$q^* = m_f H_f + m_{air} H_{air} - m_g H_g + m_f(\Delta H_c) \tag{7.4}$$

where q^* is the rate of heat lost through the walls (energy/time), and m and H are the mass flows (mass/time) and specific enthalpies (energy/mass) of various constituents (subscripted f—fuel/waste mixture, air—combustion and excess air, g—effluent gas mixture), and ΔH_c is the specific enthalpy of combustion (energy/mass of fuel-waste mixture). Heats of formation of fuel mixtures (even without being mixed with wastes) are not readily available, since these represent chemical mixtures. However, this enthalpy balance may nevertheless be completed by use of the "net heating value" (NHV) for the fuel-waste mixture. The NHV is defined as the heat (enthalpy) released by the combustion process with all reactants used and all products produced in their standard state (except that water is released as the vapor). This implicitly incorporates the heat of formation and the enthalpy of formation of products and reactants at their standard states. The term δH is the additional enthalpy required to raise the temperature from the standard state to temperature T ("correction enthalpy"). With these changes, Equation 7.4 can be rewritten as:

$$q^* = m_{air}\delta H_{air} - m_g \delta H_g + m_f(\text{NHV}) \tag{7.5}$$

To compute the enthalpies of the input and output streams it is necessary to know the input and output temperatures. The input temperature may be that of the surroundings (e.g., ambient air), or the input streams may be preheated by contact with the hot exhaust gases. The output temperature is generally assumed to be equal to that of the temperature during combustion itself. The enthalpy of any constituent at a given temperature may be regarded as the sum of its enthalpy under standard conditions (25°C, 1 atm) and a correction for temperature change due to the integration of the heat capacity over the temperature difference. For many gaseous compounds, this temperature correction may be written in the following form (Theodore and Reynolds, 1987):

$$\delta H = a(T_1 - T_0) + (b/2)(T_1^2 - T_0^2) - c[(1/T_1) - (1/T_0)] \tag{7.6}$$

where T_0 is the standard state temperature, and T_1 is the temperature to which the correction is required—both in absolute, or kelvin, units. Table 7-3 summarizes the values of a, b, and c in this equation, along with standard-state specific enthalpies for a variety of major gaseous compounds likely to be present in the feed or exhaust gas of combustion systems. Using this table, the enthalpy correction of a gaseous constituent at a given temperature may be computed. Also given in the table are free energies

TABLE 7-3 SPECIFIC ENTHALPIES, FREE ENERGIES AND TEMPERATURE CORRECTION FACTORS FOR VARIOUS COMPOUNDS. (Adopted from Theodore and Reynolds, 1987; Perry and Chilton, 1973.)

	kJ/kg				
Compound	a	$b \times 1000$	c	H_{std}	G_{std}
Gaseous Compounds					
Ammonia	1.750	1.4767	−9106.4	−2697.4	−960.6
Carbon dioxide	1.005	0.1997	−1046.0	−8943.5	−8773.1
Nitrogen	1.021	0.1345	−1793.1	0.0	0.0
Oxygen	0.936	0.1308	−6537.5	0.0	0.0
Sulfur dioxide	0.722	0.1229	−12029.0	−4637.7	−4686.1
Water	1.697	0.5718	0.0	−13434.8	−12699.6
Hydrogen chloride	0.720	0.1423	3443.6	−2532.6	−2614.6
Chlorine	0.522	0.0094	−4012.9	0.0	0.0
Carbon monoxide	1.015	0.1464	−1643.7	−3947.3	−4902.5
Nitrous oxide	1.038	0.1959	−19398.5	219.6	2370.9
Nitric oxide	1.131	0.1480	−2252.9	1294.0	3329.5
Nitrogen dioxide	0.916	0.2074	−15189.7	1772.7	1115.6
Nitrogen tetroxide	0.912	0.4320	−16190.3	−12292.8	1064.0
Liquids					
Chloroform	0.312	0.9528	−4697.6	−848.4	−574.2
n-butyl chloride	0.146	3.8763	−15025.3	−1593.0	−419.5
Chlorobenzene	−0.042	3.5582	−15776.0	461.0	881.8
Hexachlorobenzene	0.316	1.1998	−6069.5	−119.0	155.2

of formation for these compounds. The use of G_{std} will be discussed in a later section. It should be noted that the heat capacity terms in Table 7-3 pertain to all compounds in their gaseous state. Hence, if the enthalpy of formation of a gaseous compound that is present as a liquid is desired, it is also necessary to have data on the enthalpy of vaporization of that compound.

The net healing value of the fuel : waste mixture may be computed by addition of the net heating values of the reactants. If w is the ratio of mass of fuel per mass of waste combusted, then the overall NHV (per unit mass of *waste* burned) is obtained from the net heating values of the individual constituents by mass averaging:

$$\text{NHV} = (w\, \text{NHV}_{\text{fuel}} + \text{NHV}_{\text{waste}}) \tag{7.7}$$

The net heating value per mass of *mixture* burned is given by the above divided by $(1 + w)$. NHV_{fuel} can be obtained directly from Table 7-2. The net heating value of the waste ($\text{NHV}_{\text{waste}}$) may be obtained either from experimental determination, from knowledge of the chemical composition of the material (and component addition), or from use of the empirical Dulong correlation which is (Theodore and Reynolds, 1987):

$$\text{NHV}_{\text{waste}} = 32{,}463\, m_c + 104{,}346(m_H - m_O/8) - 1762\, m_{Cl} + 10{,}435\, m_S \tag{7.8}$$

with NHV in kJ/kg of waste, and where the m's are the mass fractions of the indicated element in the waste.

Example 7-2

Consider the waste discussed in Example 7-1, with an empirical composition of $C_{0.5}H_{0.2}N_{0.05}S_{0.01}O_{0.15}Cl_{0.09}$. This is to be co-combusted with #6 fuel oil. What ratio of fuel to waste should be used such that the net heating value of the mixture is 25,000 kJ/kg?

Solution

First, NHV_{waste} must be computed using the Dulong equation. The formula weight of the waste is 12.8105 g/mole as noted above. For each element, its contribution to the formula weight divided by the formula weight is the mass fraction. Thus, for the components in the Dulong equation, one obtains:

$$m_C = 0.468365$$

$$m_H = 0.015612$$

$$m_O = 0.187346$$

$$m_{Cl} = 0.249053$$

$$m_S = 0.024979$$

Therefore, from Equation 7.8:

$$\begin{aligned} NHV_{waste} &= 32{,}463 \times 0.468365 + 104{,}346\,(0.015612 - 0.187346/8) \\ &\quad - 1762 \times 0.249053 + 10{,}435 \times 0.024979 \\ &= 14{,}212 \text{ kJ/kg} \end{aligned}$$

Now, from Table 7-2, NHV_{fuel} for #6 oil is 45,200 kJ/kg. By the problem conditions, and application of Equation 7.7, the desired quantity is w:

$$25{,}000 = (45{,}200\, w + 14{,}212)/(1 + w)$$

$$w = 0.534$$

Thus, for every kg of waste, 0.534 kg of fuel must be burned to produce the desired net heating value of the mixture.

If a fuel : waste ratio is specified along with the fraction of excess air that is to be used, the relative masses of reactants and products are fixed by the stoichiometric combustion Equation 7.1. The NHV is computed using the procedure outlined in Example 7-2. The enthalpies for the reactant and product gases may each be expressed as functions of temperature using the information in Table 7-3, or analogous data. Providing that the rate of heat loss from the combustion process by transfer through the walls is known, the temperature in the process can be obtained by solving the resulting equation.

Heat loss through insulating materials is generally described by use of thermal diffusivities for solid materials. The rate of heat transport through such materials is given by the product of the area through which transfer occurs, an overall heat transfer coefficient, and the temperature differential through the surface. It is beyond the scope of this chapter to elaborate on such computations, which are relatively straightforward (Perry and Chilton, 1973). However, empirical experience suggests that, in incinerators, the rate of heat loss by such processes is no more than 5% of the heat produced by combustion (Bonner et al., 1981). Thus, if η is the fraction of such loss, then the equation can be modified by setting $q^* = \eta m_f(\text{NHV})$, and therefore:

$$m_{\text{air}}\delta H_{\text{air}} - m_g \delta H_g + m_f(\text{NHV})(1 - \eta) = 0 \tag{7.9}$$

If $\eta = 0$, the process occurs without heat loss, and is termed "adiabatic." The temperature obtained during adiabatic combustion sets an upper limit to the temperature that can be obtained at a given ratio of fuel and air to waste. It is also apparent, from this equation, that the ratio of flows, and not their absolute values, govern the temperature attained, provided that the fraction of heat loss is set (this might not be true, however, since heat loss is proportional to a temperature difference). The solution to this equation is illustrated by Example 7-3.

Example 7-3

Consider again the waste described in Examples 7-1 and 7-2. This is to be combusted using a fuel ratio (#6 fuel oil) of 0.534 kg fuel per kg waste and 100% excess air. Compute the temperature expected if the process were to occur (a) adiabatically or (b) with a 5% heat loss. The inlet air is preheated to 75°C (348.14 K).

Solution

The conditions of 0.534 kg fuel/kg waste represent a molar ratio of 1.203 : 1 fuel : waste. The overall elemental composition is computed by direct addition, as in the following table. The stoichiometric coefficients are as indicated. From this information, Φ is computed (for use in Equation 7.1) as 0.100466.

From Equation 7.1, applying the composition of the fuel-waste mixture, and at an excess air ratio of 1.0 (r = 2), the following stoichiometric coefficients are computed, and by application of molecular weights, the resulting mass required or produced per mass of waste. Also, in this table, the total masses of reactants and products are computed—this latter procedure provides a check on the computation, since there must be a conservation of mass during the combustion process.

	Mixture	
	Moles	Mole Fraction
C	0.995588	0.451947 = z
H	0.901281	0.409137 = y
N	0.05	0.022697 = x
S	0.016014	0.007269 = w
O	0.15	0.068092 = v
Cl	0.09	0.040855 = u

Mixture molecular weight = 8.921 g/mol

Component	Moles	Mass
Reactants		
Mixture	1	8.921
O_2	1.218	38.986
N_2	4.583	128.329
Products		
CO_2	0.452	19.886
H_2O	0.201	3.617
HCl	0.0073	0.258
SO_2	0.0681	4.358
N_2	4.595	128.647
O_2	0.609	19.493
Mass bal		
Reactants	176.236	
Products	176.258	

The net heating value (NHV) of the fuel:waste mixture is computed first by observing, from Table 7-2 that the NHV for #6 fuel oil is 45,200 kJ/kg. Using the waste composition and the Dulong correlation, the NHV for waste is 14,212 kJ/kg. Therefore the NHV for the mixture can be expressed as either 24,999 kJ/kg of mixture (i.e., (0.534 × 45,200 + 14,212)/1.534) or 38,349 kJ/kg of waste processed. The other reactants are oxygen and nitrogen. Using the enthalpy data at the inlet temperature (Table 7-3), the enthalpies of oxygen and nitrogen are, respectively, 71.15 and 81.40 kJ/kg. Multiplying each of these by the mass reactant used per mass of mixture combusted, it is found that the contribution of the gases to the input enthalpies is 26,031 kJ/kg of mixture. The sum of the product enthalpies at the exit (combustor) temperature must equal this input enthalpy.

The product enthalpies are functions of the exit temperature. Let this be T_1, and let $T_0 = 298.14$ (i.e., the "standard" temperature in Equation 7.6). Then each of the coefficients of the excess enthalpy function (Equation 7.6) are multiplied by the mass of the product per mass of mixture combusted, and the results are summed. The process is illustrated in the following table. Based upon this, the combustion temperature is the root (T_1) of the following equation:

$$26{,}031 = 13.973(T_1 - T_0) + 0.00103\,(T_1^2 - T_0^2) - 33{,}601.6(1/T_1 - 1/T_0)$$

This equation can be solved by trial and error or by iteration (e.g., Newton–Raphson). By such methods, the root is found to be 1895.3 K or 2951.9°F. This should be compared to the heat tolerance of the refractory materials used in the combustor. The following table lists tolerance temperatures for many common refractory materials.

	Multiplier		
Products	$(T_1 - T_0)$	$(T_1^2 - T_0^2)$	$(1/T_1 - T_0)$
CO_2	1.560	0.00015	−1623.7
H_2O	0.479	0.00008	0.0
HCl	0.014	0.00000	69.3
SO_2	0.246	0.00002	−4092.1
N_2	10.249	0.00068	−18007.3
O_2	1.425	0.00010	−9947.8

For the situation in which a 5% loss due to heat transfer through the walls is assumed, the computations proceed identically as above except that the input enthalpy must be reduced by 5%. In other words, the input enthalpy is $0.95 \times 26{,}031$, or 24,729 kJ/kg. Therefore, the following equation must be solved for the unknown T_1:

$$24{,}729 = 13.973(T_1 - T_0) + 0.00103\,(T_1^2 - T_0^2) - 33601.6(1/T_1 - 1/T_0)$$

The root of this equation is 1822.3 K or 2820.5°F, which is 73 K lower than the adiabatic case—it is anticipated that the loss of heat reduces the combustion temperature.

This computation can also be repeated at different fuel:waste ratios and excess air ratios. The results from such computations are summarized in Figure 7-2. It can be seen that, in general, as excess air increases, the temperature decreases, and as the amount of fuel increases, the temperature increases.

Once the combustion temperature is known, for any particular flow rate of feed, the combustor volume required to achieve a given residence time can be computed by assuming the ideal gas law for the product gases. The required residence time to achieve a given degree of organic destruction can be computed from kinetic considerations noted later. The computation of residence times is illustrated in the following example problem.

Example 7-4

For the waste stream considered in Example 7-3, one metric ton per day of material is to be combusted. If the 5% heat loss is assumed, compute the combustor volume required to give a residence time of three seconds in the incinerator.

Solution

If one metric ton/d (1,000 kg/d) of waste are combusted, the feed of the fuel: waste mixture will be 1534 kg/d. The products will be 5.93 moles per mole of the mixture

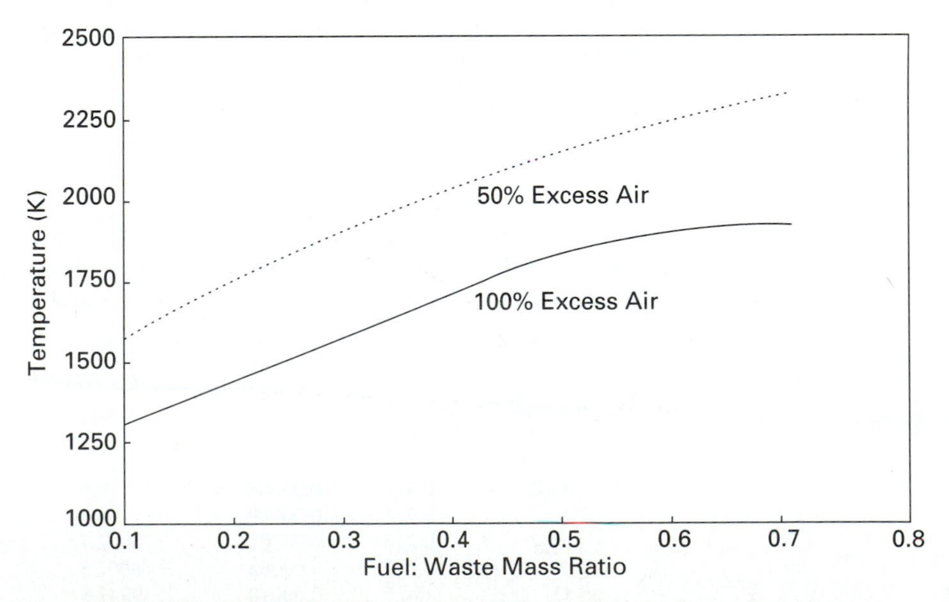

Figure 7-2 Relationship between fuel, excess air, and combustion temperature.

combusted. The formula weight of the mixture is 8.9207 g/mole. Therefore the moles per day of fuel: waste mixture combusted is given by:

$$10^6 \text{ g/d} / 8.9207 \text{ g/mole} = 1.12 \times 10^5 \text{ moles/day}$$

Hence, the reaction products have a molar flow rate of $5.93 \times 1.12 \times 10^5$ moles/day = 6.64×10^5 moles/day. This can also be expressed as 7.69 moles/s. The required combustor volume must contain 3 seconds of this flow—or 23.07 moles:

The ideal gas law, rearranged for volume, is as follows:

$$V = nRT/P$$

The pressure within the combustor is assumed to be at one atmosphere. From Example 7-3, the temperature at these conditions is 1895.3 K. Therefore the required volume is:

$$\begin{aligned} V &= 23.07 \text{ moles } (0.082 \text{ L-atm/mol-K}) (1895.3 \text{ K})/1 \text{ atm} \\ &= 3585 \text{ L} \\ &= 3.585 \text{ m}^3 \end{aligned}$$

If a cubical chamber were used, this would be 1.53 m on each side; however it is desirable, for fluid flow characteristics, not to use such a system (discussed below).

EQUILIBRIA AND KINETICS OF COMBUSTION REACTIONS

The preconditions for the stoichiometric combustion reaction to go completely as written in Equation 7.1 are that the equilibria for the combustion is favorable and that sufficient time is allowed to permit any kinetically limited reactions to go to completion. Hence it is required that the combustion equilibrium and rate constants be estimated.

Equilibrium Constant Estimation

From fundamental thermodynamic relationships, the equilibrium constant for a reaction (K) is related to temperature by the following equation:

$$d \ln K/dT = \Delta H^\circ / RT^2 \tag{7.10}$$

where ΔH° is the net enthalpy of the reaction and R is the ideal gas constant. However, the quantity ΔH° is temperature dependent. If $H_{\text{std},i}$ and $\delta H_i(T)$ are the standard enthalpies of formation of reactant or product i and the enthalpy correction for temperature of the reactant or product (to a temperature T), then the value of the net enthalpy of reaction is given by:

$$\Delta H^\circ(T) = \sum_i z_i MW_i [H_{\text{std},i} + \delta H_i(T) + \Delta H_{\text{vap},i}] \tag{7.11}$$

where z_i is the stoichiometric coefficient of the reactant or product (with a positive sign for products and a negative sign for reactants), MW_i is the molecular weight of that component, and $\Delta H_{\text{vap,i}}$ is the enthalpy of vaporization of component i (if the standard state is a liquid; if the standard state is a gas then this enthalpy is zero).

Equation 7.10 can be integrated from standard conditions (25°C) to some particular temperature at which the equilibrium constant is sought (which would generally be that of the incinerator) yielding the following equation:

$$\ln K = \frac{1}{R}\int \frac{\Delta H^o(T)}{T^2}\,dT \tag{7.12}$$

The constant of integration in this equation is evaluated as the value of ln K under standard conditions, which can be obtained from the free energies of formation:

$$-RT\ln K(25°\text{C}) = \sum_i z_i MW_i G_{\text{std},i} \tag{7.13}$$

If the functional form of the enthalpy versus temperature function (Equation 7.6) is substituted into Equation 7.11, and the result is integrated according to Equation 7.12, the following results:

$$\ln K(T) = \frac{1}{R}\left\{ \sum z_i MW_i \left[-\left(\frac{1}{T} - \frac{1}{T_0}\right)\left(H_{\text{std},i} + H_{\text{vap},i} - a_i T_0 - \frac{b_i T_0^2}{2} + \frac{c_i}{T_0}\right) + a_i \ln\left(\frac{T}{T_0}\right) + \frac{b_i}{2} + 2c_i\left(\frac{1}{T^2} - \frac{1}{T_0^2}\right)\right]\right\} + \ln K(25°\text{C}) \tag{7.14}$$

This equilibrium constant can now be related to the composition of the gas using a mass action expression. If P_i is the partial pressure (in atmospheres) of component i in the gaseous mixture (which has a signed stoichiometric coefficient of z_i), then if gas fugacity coefficients are equal to unity, the following may be written:

$$K = \prod_i P_i z_i \tag{7.15}$$

By setting this product equal to the numerical value of the equilibrium constant computed in Equation 7.14, the composition at equilibrium can be computed to ascertain whether the desired composition is achieved.

Example 7-5

Chloroform is to be combusted along with a supplemental fuel at a temperature of 1,200°C. If the final gas has an oxygen partial pressure of 0.1 atm, and a water vapor partial pressure of 0.001 atm, and if the products are carbon dioxide, water vapor, and hydrogen chloride, compute the residual concentration if the initial partial pressure of chloroform (after vaporization at the incinerator nozzle) is 0.01 atm. The enthalpy of vaporization of chloroform at 25°C may be estimated as 100 kJ/kg. The free energy of vaporization of chloroform may be assumed equal to the enthalpy of vaporization.

Solution

First the stoichiometric combustion reaction for chloroform must be computed. If hydrogen chloride (rather than Cl_2) is the product, then water must appear as a reactant (it is supplied from the combustion of a supplemental fuel). Then, the following balanced reaction may be written (molecular weights are written below each component):

$$\begin{array}{ccccccccc} CHCl_3 & + & H_2O & + & \frac{1}{2}O_2 & \rightarrow & 3\ HCl & + & CO_2 \\ 119.35 & & 18 & & 32 & & 36.45 & & 44 \end{array}$$

From the stoichiometric equation, the value of K(25°C) can be computed via Equation 7.13 and the data in Table 7-3 as follows (note that the term for chloroform is

$-574.2 + 100$ due to the vaporization):

$$
\begin{aligned}
-RT \ln K(25°C) &= -1(119.35)(-474.2) - 1(18)(-12{,}699.6) \\
&\quad -(1/2)(32)(0) + 3(36.45)(-2614.6) + 1(44)(-8773.1) \\
&= -443{,}330 \text{ kJ/kg-mol} \\
&= -443.3 \text{ kJ/mol} \\
\ln K(25°C) &= 443{,}330/(298.14 \times 8.314 \text{ J/mol-k}) \\
&= 178.85
\end{aligned}
$$

Note that this value is quite high, indicating that, even at standard conditions, the oxidation of chloroform (and, in general, most organic compounds) is highly favored thermodynamically.

The enthalpy correction terms are computed in the following:

$$
z_i MW_i[-(1/T - 1/T_0)(H_{\text{std},i} + H_{\text{vap},i} - a_i T_0 - b_i T_0^2/2 + c_i/T_0) + a_i \ln(T/T_0) + b_i/2 + 2c_i(1/T^2 - 1/T_0^2)]
$$

$CHCl_3$ $\quad -1(119.35)[-(-2.675 \times 10^{-3})(-848.4 + 100 - 0.312(298.14) - 0.0009528(298.14^2)/2 - 4697.6/298.14) + 0.312 \ln(1200/298.14) + 0.00953/2 + 2(-4697.6)(-1.056 \times 10^{-5})]$
$= 223.43$ kJ/kg-mol K

H_2O $\quad -1(18)[-(-2.675 \times 10^{-3})(-13434.8 - 1.697(298.14) - 0.0005718(298.14^2)/2 - 0/298.14) + 1.697 \ln(1200/298.14) + 0.0005718/2 + 2(0)(-1.056 \times 10^{-5})]$
$= 629.93$ kJ/kg-mol K

O_2 $\quad -(1/2)(32)[-(-2.675 \times 10^{-3})(-0 - 0.936(298.14) - 0.0001308(298.14^2)/2 - 6537.5/298.14) + 0.936 \ln(1200/298.14) + 0.0001308/2 + 2(-6537.5)(-1.056 \times 10^{-5})]$
$= -9.933$ kJ/kg-mol K

HCl $\quad 3(36.45)[-(-2.675 \times 10^{-3})(-2532.6 - 0.720(298.14) - 0.0001423(298.14^2)/2 + 3443.6/298.14) + 0.72 \ln(1200/298.14) + 0.0001423/2 + 2(3443.6)(-1.056 \times 10^{-5})]$
$= -498.68$ kJ/kg-mol K

CO_2 $\quad (44)[-(-2.675 \times 10^{-3})(-8943.5 - 1.005(298.14) - 0.0001997(298.14^2)/2 - 1046/298.14) + 1.005 \ln(1200/298.14) + 0.0001997/2 + 2(-1046)(-1.056 \times 10^{-5})]$
$= -910.42$ kJ/kg-mol K

The overall enthalpy correction term is the sum of the above, that is, -565.67 kJ/kg-mol K or -565.67 J/mol-K. Therefore, the equilibrium constant at 1200 K is computed as:

$$
\begin{aligned}
\ln K &= -565.67/8.314 + 178.85 \\
&= 110.8
\end{aligned}
$$

Hence,

$$
K = 1.31 \times 10^{48} = P_{CO_2}(P_{HCl})^3/P_{CHCl_3}P_{H_2O}(P_{O_2})^{1/2}
$$

If x is the residual partial pressure of chloroform, then from the problem conditions, the following are true:

$$P_{CO_2} = 0.01 - x$$
$$P_{HCl} = 3(0.01 - x)$$
$$P_{CHCl_3} = x$$
$$P_{H_2O} = 0.001$$
$$P_{O_2} = 0.01$$

Therefore, the following equation must be solved for x:

$$1.31 \times 10^{48} = 27(0.01 - x)^4/[x(0.001)(0.01)^{1/2}]$$
$$4.85 \times 10^{42} = (0.01 - x)^4/x$$

The result is $x = 2.06 \times 10^{-51}$ atm. This indicates that, if equilibrium were to be attained, virtually all of the feed material would be oxidized. However, in the combustion of organic compounds, kinetic limitations generally govern the degree of conversion. Nonetheless, equilibrium information can be used to provide broad insights into the nature of a combustion process.

Kinetics

Very frequently the size of an incinerator will be controlled by the necessity to provide sufficient residence time for the destruction of organics, and possibly carbon monoxide, to proceed to a satisfactory extent. In order to determine this, it is necessary to have expressions for the rates of the various combustion processes. From a mechanistic point of view, the destruction of organics, particularly chlorinated organics, during incineration occurs by a complex series of reactions that may involve intermediates of greater thermal stability than the parent molecule (Senkan et al., 1981). However, empirical evidence suggests that the degree of destruction of organic materials may be correlated by relatively simple rate expressions.

The most useful correlating framework for destruction kinetics of organic compounds at present appears to be that of Lee et al. (1982). For a plug flow reactor at steady state, or a batch system, the ratio of exit (C) to inlet (C_0) concentrations of a particular organic compound can be given by a first-order reaction law incorporating a delay time as follows:

$$\ln(C/C_0) = -k(\Theta - \Theta_1) \tag{7.16}$$

where k is a compound and temperature specific rate constant, and Θ_1 is a compound and temperature specific lag time. The dependencies of k and Θ_1 on temperature are given by the following relations:

$$\Theta_1 = x_1 - x_2 T \tag{7.17}$$

$$k = A \exp(-E/RT) \tag{7.18}$$

where T is the absolute temperature, and x_1, x_2, and A are compound (but not temperature) specific. Some values of these parameters are shown in Table 7-4. The validity of these rate parameters is at temperatures above the indicated lower limit,

TABLE 7-4 KINETIC CONSTANTS FOR ORGANIC DESTRUCTION. (Adapted from Theodore and Reynolds, 1987.)

Compound	A (s^{-1})	E(kJ/mol)	x_1 (s)	x_2 (s/K)	Lower temp (K)
Acrolein	3.3×10^{10}	150.31	0	0	700
Benzene	7.4×10^{21}	401.51	2.8425	0.000989	964
Chlorobenzene	1.3×10^{17}	320.71	1.3335	0.000444	1005
Methyl chloride	7.3×10^{8}	171.24	1.6372	0.000467	1089
Toluene	2.3×10^{13}	236.55	1.4808	0.000512	964
Vinyl chloride	3.6×10^{14}	265.02	0	0	950

but below about 1,250°K. At higher temperatures other reaction mechanisms are believed to be operable.

Given a rate constant, the degree of destruction in a given incinerator can be estimated assuming plug flow performance. An illustration of this is given below.

Example 7-6

In a given waste stream, benzene is one of the POHC's to be destroyed (to the required "four nines" level). If a temperature of 1,050 K is to be used, compute the required residence time under plug flow conditions.

Solution

For benzene, from Table 7-4, we have $A = 7.4 \times 10^{21} s^{-1}$, $x_1 = 2.8425$ s, and $x_2 = 0.000989$ s/K. From Equation 7.16, the lag time is computed:

$$\Theta_1 = 2.8425 - 0.000989\,(1050) = 1.804 \text{ s}$$

From Equation 7.15, the rate constant is computed:

$$k = 7.4 \times 10^{21} \exp(-401{,}510/(1050 \times 8.314)) = 78.4 \text{ s}^{-1}$$

Therefore, to achieve the required destruction the following must be true:

$$\ln(0.0001/1) = -78.4\,(\Theta - 1.804)$$

$$\Theta = 1.92 \text{ s}$$

From the analysis of the kinetics of destruction according to this formulation, it has been found possible to correlate the time required for destruction of an organic material to the residual time, and to other molecular properties (Cooper and Alley, 1986; Theodore and Reynolds, 1987). The temperatures (in °F) required for 99, 99.9 and 99.99% destruction are given by the equations of the following form:

$$T_{\text{DRE}} = \sum a_{i,\text{DRE}} W_i \tag{7.19}$$

where W_i is a property of the molecule or the incinerator, and $a_{i,\text{DRE}}$ is a numerical coefficient that is a function of the destruction efficiency and the property—these coefficients, along with overall estimates of the error of the regression, are given in Table 7-5. This correlation may be used to select compounds which are most likely to be the hardest to decompose in a given incinerator. On this basis, POHC's likely to be limiting for the design of the incinerator can be selected.

TABLE 7-5 CORRELATIONS FOR DESTRUCTION EFFICIENCY. (After Cooper and Alley, 1986; Theodore and Reynolds, 1987.)

		Coefficient of W_i											Standard
Destruction	Constant	1	2	3	4	5	6	7	8	9	10	11	error (°F)
99%	577	−10.0	110.2	67.1	72.6	0.586	−23.4	−430.9	85.2	−82.2	65.5	−76.1	19.8
99.9%	594	−12.2	117.0	71.6	80.2	0.592	−20.2	−420.3	87.1	−66.8	62.8	−75.3	20.9
99.99%	605	−13.8	122.5	75.5	85.6	0.597	−17.9	−412.0	89.0	−55.3	60.7	−75.2	22.1

W_1 = number of carbon atoms in molecule
W_2 = aromatic (0 – no; 1 – yes)
W_3 = presence of carbon-carbon double bonds (0 – no; 1 – yes)
W_4 = number of nitrogen atoms in molecule
W_5 = autoignition temperature of organic
W_6 = number of oxygen atoms in molecule
W_7 = number of sulfur atoms in molecule
W_8 = hydrogen/carbon mole ratio
W_9 = allyl compound (0 – no; 1 – yes)
W_{10} = chlorine bond to a carbon which is double bonded to another carbon (0 – no; 1 – yes)
W_{11} = ln (residence time in seconds)

Turbulent Conditions during Incineration

The prior computations assume that the flow conditions in the incinerator are plug flow—in other words, that each and every amount of material has a residence time in the system exactly equal to the mean residence time. Any deviation from this assumption will result in some material having a lower residence time, and therefore, a lower degree of destruction, thus diminishing performance. It is therefore important to consider the degree to which a design is capable of achieving plug flow behavior.

The achievement of turbulence in incinerators is generally considered primarily a function of the velocity of the hot gases as they move through the combustion chamber. If the velocity of these gases is in excess of 10 – 15 ft/s, sufficient turbulence to achieve plug flow may be assumed. In the absence of this velocity, mechanical devices (baffles, cyclone inducers) to promote mixing may be necessary (Bonner et al., 1981).

Formation of NO_x, CO, and Acid Gases

As noted from the stoichiometric Equation 7.1, the combustion process will result in the formation of hydrogen chloride, carbon dioxide, and nitrogen in the gaseous phase. However, additional reactions or kinetic limitations may result in the formation of carbon monoxide or nitrogen oxides, which may have unwanted effects. Furthermore, the release of carbon monoxide is often regarded as an indicator of incomplete combustion.

The extent to which carbon monoxide is produced during a given combustion process is difficult to predict. However, for a plug flow system, the investigations of Howard et al. (1973) lead to the following model for carbon monoxide destruction:

$$\ln f = -1.3 \times 10^{14} \exp(-15100/T)\{O_2\}^{1/2}\{H_2O\}^{1/2}\,\Theta \qquad (7.20)$$

where f is the fraction carbon monoxide remaining, Θ is the residence time in seconds, and the braces {} denote concentrations in mol/cm^3. In certain circumstances, the incinerator design (in terms of temperature and excess oxygen) may be controlled by the destruction of carbon monoxide rather than the destruction of organic matter per se. Generally the term combustion efficiency is defined as the ratio of carbon dioxide to the sum of carbon monoxide plus carbon dioxide in the stack gases (Lee and Huffman, 1989a). Continuous monitoring of carbon monoxide in stack gases is generally required under the RCRA incinerator standards; although there is a qualitative relationship between such emissions and DRE, the quantitative correlations are poor (Oppelt, 1987).

Example 7-7

An incinerator is designed to operate at 1,100 K with a residence time of 2 seconds. Based on combustion stoichiometry, the off-gas oxygen and water mole fractions are 0.07 and 0.05, respectively. Compute the destruction efficiency of carbon monoxide in this incinerator.

Solution

Applying the ideal gas law, the molar density of the combustion gases at the chosen temperature is computed by:

$$\begin{aligned}(n/V) &= P/RT \\ &= (1\ \text{atm})/(82.0567\ \text{cm}^3\ -\text{atm/mol-}K)(1{,}100\ K) \\ &= 1.108 \times 10^{-5}\ \text{mol/cm}^3\end{aligned}$$

Therefore the molar concentrations of water vapor and oxygen are given as:

$$\begin{aligned}\{O_2\} &= 0.07\ (1.108 \times 10^{-5}\ \text{mol/cm}^3) \\ &= 7.76 \times 10^{-7}\ \text{mol/cm}^3\end{aligned}$$

$$\begin{aligned}\{H_2O\} &= 0.05\ (1.108 \times 10^{-5}\ \text{mol/cm}^3) \\ &= 5.54 \times 10^{-7}\ \text{mol/cm}^3\end{aligned}$$

Substitution of the information in Equation 7.20 yields:

$$\begin{aligned}\ln f &= -1.3 \times 10^{14} \exp(-15100/1100)[7.76 \times 10^{-7}\ \text{mol/cm}^3]^{1/2} \\ &\quad [5.54 \times 10^{-7}\ \text{mol/cm}^3]^{1/2}\ (2) \\ &= -186\end{aligned}$$

This indicates that the destruction of carbon monoxide would be virtually complete. In practice, any carbon monoxide in the emission would be the result of nonideal flow patterns (e.g., short circuiting).

As indicated by Equation 7.1, sulfur dioxide and hydrogen chloride may be produced during combustion, depending upon the sulfur and chlorine content of the waste : fuel mixture. In addition, depending upon the temperature and mixing conditions, nitrogen oxide gaseous (NO_x) may also be produced.

The formation of nitrogen oxides results from a combination of kinetic and equilibrium factors. There are two distinct routes to formation. In the first, termed fuel NO_x, the nitrogen present in the fuel : waste mixture reacts during combustion to form nitrogen oxides. In the second, termed thermal NO_x, there is a direct gas phase reaction between nitrogen and oxygen resulting in byproduct formation.

The rate of formation of fuel NO_x is still not thoroughly understood. Empirically, however, from 10 to 60% of the nitrogen present in a fuel : waste mixture can be expected to be converted into nitrogen oxides (U.S. EPA, 1983). The formation is dependent on the degree of excess air, and tends to increase as excess air increases, and is also increased with increasing mixing; however, temperature has little influence on fuel NO_x formation (Cooper and Alley, 1986).

Formation of thermal nitrogen oxides is more directly a function of the oxygen and nitrogen concentrations in the highest temperature combustion zone. Temperature increases up to 1,990°C increase nitrogen oxide concentrations; above this temperature decomposition reactions to higher nitrogen oxides result (Cooper and Alley, 1986).

Products of Incomplete Combustion

During the combustion of organic materials, it is possible that other compounds may form and, if combustion is incomplete, may be released into the stack gases. These compounds are generally termed "products of incomplete combustion" (PIC's). Although there is no formal definition of PIC's that are used, a common definition is that these are compounds listed in 40 C.F.R. Part 261 Appendix VIII but not present in the fuel or the waste itself (Oppelt, 1986). While there are no current regulatory requirements on PIC formation, there is continuing concern for the potential health risk associated with PIC emissions.

These compounds may result by several mechanisms (Oppelt, 1987):

- they may be fragments of regulated POHC's or other compounds present in the feed or fuel itself.
- they may be produced by radical–molecule recombination reactions in the combustion zone.
- they may originate from contaminants present in the air stream, or in water used as scrubber water.

In a review of performance of full-scale incinerators, it has been found to be useful to describe organic emissions in stack gases (PIC's and POHC's) using an emission factor of mass production per unit combustion energy input. On this basis, the bulk of organic emission occurs in the 10 to 100 ng/kJ range, and tends to be higher for volatile compounds than for semivolatile compounds (Oppelt, 1987).

The regulatory significance of PIC formation is unclear. U.S. EPA has generally considered that the formation of PIC's is controlled indirectly by the requirement for POHC destruction (Oppelt, 1986). While it has often been proposed that combustion monitoring of either carbon monoxide or total unburned hydrocarbons can function as surrogates for analysis of hazardous organic emissions, field tests of such techniques have been unpromising (Oppelt, 1987).

PHYSICAL CONFIGURATIONS

A variety of physical devices may be used for incineration. These depend on the physical state of the waste to be incinerated (e.g., liquid, solid, bulk material), the nature of the mixing process during combustion, and the system residence time.

Liquid Injection

In liquid injection systems, a variety of techniques can be used to intimately mix combustion air, fuel and waste, and to form finely dispersed droplets. Generally, the maximum viscosity for a fuel: waste mixture is limited to approximately $1-2$ cm^2/s; however, preheating up to 200°C can be used to maintain this value (Bonner et al., 1981). Liquid injection systems are the most frequently used for hazardous waste processing (Oppelt, 1987).

In one type of atomizer, a rotating cup is used to generate a centrifugal spray of droplets (rotating cup atomization). Other types of atomizers rely upon pressurization of only the fuel: waste mixture. In other systems, air (or steam) is used as a secondary fluid, either mixed internally prior to atomization, or fed concurrently and mixed after atomization. In the single fluid and internal mix systems, suspended solids concentrations must be essentially zero. The rotary cup system can handle solids concentrations of up to 20% as long as their size is less than 35 mesh. The external air system, at low pressure, can handle up to 30% solids at up to 100 mesh. Higher pressure air or stream can be used for solids concentrations up to 70% (Bonner et al., 1981).

Multiple injectors are used in incinerators to achieve a greater ability to handle varying flow rates as well as to maintain adequate mixing patterns in the combustion chamber. A single burner is wall-mounted within a somewhat shielded space to control combustion conditions (Figure 7-3). The relative location of burners to each other, and with respect to the combustion chamber is also quite important. Combustion chambers can be classified as axial firing, radial firing, or tangential firing. In axial-fired combustors, burners are located at the ends of the chamber. In radial firing, burners are located at the periphery of the combustor with flames directed inward. While both of these configurations are relatively easy to construct, they are relatively inefficient in their utilization of the combustion volume. The tangentially fired systems involve location of burners along the walls at angles to encourage the formation of a swirl or vortex flow in the combustor, and achieve a greater degree of mixing. In any event, careful location of burners is required to minimize flame impingement on the refractory walls (Theodore and Reynolds, 1987). Typical design features of liquid injection incinerators are given in Table 7-6.

Rotary Kilns

Rotary kiln incineration systems are the second most frequently used for hazardous waste processing (Oppelt, 1987), although they tend to be larger than liquid injection systems (Oppelt, 1987). Originally this system was designed for thermal processing

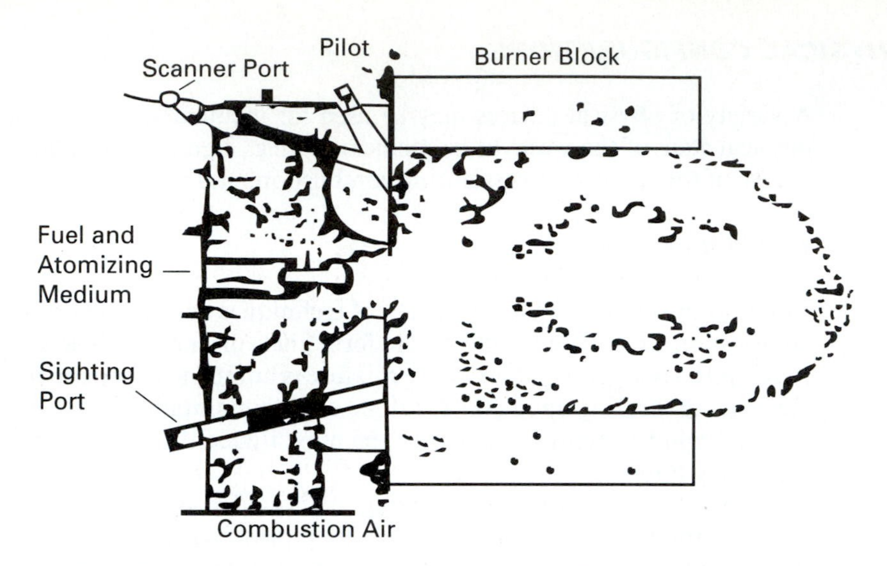

Figure 7-3 Schematic of single burner installation (Theodore and Reynolds). (Reprinted from Theodore and Reynolds (1987), with permission. Copyright John Wiley and Sons, Inc.)

of lime and cement. The central feature of a rotary kiln incinerator is a long cylinder that is slightly inclined from the vertical and rotates at slow speeds (0.2–1 inch/second at the periphery). Solids, sludges, and drummed materials may be fed directly into this system, while nozzles can be used to inject pumpable materials and fuel in a conventional manner. The slow rotation of the kiln allows control of the solids' residence time in the system. As the solids progress through the kiln, volatilization of organics occurs and is succeeded by combustion in the gas phase. Figure 7-4 is a schematic of a rotary kiln incinerator.

The residence time in a rotary kiln incinerator of solid material may be estimated from the following empirical equation of Manson and Unger (1979):

$$\Theta = 0.19\ L/DSN$$

where Θ is the residence time in minutes, L and D are the kiln length and diameter in meters, S is the kiln slope (m/m) and N is the rotational velocity (rpm). Typical ranges are $L/D = 2-10$, and $S = 0.03-0.09$ (Bonner et al., 1981).

In the design of rotary kiln systems, a limiting factor may be the time required for thorough volatilization of solids ("burnout"). Empirical experience has suggested that 0.5 s for propellants, 5 minutes for wooden boxes, 15 minutes for refuse, and 60 minutes for railroad ties are satisfactory (Manson and Unger, 1979). As

TABLE 7-6 DESIGN VARIABLES FOR LIQUID INJECTION INCINERATORS. (After Theodore and Reynolds, 1987.)

Combustion chamber residence time	0.5–2 s
Combustion chamber temperature	700–1650°C
Maximum feed rate (organic wastes)	5.7 m^3/hr

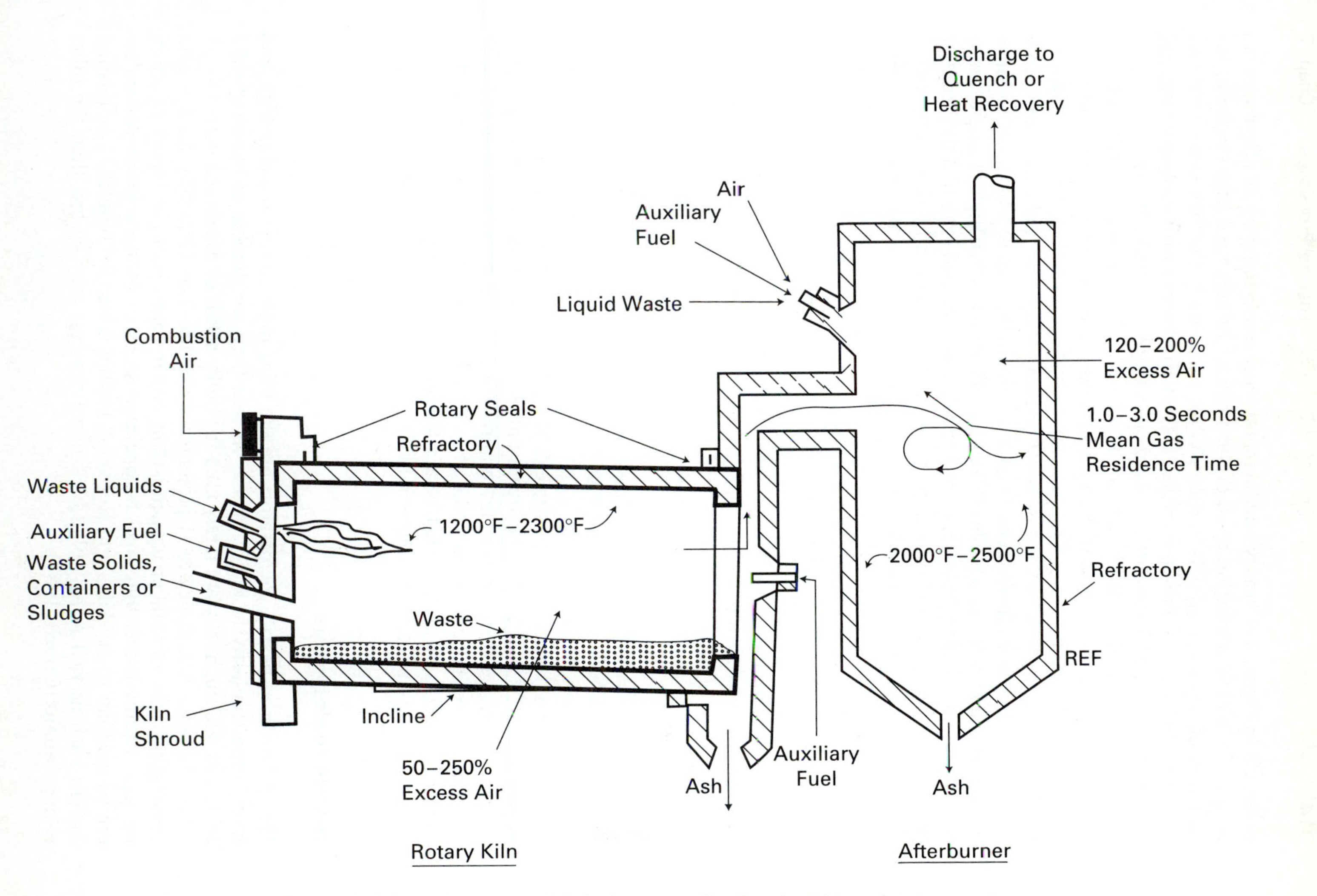

Figure 7-4 Schematic of a rotary kiln incinerator. (After Oppelt, 1987; reprinted by permission of the Air and Waste Management Association.)

indicated in Figure 7-4, rotary kiln incinerators generally employ an afterburner to ensure complete combustion of all evolved gases. To achieve a high degree of destruction of organic materials in rotary kilns, particularly the "six nines" level required for PCB destruction, it is critical that an afterburner with adequate time and temperature be used to treat the evolved gases (Tessitore et al., 1987).

The design and operation of rotary kiln incinerators is complicated by the need to maintain provision for introduction of solid materials in the feed end while minimizing the exhaust of volatilized, uncombusted, hazardous gases. Generally, the kiln itself is maintained at a negative pressure relative to ambient to minimize emissions. However, the feed end must have high temperature seals to allow introduction of batches of solids while preventing off gassing. With highly volatile materials, transient "puffing" may cause leakage through the end seals and thus cause a potential safety and environmental problem (Theodore and Reynolds, 1987). Careful consideration is therefore required in the design of the front-end solids handling systems for rotary kiln incinerators.

Rotary kiln processes are routinely used in the manufacture of a number of bulk mineral products, including cement and lime. The fuel value of hazardous wastes may be beneficially used while obtaining acceptable removal of POHC's and other criteria air pollutants. Mournighan and Branscome (1988) reported on a series of nine tests in full-scale lime or cement kilns. Well over four nines DRE was achieved, and emissions of sulfur oxides, nitrogen oxides, particulates, and HCl were acceptable provided that total chlorine inputs did not exceed 0.6% of the throughput of the kiln.

It is also possible to design and operate small rotary kilns of sufficient portability to permit transport to Superfund sites for on-site cleanup. These systems include secondary combustion chambers to ensure organic destruction and air pollution control devices (scrubbers). Freestone et al. (1987) used such a unit, designed for a heat release of 15 GJ/hr for the treatment of dioxin-containing soils at a site in Missouri. Greater than the six nines required for dioxin removal was achieved, with adequate removal of criteria air pollutants. Based on this work, it was suggested that modular systems of up to 80 GJ/hr are feasible for short-term (1 – 3 year) use at Superfund sites, after which they can be moved to other sites for subsequent use.

Hearth Incinerators

Hearth, or fixed-hearth, incinerators are the third most common technology, and have some similarity to kiln processes in that they are capable of handling solid materials, although they are rarely used for bulky solids or drummed materials. A schematic of a hearth device is presented in Figure 7-5. Generally fixed hearth installations are of smaller capacity than either rotary kiln or liquid injection devices. Furthermore, they may be operated so as to produce lower particulate emissions than rotary kilns and so overall capital costs (for both the combustion system itself as well as air pollution control devices) may be less than the other configurations (Oppelt, 1987).

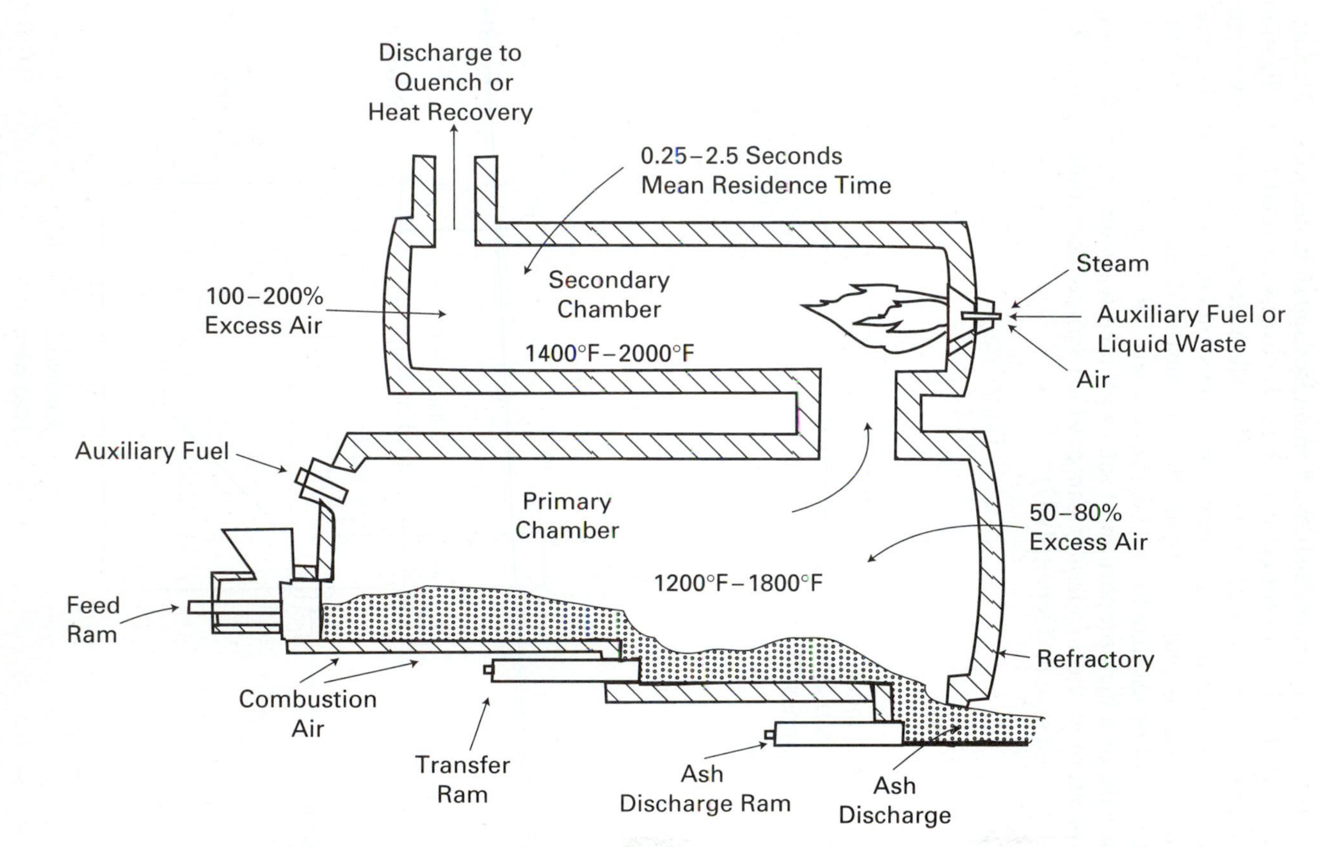

Figure 7-5 Schematic of fixed hearth incinerator. (After Oppelt, 1987; reprinted by permission of the Air and Waste Management Association.)

Fluidized Bed Systems

The last category of incinerator in widespread use is the fluidized bed combustor (Figure 7-6). This device can be used on sludges and shredded solid materials. A bed of sand or alumina is kept circulating by the excess air used for fluidization. The mixing of air, fuel, and waste in the sand bed allows for a control of the residence time and for good heat transfer. Volatile materials are emitted from the bed and combusted in the gas phase overlying the combustor. Generally, solids fed to such a system must be less than two inches in diameter. The presence of finely divided solid materials, or a high proportion of salts with melting points below the operational temperature of the incinerator can produce serious operational problems in the fluidized bed system (Oppelt, 1987).

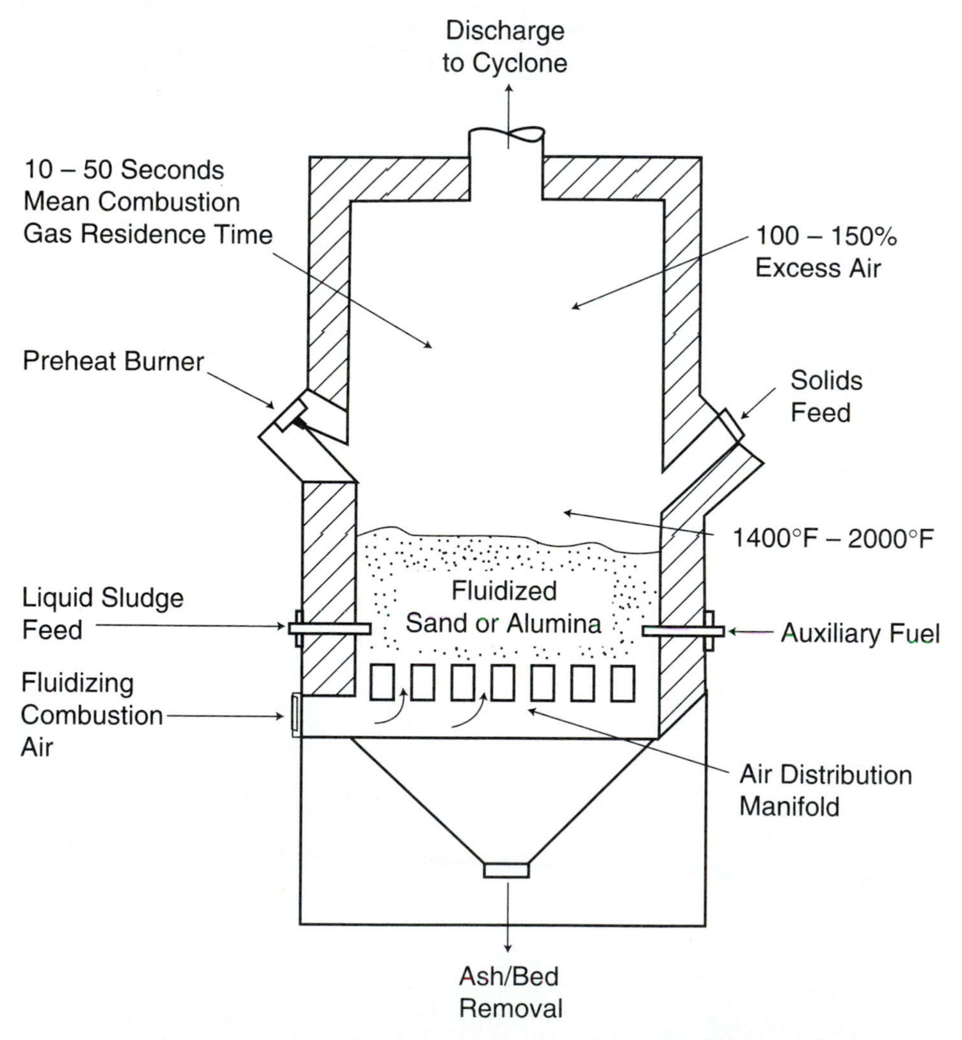

Figure 7-6 Schematic of fluidized bed combustor. (After Oppelt, 1987; reprinted by permission of the Air and Waste Management Association.)

Miscellaneous Systems

A variety of additional incineration systems have been proposed, although none of these has yet achieved the degree of utilization of those previously described. Two of the most promising technologies are the infrared incineration system and the plasma arc furnace.

Infrared incineration is a technology suited for the treatment of soils and particulate matter at Superfund sites, and can be assembled in the field. Liquids can also be incinerated if they are coated onto sand or similar solids. Solids are screened to pass a one-inch screen. The resulting material is placed onto a conveyor at a controlled rate at a depth of 1 – 1 1/2 inches. Above the conveyor, electric infrared heating elements are installed to heat the waste to 870 – 1040°C. Combustion air is introduced along the sides. Rakes are also mounted above the belt to achieve turning. The overall residence time of solids in this primary combustion chamber may be up to 25 minutes. The evolved gases are collected and taken to a secondary combustion chamber operated at a gas residence time of 2 seconds at 1,200°C, with the addition of secondary combustion air and fuel. In tests performed to date, including two demonstrations at Superfund sites, acceptable DRE's and air emissions were obtained (Lee and Huffman, 1989b; Wall, 1989).

Plasma arc furnace technology is at a less advanced stage of demonstration, but holds promise for achieving substantially higher destruction efficiencies than other technologies, and thus are suited for treatment of highly concentrated materials, such as PCB-laden capacitors. A plasma is produced by high energy electric discharge into a gas which converts all materials into their atomic state, thus potentially minimizing byproduct formation. Test burns using PCB oils showed six to eight nines destruction (Lee and Huffman, 1989b). A capacitor destruction facility has been designed for Model City, New York, which would operate at 1.5 – 2.5 MT of material per hour. Plasma arc systems do not require supplemental fuel and only require limited supplemental air; however, materials durability and complexity of operation remain issues of concern (Lee and Huffman, 1989b).

AIR POLLUTION EQUIPMENT

Ordinarily, the thermal destruction process can be relied upon to achieve the DRE required for the organic constituents subject to regulation (POHC's). However, to achieve compliance with both the particulate and the hydrogen chloride emissions standards, air pollution control devices are also necessary. Most facilities use one of the following treatment trains for air pollution control (Theodore and Reynolds, 1987):

- Venturi scrubber (particulate removal) followed by a packed tower absorber (gas removal)
- ionizing wet scrubber (particulate removal) followed by a packed tower
- dry scrubber (gas removal) followed by a baghouse (for particulate removal) or an electrostatic precipitator (particulate removal)

The design and specification of these devices are beyond the scope of this chapter, but they are well described elsewhere (e.g., Cooper and Alley, 1986).

It should be noted that all of these devices result in the formation of a residue. Baghouses and electrostatic precipitators produce a solid material which is removed from the stack ("flyash"). This material, if it results from the combustion of a specifically listed hazardous waste remains hazardous and must be disposed accordingly. If the ash results only from the combustion of wastes hazardous by characteristic, then, depending upon the leaching properties of the waste under the EP or TCLP procedures, it may or may not be hazardous. Furthermore, some ashes from incineration of specific materials are themselves specifically listed wastes under RCRA. For example, ash from treatment of soils contaminated with certain chlorophenolics or chlorobenzenes are hazardous (RCRA code F028).

CONDUCT OF A TRIAL BURN

The lack of availability of inexpensive monitoring equipment for the detection of POHC's prevents the enforcement of a destruction efficiency (DRE) on the basis of self-monitoring, as is the case in many other aspects of pollution control. Instead, prior to the combustion of a waste stream, a test-burn must be performed.

As part of the application for a permit (or an amendment to a permit to burn an additional waste stream), plans for a test burn of waste streams must be drawn up. The objective of this test burn is to demonstrate the range of operating conditions over which acceptable DRE of POHC's can be achieved. If the test burn plan is accepted, then a temporary permit is issued during which time (not to exceed 30 days), the define plan can be executed. Typically, this test burn consists of three phases (Theodore and Reynolds, 1987):

- mechanical shakedown using nonhazardous feed (or previously permitted material)
- trial burn phase itself
- work up and analysis of data

If acceptable performance has been demonstrated during the test-burn then a regular permit for the treatment of the particular waste stream(s) will be granted. The entire process can be time consuming, requiring 30–60 days for completion of all analyses (Theodore and Reynolds, 1987). The cost for preparation of a trial burn plan is typically \$10,000–\$20,000, with \$50,000–\$100,000 being required to conduct the sampling and analysis phases (Freeman, 1989).

Activities to be completed during the trial burn include the following (Bonner et al., 1981):

- site survey
- preparation of sampling and analytical equipment

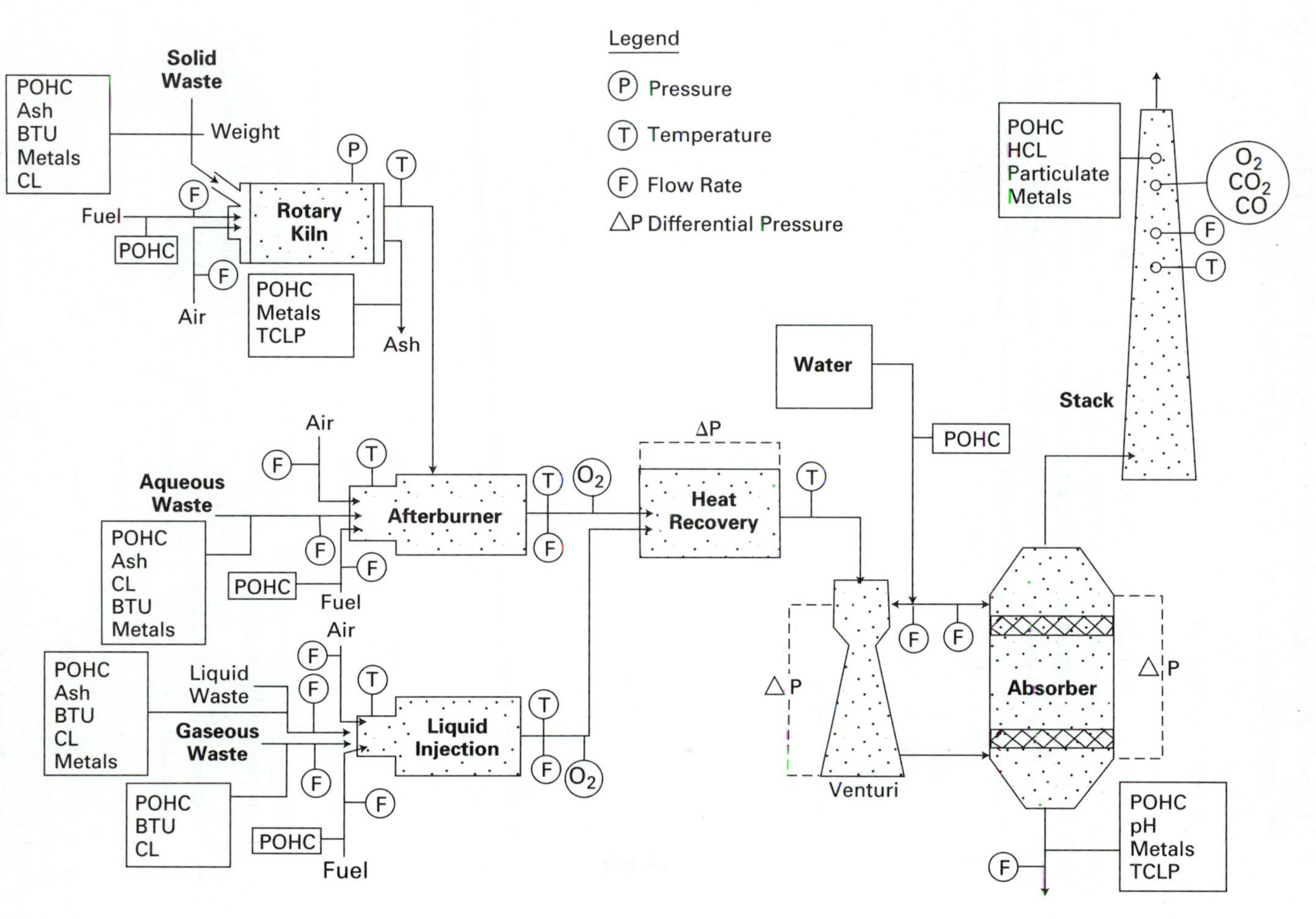

Figure 7-7 Sampling points of potential use in a trial burn. (Reprinted by permission of the Air and Waste Management Association.)

- set up of equipment and takedown
- stack sampling, generally at least three tests with 1 – 1 1/2 hours per test.
- sample analysis, generally at an off site laboratory
- equipment cleanup
- report preparation

There are a number of potential control points that must be monitored to determine performance acceptability. Figure 7-7 illustrates the possibilities. Additionally, the combustion air itself (Lee, 1987) may contain organic materials which need to be accounted for in assessing destruction efficiencies and in designing the sampling protocol. A typical analytical protocol for organic measurements during a trial burn is shown in Table 7-7. In addition, stack gas flows, particulate concentrations, hydrogen chloride concentrations, carbon monoxide, and oxygen concentrations would also be measured (Theodore and Reynolds, 1987). The details of the conduct of the trial burn would be specified in the temporary operating permit. In general, three replicate burns at a given condition are desirable to estimate process and waste variability.

TABLE 7-7 ORGANIC SAMPLING MEASUREMENTS FOR A TRIAL BURN (Oppelt, 1987)

Sample	Frequency for each run	Sampling method	Analysis parameters
Liquid waste feed	Grab sample every 15 minutes	S004	Volatile and semivolatile POHC's, chloride, ash, ultimate analysis, viscosity, high heating value
Solid waste feed	Grab sample from each drum	S006, S007	Volatile and semivolatile POHC's, chloride, ash, high heating value
Chamber ash	Grab sample at completion of a set of replicate runs	S006	Volatile and semivolatile POHC's, TCLP
Stack gas	Composite	MM5 (3 hr)	Semivolatile POHC's, particulates, water vapor, HC1
	Three pairs of traps (40 min/pair)	VOST (2 hr)	Volatile POHC's
	Composite in Tedlar gas bag	S011	Volatile POHC's (only if VOST samples nonquantifiable)
	Composite in Mylar gas bag	M3 (1 – 2 hr)	CO_2 and O_2
	Continuous (3 hr)	Continuous plant monitor	CO

Notes:
VOST—volatile organic sampling train
MM5—EPA modified method 5
M3—EPA method 3
Methods described in U.S. EPA (1982), Gorman et al. (1985), Harris et al. (1984).

OPERATIONS AND CONTROL

The trial burn ensures that, under proper operating conditions, the incinerator as designed can achieve the necessary reductions in organic material, while maintaining satisfactory emission levels. However, due to variabilities in feed and fuel composition, and other factors, it is necessary to have safeguards on the incinerator to ensure the continuous satisfactory operating performance. As noted above, it is not practicable, at present, to monitor individual organic materials in the stack gases on a continuous basis. Therefore, a variety of more indirect measurements are often used. The following types of emissions monitoring currently exist (Oppelt, 1987):

- paramagnetic and electrocatalytic oxygen analyzers
- nondispersive infrared analysis for CO and CO_2
- chemiluminescence for NO_x
- flame photometry, pulsed fluorescence, or nondispersive ultraviolet analysis for SO_2

Despite the availability of this instrumentation, practical experience to date has failed to find a single set of monitoring conditions which can be quantitatively used to

TABLE 7-8 CHARACTERISTICS OF COMMON REFRACTORY MATERIALS
(Modified from Bonner et al. (1981.)

Type	Typical composition	Fusion temperature (°C)	Resistant to	Degraded by
Silica	95% SiO_2	1700	HC1, NH_3, acid slags	Basic slags, A1, Na, Mg, F_2, $C1_2$, H_2 (>1,400°C)
High-duty fireclay	54% SiO_2 40% $A1_2O_3$	1710	Most acids, slag conditions	High-lime slags, other bases at high temperature
Super-duty fireclay	52% SiO_2 42% $A1_2O_3$	1740	HC1, NH_3, most acids	Basic slags, Na, Mg, F_2, $C1_2$, H_2 (>1,400°C)
Acid resistant (Type H)	59% SiO_2	1670	Excellent for most acids; bases in moderate concentration	HF, H_3PO_4
High alumina	50–85% $A1_2O_3$	1760–1870	HC1, NH_3, SO_2	Basic slags, Na, Mg, F_2, H_2 Cl_2, (>1,400°C)
Extra high alumina	90–99% $A1_2O_3$	1650–2010	HC1, HF, NH_3, SO_2, S_2, HNO_3, H_2SO_4, $C1_2$	Na, F_2 (>980°C)
Mullite	71% $A1_2O_3$	1810	HC1, SO_2, NH_3	Na, F_2, $C1_2$, H_2 (>1,400°C)

Note: A safety factor of at least 120°C between the fusion temperature and the operating temperature is advisable.

correlate organic destruction efficiency, although some (Lee, 1987) have suggested that CO monitoring can be used as a conservative indicator of process performance.

In addition, flow monitoring and temperature probes can be used to determine the existence of adequate time and temperature in the incineration system. These can be coupled to the feed systems to take certain precautions if an "off spec" operating condition occurs. For example, in the case of rotary kiln incineration, the following actions might be taken (Theodore and Reynolds, 1987):

- termination of feed and fuel supply if flameout of auxiliary burners occurs
- termination of waste feed if low temperatures are sensed either at the kiln outlet or the afterburner outlet
- reduction in fuel and/or feed if high temperatures are sensed either at the kiln or afterburner outlet (to minimize refractory damage)

One potential operating problem faced by all incineration systems is the resistance of materials in contact with the high temperature gaseous and slag materials. These materials may be sensitive to particular chemical components of the waste. Table 7-8 can be used as a guide to the chemical and thermal resistance of available refractories.

PROBLEMS

7.1. Waste biological solids from the treatment of a hazardous waste have an empirical formula of $C_5H_7O_2NCl_{0.02}S_{0.01}$. This is to be combusted with #6 fuel oil at a 1 : 1 weight ratio and a 50% excess air level. Compute the stoichiometric combustion equation and the composition of the gas phase products.

7.2. Consider the combustion process in problem 7.1. What should be the ratio of fuel oil to waste (on a mass basis) if the mixture is to have a net heating value of 25,000 kJ/kg?

7.3. A combustion process has an air : fuel-waste mixture of 5 : 1 (mass : mass) and a product gas phase composition (in mass percent) as follows:

CO_2	12%
H_2O	2%
HCl	0.1%
SO_2	2.5%
N_2	70%
O_2	13.4%

If the net heating value of the fuel-waste mixture is 20,000 kJ/kg, estimate the combustion temperature under adiabatic and more realistic (5% loss) conditions.

7.4. An incinerator combustion box has a volume of 50 m^3. If the incinerator is to be operated at 1,200°C, what is the maximum molar flow rate of products such that the residence time is at least 2.5 seconds?

7.5. An incinerator combustor is to be lined with silica. If the operating temperature is to be 200°C below the fusion temperature, determine whether four nines destruction can be achieved in three seconds residence time in a plug flow system for the following compounds: acrolein, methyl chloride, toluene, vinyl chloride.

REFERENCES

AMEND, L. J., AND LEDERMAN, P. B. "Critical Evaluation of PCB Remediation Technologies." *Environmental Progress* 11(3): 173–177 (1992).

BONNER, T., DESAI, B., FULLERKAMP, J., HUGHES, T., KENNEDY, E., MCCORMICK, R., PETERS, J., AND ZANDERS, D. *Engineering Handbook for Hazardous Waste Incineration,* prepared under U.S. EPA Contract 68–03–2550 (1980). Republished as *Hazardous Waste Incineration Engineering.* Noyes Data Corporation, Park Ridge N.J. (1981).

CLAUSEN, C. A., COOPER, C. D., HEWETT, M., AND MARTINEZ, A. "Enhancement of Organic Vapor Incineration by Using Ozone." *Journal of Hazardous Materials* 31(1): 75–87.

COOPER, C. D., AND ALLEY, F. C. *Air Pollution Control: A Design Approach.* PWS Publishers, Boston (1986).

FREESTONE, F., HAZEL, R., WILDER, I., BRUGGER, J., YEZZI, J., MILER, R., PFROMMER, C., HELSEL, R., AND ALPERIN, E. "Evaluation of On-Site Incineration for Cleanup of Dioxin-Contaminated Materials." *Nuclear and Chemical Waste Management* 7, 3–20 (1987).

FREEMAN, H. M., *Standard Handbook of Hazardous Waste Treatment and Disposal.* McGraw-Hill, New York (1989).

GORMAN, P., ET AL. "Practical Guide—Trial Burns for Hazardous Waste Incineration." U.S. EPA, NTIS PB 86–190246/AS (1985).

HARRIS, J. C., LARSEN, D. J., RECHSTEINER, C. E., AND THRUN, K. E. "Sampling and Analysis Methods for Hazardous Waste Combustion." U.S. EPA, EPA-600/8–84–002 (1984).

HOWARD, J. B., WILLIAMS, G. E., AND FINE, D. H. "Kinetics of Carbon Monoxide Oxidation in Postflame Gases." 14th Symposium on Combustion, p. 975 (1973).

KOTHANDARAMAN, S., AHLERT, R. C., VENKATARAMANI, E. S., AND ANDREWS, A. T. "Supercritical Extraction of Polynuclear Aromatic Hydrocarbons from Soil." *Environmental Progress* 11(3): 220–222 (1992).

LEE, C. C., AND HUFFMAN, G. L. "Incineration of Solid Waste." *Environmental Progress* 8, 3, 143–151 (1989a).

———. "Innovative Thermal Destruction Technologies." *Environmental Progress* 8, 190–199 (1989b).

LEE, K. C. "Prepared Discussion." *Journal APCA* 37, 9, 1011–1017 (1987).

LEE, K. C., MORGAN, N., HANSON, J. L., AND WHIPPLE, G. "Revised Predictive Model for Thermal Destruction of Dilute Organic Vapors and Some Theoretical Explanation." Air Pollution Control Association Annual Meeting Paper #82–5.3 (1982).

MANSON, L., AND UNGER, S. "Hazardous Material Incinerator Design Criteria." U.S. EPA 600/2–79–198 (1979).

MOURNIGHAN, R. E., AND BRANSCOME, M. "Hazardous Waste Combustion in Industrial Processes: Cement and Lime Kilns." U.S. EPA 600/S2–87–095 (1988).

OPPELT, E. T. "Hazardous Waste Destruction." *Environmental Science and Technology* 20, 312–318 (1986).

———. "Incineration of Hazardous Waste: A Critical Review." *JAPCA* 37, 558–586 (1987).

PERRY, R. H., AND CHILTON, C. H. *Chemical Engineers' Handbook.* McGraw-Hill, New York (1973).

SENKAN, S., ROBINSON, J., BOSE, D., AND GUPTA, A. "Combustion/Incineration Characteristics of Chlorinated Hydrocarbons." Presented at the 1981 Annual Meeting of AICE.

TESSITORE, J. L., CROSS, F., AND MUNOZ, J. H. "PCB Destruction by Incineration." *Pollution Engineering* 19, 8, 70–75 (1987).

THEODORE, L., AND REYNOLDS, J. *Introduction to Hazardous Waste Engineering.* Wiley-Interscience, New York (1987).

U.S. EPA. "Test Methods for Evaluating Solid Waste, Physical/Chemical Methods," 2nd edition, SW–846 (1982).

U.S. EPA. "Control Techniques for Nitrogen Oxide Emissions from Stationary Sources," 2nd ed., EPA–450/3–83–002 (1983).

WALL, H. "Technology Demonstration Summary: Shirco Pilot-Scale Infrared Incineration System at the Rose Township Demode Road Superfund Site." U.S. EPA 540/S5–89–007 (1989).

ZEMANSKY, M. W., ABBOTT, M. M., AND VAN NESS, H. C. *Basic Engineering Thermodynamics,* 2nd edition. McGraw-Hill, New York (1975).

8

Waste Elimination Options

INTRODUCTION

Under the Hazardous and Solid Waste Amendments of 1984, Congress has stated:

> " . . . that it is to be a national policy of the United States that, where feasible, the generation of hazardous waste is to be reduced or eliminated as expenditiously as possible. Waste nonetheless generated should be treated, stored, or disposed of so as to minimize the present and future threat to human health and the environment."

The diminishing number of available disposal facilities, the increasing cost of disposal, and the strict joint, several, and perpetual liability imposed under CERCLA have also acted as incentives for generators to seek to reduce the amount of waste that they generate. Also under the 1984 amendments, generators must certify on each waste manifest that they have a program "to reduce the volume or quantity and toxicity of such waste to the degree determined by the generator to be economically practicable."

EPA has further stated that such a waste minimization program must generally include six elements:

- top management support
- characterization of waste generation
- periodic waste minimization assessments
- cost allocation system to fully load disposal costs back to the production unit
- encourage technology transfer
- program evaluation

The term waste minimization is ordinarily defined as "The reduction, to the extent feasible, or waste that is generated or subsequently treated, stored or disposed of. It includes any source reduction or recycling activity undertaken by a generator that results in either the reduction of the total volume or quantity of waste or the reduction in the toxicity of waste (or both) as long as the reduction is consistent with the goal of minimization of present and future threats to human health and the environment" (Stephan and Atcheson, 1989).

In addition to the requirements under legislation, industry has a number of other incentives to pursue waste minimization activities. These include (National Research Council. 1985):

- high costs of disposal of hazardous wastes
- prospects for liability associated with remedial action at disposal sites
- risks of third party liability
- potential for adverse public relations

There appears to be a widespread perception that substantial waste reduction could be achieved by implementation of best available technology. In a survey of industries, OTA (1986) reported that 15% of respondents believed that best available technology could achieve 50–75% weight reduction and 30% of the respondents believed that such technology could achieve 25–50% reduction.

When the economics of waste reduction are considered, it is often found that the process or raw materials changes that are required may be more than compensated for by reduction in disposal costs, potential liabilities, and regulatory expenses (for example, of being a large versus a small quantity generator). Formal economic analysis of such changes are presented by McHugh (1990).

STRATEGIES FOR MINIMIZATION

The Office of Technology Assessment has reviewed over 300 case studies of waste reduction, and has classified these into five generic categories. These are: improving routine plant operations, altering production technology, recycling waste back into production, changing raw materials, and redesigning or reformulating the product itself (Hirschorn, 1988). The applicability of these techniques is dependent on industry, as shown in Table 8-1. Examples of each of these strategies will be given below. However, in general, a waste minimization audit should be conducted for each individual facility. The audit process will be discussed in a subsequent section.

Improving Plant Operations

Improving the management or housekeeping practices of a manufacturing facility may reduce waste generation in a variety of ways. Examples of such changes include (OTA, 1986):

- improvements in ancillary plant operations such as better predictive and preventive maintenance;

TABLE 8-1 POTENTIAL FOR WASTE REDUCTION BY INDUSTRY TYPE. (OTA, 1986.)

Industry characteristic	Example industries	Operations changes	In-plant recycling	Process changes	Input changes	End product changes
Mature, high-volume product	Rubber Petroleum Commodity chemicals Paper Lumber	+	+	−	−	−
Stringent product specifications; high-profit product	Pharmaceuticals Weapons Robotics Specialty chemicals	+	−	−	−	−
Frequently changing product mix; high tech	Electronics Medical equipment	+	+	+	+	−
Job shop processing	Electroplating Printing	+	+	+	+	−
	Foundries Machine shops	+	+	−	+	−
Changing production technology for commodity goods	Steel Nonferrous metals Textiles	+	+	+	−	−
Large-scale manufacture of consumer goods	Automobile Appliances Paints Consumer electronics	+	+	+	+	+

- better handling of materials to reduce fugitive emissions, leaks, and spills;
- changes in methods of cleaning equipment to avoid use of hazardous materials (such as solvents);
- better monitoring of process equipment for corrosion, vibration and leaks;
- more automation of processing;
- separation of waste streams to facilitate in process recovery and volume reduction;
- use of covers on tanks to reduce volatilization losses;
- use of sensors to detect and prevent nonroutine releases.

Some specific examples of plant operations improvements are the following:

- to minimize dragout of cyanide solutions in electroplating, installation of drain boards is estimated to have a pay back period of under 2 months (Freeman, 1988)

- reduction of dragout by use of low pressure air blowoff devices or intermediate dragout tanks (Hunt, 1988)
- periodic maintenance of fittings and seals to minimize intrusion of oil and hydraulic fluid into metal working fluids (Hunt, 1988)
- when solvents part cleaning processes are used, an aqueous prewash may increase the lifetime of the solvent baths (Toy, 1988)
- improving rinse processes by use of countercurrent rinse baths and/or improved mixing within rinse tanks (Hunt, 1988)
- for tank cleaning, an initial small volume wash may be used and the concentrated spent liquid recycled for process use (OTA, 1986)
- replacement of common baghouses by separate baghouses over individual process lines might permit recovery of dusts for reuse in the process (OTA, 1986)
- optimal scheduling of batch manufacturing processes may minimize the necessity for tank dumps or cleanouts (OTA, 1986)
- manual scraping of paint tanks prior to cleanout, use of teflon coated tanks or minimization of surface area : volume ratio can reduce per batch paint waste for disposal (Lorton, 1988)
- replacement of caustic cleaners by nontoxic alkaline cleaning solutions (Lorton, 1988)

In many cases, training procedures can be modified to stress to operating personnel the importance of waste minimization and the necessity for proper process supervision and quality control (to minimize off-spec products which must be discarded).

Altering Process Technology

Changes in process technology include modernization, modification or better control of process equipment (OTA, 1986). Such process modifications often require technological development and may be capital intensive; hence they may be more suitable for use when a manufacturing operation is to be redesigned or built rather than as a retrofit. However, they may offer substantial opportunities for cost reduction.

Specific examples of alterations in process technology leading to waste reduction are:

- replacement of chromic acid pickling of copper and copper alloys by a hydrogen peroxide/sulfuric acid based pickling process (Steward, 1981)
- replacement of ammonia based etchant solutions in printed circuit board manufacture by a hydrogen peroxide/sulfuric acid-based etchant facilitates subsequent recovery of copper sulfate by refrigerated crystallization (Foecke, 1988)
- replacement of cyanide-based solutions in electroplating operations (NRC, 1985)

- replacement of open cycle water-based cooling, which results in a chemical-laden blowdown stream, by closed-cycle refrigeration processes (OTA, 1986)
- paint stripping by plastic media blasting rather than by solvent stripping (OTA, 1986). This can also be used for alternative stripping of prior coatings in plating shops (Freeman, 1988).
- replacement of solvent evaporative drying by hot air drying of printed circuit boards (Foecke, 1988).
- Du Pont found that use of a new catalyst for the production of nylon resulted in the reduction of hazardous waste, as well as nonhazardous aqueous waste (Hollod and McCartney, 1988).
- A mercury laden waste is produced at mercury cell chloralkali manufacturing plants. The replacement of this technology by membrane cell processes, while highly capital intensive, can eliminate this waste stream (RCRA code K071). In one study, a payback period of approximately two years was computed for this substitution (Drabkin and Rissmann, 1988).

Recycle/Recovery/Reuse

There are several types of recycle/recovery/reuse strategies that may be explored for waste minimization. OTA strictly defines this approach to be in-process recycling; that is, the recovery of material for reuse at the same facility at which it is generated. However, this is too restrictive. It is more useful to include the following classification of recycling opportunities (Noll et al., 1985):

- direct recycle for primary (generator) use; this is the approach emphasized by OTA.
- use by a second industry as a raw material (secondary recycle).
- energy recovery.
- utilization in pollution control systems.

It may be necessary, to facilitate recycling, to employ unit operations for material concentration or purification. The processes that may be used include those discussed in the previous chapters on physical, chemical, biological, or thermal treatment, but they may also include processes highly specific to the industry that has generated the waste.

When dragout recovery operations are practiced, the recovered solutions can be returned to the original process. Rinsewaters from various operations may be suited for use as makeup water for the manufacturing process. In polymer manufacture, the grinding pumpout solution from polymer manufacture can be returned for reuse in the original process rather than disposed (Hollod and McCartney, 1988).

Potential examples for downward (secondary) recycle are the use of waste solvents generated in electronics as raw materials in paint manufacture and the use of paint sludges as sealant materials (NRC, 1985). The availability of waste exchanges can facilitate such transfers.

A variety of techniques exist for the recovery of metals from solutions generated in metal-working industries. These include electrolytic recovery, solvent extraction, floatation, precipitation (including coprecipitation), ion exchange, and membrane separation. Pyrometallurgical processes, acid solubilization, and biologically assisted leaching can be used to recover metals from solid wastes (Brooks, 1986).

In addition to metals recovery, a number of other specific processes have been identified as suitable for materials recovery:

- waste oil rerefining (NRC, 1985)
- solvent recovery by distillation (NRC, 1985; Foecke, 1988; Nunno et al., 1988)
- recovery of iron salts from steel pickle liquor (NRC, 1985)
- use of sodium borohydride reductive precipitation for metal recovery from printed circuit-board sludge wastes (Nunno et al., 1988).
- retorting of sludges for mercury cell chloralkali plants for mercury removal and recovery (Drabkin and Rissmann, 1988).

In addition, it has been found technically feasible to hydrogenate chlorinated hydrocarbon wastes, including PCB's, to produce a usable hydrocarbon fuel and inorganic solids. Although the economics of this process need further work, it appears to have a positive energy balance (Kalnes and James, 1988).

The recovery of energy value from organic-laden wastes is primarily accomplished through thermal processes. Cement or brick kilns, or steel mills, have been identified as potentially useful combustion locations at which fuel value of wastes can be beneficially utilized. These processes should be considered as analogous to the thermal destruction processes discussed previously with the additional benefit that the heating value of the waste stream is used to replace an energy input in the form of natural gas, oil, or coal.

In several cases, a waste produced by industry may, with minimal processing, be suitable for recycle to a pollution control facility. The major example of this is the use of pickle liquor as a source for iron salts which may be used to precipitate phosphate at minicipal sewage treatment plants (Noll et al., 1985). In addition, an acid or alkaline waste with minimum concentrations of other components may be potentially useful to effect neutralization either at the original generator or at another facility.

Changing Raw Materials

The option of changing raw materials used in a process is difficult to distinguish from a process modification per se. In addition to options discussed previously, the following additional strategies for minimization using this approach have been identified:

- replacement of solvent-based inks or paints by water-based inks or paints (NRC, 1985; OTA, 1986)

- replacement of tap water by de-ionized water in the formulation of process solutions that will be treated by chemical precipitation can reduce hazardous sludge generation by the minimization of precipitation of hardness cations in the feed water (Foecke, 1988)

Finally, mercury-based fungicides and bactericides in paint often render resulting paint waste hazardous. Research is underway on the practicality of replacing these compounds by tributyltin and methylenedianiline (Lorton, 1988).

Product Reformulation

This is perhaps the most difficult option, and often requires extensive marketing efforts among the product end users. In the manufacture of a certain industrial adhesive, Monsanto originally used a filtration process to remove particulates that resulted in a hazardous waste of filter cake and spent filter media. The elimination of the filtration step, and retention of particulates in the final product, did not alter product quality, but it was necessary to conduct a multi-year marketing effort to overcome customer resistance to the altered product (OTA, 1986).

WASTE AUDITS

A waste minimization audit, or more concisely, waste audit, is a "systematic, periodic internal review of a company's processes and operations designed to identify and provide information about opportunities to reduce waste" (OTA, 1986). To implement a significant degree of waste reduction, audits are necessary to develop site specific options that are not obvious based on review of generic case studies such as those noted previously. A waste audit consists of several steps, each with given activities and products. These are tabulated in Table 8-2.

TABLE 8-2 STAGES OF A WASTE AUDIT. (OTA, 1986; Drabkin, 1988; Drabkin et al., 1988.)

Stage	Activities	Products
Initiation/planning	Organize task force Get management commitment	Establish goals
Pre-audit	Identify waste and process flows Prioritize targets for detailed study	Inspection agenda Process descriptions Assessment team assignments
Audit	On site inspection(s) Generation of options (brainstorming) Selection for feasibility study	Lists of candidate options
Feasibility analysis	Technical evaluation of options Economic evaluation of options	Estimated costs and benefits Profitability analysis Final report
Implementation	Put recommendations in place Evaluate performance	Identify needs for further study

Pre-Audit Phase

During the pre-audit phase, wastes are prioritized for attention. This prioritization is generally based on composition, quantity, degree of hazard, method, and cost of disposal (and perhaps future liability resulting therefrom), potential for minimization, and recycle and compliance status (Drabkin et al., 1988). Information for this phase can come from many sources, including the following (Lorton et al., 1988):

- hazardous waste manifests and biennial reports
- previous audit reports
- emission inventories
- waste assays
- permits (e.g., NPDES)

Depending upon the size of the facility and the level of complexity desired in the audit, the identification of waste streams can proceed to several levels of detail. These have been described as (OTA, 1986):

Level I: analyze waste streams to get order of magnitude information. This is the appropriate level of detail for many small businesses.

Level II: inventory all wastes with identification of variability over time.

Level III: conduct a mass balance on all hazardous materials/wastes to determine mass flows of material discharged. This requires an intensive effort, and may only be justifiable for larger plants or larger volume waste streams.

In addition, during the pre-audit phase, the following additional information should be gathered (Lorton et al., 1988):

- process, equipment, and facility design information
- environmental reports, assays, manifests, documents, and permits
- operating cost information
- policy and organizational information

Often, particularly in small industries such as electroplaters, process documentation sufficient to conduct an audit is not available. However, this information may be obtainable from process, equipment, and chemical vendors, and from the open literature (Freeman, 1988).

During this phase, the foci of the site inspections should be identified, and based upon this identification the audit team members selected. The team should include persons with hands-on process familiarity, such as first-line operators and process supervisors. While an overall audit task force must have broad experience, the teams assigned for the analyses of specific processes must have focused and detailed experience (Lorton et al., 1988).

Audit Phase

The formal audit site visit should generally include a walk-through of the plant following the process line from the receipt and storage of raw materials to the shipping of product. During the audit itself, the following key guidelines should be followed (Lorton et al., 1988):

- have an agenda which has been prepared in advance;
- schedule the visit to coincide with the particular operation(s) of interest. This is of particular importance in plants with substantial batch-to-batch variability.
- monitor operations at various times during a shift;
- interview operators, foremen and supervisors, and, in particular, assess awareness of waste generation aspects of the processes;
- observe housekeeping aspects of the facility;
- review organizational structure for relationships among facilities;
- assess administrative controls.

Following the actual physical inspection, the audit phase includes a "brainstorming" session in which candidate minimization options are generated. The objective of this phase is not to judge feasibility, but to enumerate a wide range of possibilities from which the most promising may be selected. In addition to using information from the physical inspection and personal knowledge of the assessment team members, additional information sources for generating options are trade associations, published literature, conference proceedings, literature from equipment vendors, information from state, federal, and local environmental agencies, and knowledge of outside consultants (Drabkin, 1988).

After potential options are enumerated, they can be winnowed down to a list of most promising using a variety of techniques. The most widely used screening methodology is a weighted sum of ranks technique, similar to that used in environmental impact assessment. A set of criteria is identified, such as the following (Lorton et al., 1988; Hanlon and Fromm, 1990):

- reduction in waste's hazard
- reduction in treatment/disposal costs
- reduction in future liability
- reduction in safety hazards
- reduction in input material costs
- extent of prior experience in industry (maturity of technology)
- effect on product quality
- low capital cost
- low operation and maintenance costs
- short implementation period
- ease of implementation

The assessment team (or management) should assign weighting factors to each of these criteria such that a larger weight is given to the factors perceived to be the most important. Each option is then ranked according to each of the criteria. This evaluation can be performed in a consensus manner by the assessment team, or by the leader of the assessment activity. A low rank is given to more favored alternatives under given criteria. Then the weighted sum of ranks is obtained for each alternative, and the options that receive the lowest score are recommended for subsequent detailed evaluation.

Feasibility Analysis

The objective of a feasibility analysis is to conduct a technical and economic evaluation of the most preferred options identified during the audit phase and to summarize the findings in a final report of the overall assessment. Technical evaluation concentrates on the ascertainment of compatibility of the proposed options with existing facilities, and on the nature of the technology. The economic evaluation is designed to quantify the overall costs and benefits and determine, using plant-specific criteria, the attractiveness of the identified options as business decisions.

Specific questions to be addressed by the technical evaluation are (Lorton et al., 1988; Hanlon and Fromm, 1990):

- Will the option work in the proposed application?
- Has the option worked in previous applications?
- Are there sufficient resources, in terms of space and utilities (electric, water, etc.) for the proposed option, or must additional resources be procured?
- Is the option compatible with current operating procedures, work flow, and production rates?
- Is it necessary to stop production to install the option, and, if so, for how long?
- Will there be an effect on product quality, and, if so, will it be beneficial or detrimental?
- Is special expertise required to operate or maintain the new system? Does the vendor provide acceptable servicing?
- Does the system or procedure create other environmental problems?

It is important during this stage of the feasibility analysis to solicit comments from potentially impacted personnel within the manufacturing operation to determine if hidden effects of proposed options exist, and also to help secure cooperation with ultimately favored options.

During the technical evaluation of the feasibility study it is possible that some of the options generated during the audit phase will be dropped. This will reduce the work of the economic evaluation.

The economic-analysis phase of the feasibility analysis uses standard engineering economic tools to evaluate each proposed option for overall profitability. While most costs can be identified with reasonable precision, only some benefits can be thus quantified. The particular costs and benefits most amenable to economic analysis are as follows (Lorton et al., 1988):

- capital costs
- O & M costs
- reduced waste management costs
- raw material cost savings
- insurance savings
- change in utilities costs
- revenue from marketable recovered byproducts

These changes (costs and benefits) may be estimated and the economic consequences evaluated using a variety of criteria, including payback period, internal rate of return, and discounted cash flow. The overall result from this analysis has been termed a "tier zero" result (McHugh, 1990). For an alternative to be feasible it should generally show a favorable economic evaluation at this level, although less precisely quantifiable benefits may justify a somewhat lower rate of return (higher payback period) than for conventional plant investment. Tier zero analyses usually consider the consequences on an after-tax basis, so that the net result to corporate profits may be ascertained.

The "tier one" (McHugh, 1990) analysis focuses on identifying benefits associated with hidden costs. This may include reduction in reporting, testing, monitoring, and manifesting requirements, reduction in sewage surcharges, or reduction in OSHA compliance costs. These costs are relatively easy to identify, although they are considered somewhat less precise than those in the previous tier.

The "tier two" analysis identifies projected costs associated with future liabilities, such as potential cost of becoming a PRP at a future Superfund site, repairing an underground storage tank, potential fines for NPDES violations, and toxic torts actions. While there is a considerable uncertainty in evaluating the likelihood of subsequent consequences and their magnitude, the impact of these downstream liabilities can be significant, if, for example, one is a PRP at a disposal site requiring a $20 million remediation twenty years in the future.

The "tier three" analysis is the most subjective and uncertain, and incorporates the impact of a modification on corporate goodwill, public acceptance, overall corporate or director risk, and shareholder value.

Several basic concepts from engineering economics are used in these computations. If an expense or income P occurs at t years in the future, the present value (PV) of this if the interest rate is given as i (expressed on a fractional basis) is:

$$PV = \frac{P}{(1+i)^t} \tag{10.1}$$

If at various years these cash flows occur, the sum is discounted to the present (i.e., the sum of the present values for all of the successive years) is called the net present value (or present worth). The interest rate, i, which results in the net present worth for a series of years (at which both positive and negative cash flows occur) is called the internal rate of return—this must be computed by iteration, or trial and error. Fortunately, many common spreadsheet programs enable the direct computation of these quantities. Example 8-1 illustrates the computation of these quantities in a sample scenario for waste minimization.

Example 8-1

A company is considering a waste minimization strategy that will require a capital investment of $300,000. The investment will be amortized over a ten year useful life of the capital equipment at a 7% interest rate. It is anticipated that the general inflation rate over the investment life will be 5%, with a 7% inflation rate for disposal costs. The following are anticipated consequences of the changes in the first year costs:

Tier 0

Savings in raw materials	$3,000
Reduced disposal costs	$20,000
Additional operating materials	$1,000

Tier 1

Reduced reporting costs	$1,000
Reduced lab analyses	$2,000
Reduced sewage surcharges	$750

The following table summarizes the analyses of tier zero. The annual payment required for amortization is computed by the following equation:

$$\text{payment} = \frac{\text{principal} \times i}{1 - (1 + i)^{-n}}$$

where i is the interest rate (as a decimal) and n is the amortization period (in years). Hence, the annual payment is computed as $32,938. For all years beyond year 1 the annual costs or savings are escalated by the assumed inflation factor. The net of the costs for each year is then tabulated (in the column headed net).

From the annual net of benefits or costs, the net present value (NPV) at a given interest rate can be computed. If i' is the interest rate used for discounting, then the NPV of a set of cash flows A_j (net in year n) can be computed as:

$$\text{NPV} = \sum_{j=1}^{j=n} \frac{A_j}{(1 + i')^{j-1/2}}$$

The "1/2" denominator is a minor correction under the assumption that the benefit or cost accrues at midyear rather than at the start of the year.

In some circumstances, it is sufficient to compute the NPV at the competing rate at which a company is able to invest its funds. In other circumstances, the internal rate of return is desired (IRR), which can be obtained from the successive application of the previous equation to determine the value of i' that results in a zero NPV.

The tier zero analysis in Table 8-3 shows that, at both 5% and 10% discount rates, the NPV (given at the bottom two right-hand columns) is negative, indicating that there is a net cost to the proposed modification at this level. Furthermore, there is no positive value to IRR. Therefore, at this level there is no economic gain associated with the proposed change.

TABLE 8-3

Tier 0
(benefits positive)

	Capital amortization	Raw materials saved	Operating materials needed	Disposal savings	Net	NPV Interest rate 10%	5%
Year 1	($32,938)	$3,000	($1,000)	$20,000	($10,938)	($9,944)	($10,418)
2	($32,938)	$3,150	($1,050)	$21,400	($9,438)	($7,800)	($8,989)
3	($32,938)	$3,308	($1,103)	$22,898	($7,835)	($5,887)	($7,462)
4	($32,938)	$3,473	($1,158)	$24,501	($6,122)	($4,182)	($5,831)
5	($32,938)	$3,647	($1,216)	$26,216	($4,291)	($2,665)	($4,087)
6	($32,938)	$3,829	($1,276)	$28,051	($2,335)	($1,318)	($2,224)
7	($32,938)	$4,020	($1,340)	$30,015	($244)	($125)	($232)
8	($32,938)	$4,221	($1,407)	$32,116	$1,991	$929	$1,897
9	($32,938)	$4,432	($1,477)	$34,364	$4,380	$1,858	$4,172
10	($32,938)	$4,654	($1,551)	$36,769	$6,933	$2,673	$6,603
					SUMS	($26,461)	($26,571)

The tier one analysis is illustrated in Table 8-4. Costs are escalated using the inflation rate, and the net cost is found by adding the totals for tier zero and tier one for each year. The NPV for the resulting sum is computed at three discount rates: 20%, 10% and 11.53%. At the latter value, which was found by trial and error, the NPV is essentially zero, and this represents the internal rate of return.

The interpretation of this computation is that, unless the company can obtain a return on cash in excess of 11.53% over the ten-year-period, the proposed minimization strategy must be regarded as economically justifiable at the tier one level.

The tier 2 costs are much more speculative. It is assumed that continuation of the status quo waste management methods will result in a required remediation 20 years in the future (perhaps associated with a landfill at which wastes were buried). The cost of this remediation (in year 20 dollars) is assumed to be $20 million dollars, and the company has a 5% exposure—meaning that it would anticipate bearing 5% of the financial liability. Thus, it would have a $1 million liability twenty years in the future. Table 8-5 summarizes the result, and shows that the net present value of such a future liability would be far greater than the net present values associated with the analyses through tier one. Hence, even if as much as a 10% probability of such future exposure would be assumed, the overall analysis would be dominated by tier two benefits (re-

TABLE 8-4

Tier 1						NPV	
	Reporting costs	Lab analysis	Sewage	Net to		Interest rate	
Year			Surcharges	Tier 1	20%	10%	11.53%
1	$1,000	$2,000	$750	($7,188)	($6,562)	($6,854)	($6,807)
2	$1,050	$2,100	$788	($5,501)	($4,185)	($4,768)	($4,670)
3	$1,103	$2,205	$827	($3,701)	($2,346)	($2,916)	($2,817)
4	$1,158	$2,315	$868	($1,781)	($941)	($1,276)	($1,216)
5	$1,216	$2,431	$912	$267	$117	$174	$163
6	$1,276	$2,553	$957	$2,451	$899	$1,451	$1,345
7	$1,340	$2,680	$1,005	$4,782	$1,462	$2,574	$2,353
8	$1,407	$2,814	$1,055	$7,268	$1,852	$3,556	$3,206
9	$1,477	$2,955	$1,108	$9,921	$2,106	$4,413	$3,924
10	$1,551	$3,103	$1,163	$12,751	$2,256	$5,156	$4,522
					($5,342)	$1,509	$3

duced future liability). If the total future liability were to be treated as an insurance benefit, and amortized over the ten-year-period of the minimization process, the annual benefits would be $36,793 (or $3,679 if the future consequences were only 10% certain). Not shown is that, if these tier two benefits were added to the tier one results, the internal rate of return (assuming only 10% certainty of the adverse future liability) would be 48.5%—or far greater than nearly all other possible investments that could be made.

TABLE 8-5

Tier 2	
Incident costs	$20,000,000
Years in future	20
Fractional exposure	0.05
Year 0 dollars, 7%	$258,419
Annualized over 10 yrs, 7%	$36,793

This computation, in general, would be modified to reflect the tax consequences of various expenditures and to place all investment and continuing expenses on a net after-tax basis.

Implementation Phase

Following a formal feasibility analysis, a company decision to implement the proposed changes must occur. The implementation phase should be viewed as ongoing, and a periodic review of performance is desirable to verify the benefits accruing from the modifications, or, if necessary, to decide to abandon such changes (Lorton et al., 1988).

INSTITUTIONAL DISINCENTIVES

Corporate and institutional considerations may affect the acceptability and ease of implementation of waste minimization opportunities. These have been summarized by NRC (1985) and OTA (1986) and are summarized below.

It is necessary to secure the active participation of top management and to ensure than an attitude conducive to minimization is communicated throughout the corporate structure. This must be accompanied by sufficient incentives to alter behavior of plant level management and line employees. Often it may also be necessary to actively involve unionized labor. Top management may show reluctance toward waste minimization due to uncertain investment returns, the potential for production downtimes, potential problems with operations or product quality, and possible loss of proprietary information to waste reduction consultants (NRC, 1985).

Product quality is often dictated by external specifications beyond the control of an individual company. This is particularly true in the case of Department of Defense procurement and Food and Drug Administration regulated products. In such industries, it is necessary to pursue means of modifying specifications to permit use of products that have been manufactured using processes incorporating waste minimization (OTA, 1986).

PROBLEM

8.1. A waste minimization strategy will require \$1,000,000 in capital costs. If this change has a useful life of 15 years, the general inflation rate (for all costs) is 5%, and the amortization is 6.5% over the 15 years with zero salvage value, compute the minimum annual year one savings (tiers zero and one) to ensure a positive net present value at interest rates for discontinuing of 6, 8, and 10%.

REFERENCES

Brooks, C. S. "Metal Recovery from Industrial Wastes." *Journal of Metals*. July, pp. 50–57 (1986).

Committee on Reducing Hazardous Waste Generation. *Reducing Hazardous Waste Generation.* Washington, D.C., National Academy Press (1985).

DRABKIN, M. "The Waste Minimization Assessment: A Useful Tool for the Reduction of Industrial Hazardous Wastes." *Jour. APCA* 38, 12, 1530–1541 (1988).

DRABKIN, M., AND RISSMANN, E. "Waste Minimization Audit Report: Case Studies of Minimization of Mercury Bearing Wastes at a Mercury Cell Chloralkali Plant." U.S. EPA/600/S2–88/011 (1988).

———. "Waste Minimization Opportunities at an Electric Arc Furnace Steel Plant Producing Specialty Steels." *Environmental Progress* 8(2):88–96.

DRABKIN, M., FROMM, C., AND FREEMAN, H. M. "Development of Options for Minimizing Hazardous Waste Generation." *Environmental Progress* 7, 3, 167–174 (1988).

EVANOFF, S. P. "Hazardous Waste Reduction in the Aerospace Industry." *Chemical Engineering Progress* 86(4):51–61 (1990).

FOECKE, T. L. "Waste Minimization in the Electronics Products Industries." *Jour. APCA* 38, 3, 283–291 (1988).

FREEMAN, H. A. "Waste Minimization Audit Report: Case Studies of Minimization of Cyanide Waste from Electroplating Operations." U.S. EPA/600/S2–87/056 (1988).

FREEMAN, H. M., ed. *Hazardous Waste Minimization.* McGraw-Hill, New York (1990).

GHASSEMI, M. "Waste Reduction: An Overview." *The Environmental Professional* 11:100–116 (1989).

HANLON, D. AND FROMM, C. "Waste Minimization Assessments." Chapter 5 in Freeman, H. A. [ed.], *Hazardous Waste Minimization.* McGraw-Hill, New York (1990).

HIRSCHORN, J. S. "Cutting Production of Hazardous Waste." *Technology Review* 91(3):52–61 (1988).

HOLLOD, G. J., AND MCCARTNEY, R. F. "Waste Reduction in the Chemical Industry." *Jour. APCA* 38, 2, 174–179 (1988).

HUNT, G. E. "Waste Reduction in the Metal Finishing Industry." *Jour. APCA* 38, 5, 672–680 (1988).

KALNES, T. N., AND JAMES, R. B. "Hydrogenation and Recycle of Organic Waste Streams." *Environmental Progress* 7, 3, 185–191 (1988).

KAMINSKI, J. A. "Waste Reduction in the Department of Defense." *Jour. APCA* 38(9):1174–1185 (1988).

———. "Waste Reduction in the Department of Defense." *Jour. APCA* 38(8):1042–1050 (1988).

LEEMAN, J. E. "Waste Reduction in the Petroleum Industry." *Jour. APCA* 38(6):814–823 (1988).

LORTON, G. A. "Waste Minimization in the Paint and Allied Products Industry." *Jour. APCA* 38, 4, 422–427 (1988).

LORTON, G. A., FROMM, C. H., AND FREEMAN, H. "The EPA Manual for Waste Minimization Opportunity Assessments." U.S. EPA/600/S2–88/025 (1988).

MCHUGH, R. T. "The Economics of Waste Minimization." Chapter 6 in *Hazardous Waste Minimization,* H. M. Freeman (ed.), McGraw-Hill, New York (1990).

National Research Council, Committee on Institutional Considerations in Reducing the Generation of Industrial Hazardous Wastes. *Reducing Hazardous Waste Generation.* National Academy Press, Washington D.C. (1985).

NOLL, K. E., HAAS, C. N., SCHMIDT, C., AND KODUKULA, P. *Recovery, Recycle and Reuse of Industrial Waste.* Lewis Publishers, Chelsea, MI (1985).

NUNNO, T., PALMER, S., ARIENTI, M., AND BRETON, M. "Waste Minimization in the Printed Circuit Board Industry—Case Studies." U.S. EPA/600/S2–88–008 (1988).

Office of Technology Assessment. *Serious Reduction of Hazardous Waste.* OTA–ITE–317, Washington, D.C. (1986).

OMAN, D. E. "Waste Reduction in the Foundry Industry." *Jour. APCA* 38(7):932–940 (1988).

STEPHAN, D. G., AND ATCHESON, J. "The EPA's Approach to Pollution Prevention." *Chemical Engineering Progress* pp. 53–58, June (1989).

STEWARD, F. A. "Process Change to Reduce the Production of Industrial Sludges." Presented at the 1981 Annual Meeting of the American Institute of Chemical Engineers (1981).

TOY, W. M. "Waste Minimization in the Automotive Repair Industry." *Jour. APCA* 38, 11, 1422–1426 (1988).

WOLF, K. "Source Reduction and the Waste Management Heirarchy." *Jour. APCA* 38(5):681–686 (1988).

9

Systems Analysis for Regional Planning of Hazardous Waste Management Options

INTRODUCTION

Although the network of hazardous waste management has often developed in the absence of overt planning, it may be desirable in many instances to develop a regional plan. Such a plan would include decisions on optimal mixes of treatment and disposal facilities, and locations. To consider the optimization of such a system it is necessary to develop a mathematical model for the hazardous waste system and then solve it. This chapter will outline the procedures required. In the formulation of this problem, the general framework developed by Liebman (1972), Pierce and Davidson (1982), and Jennings and Sholar (1984) will be followed.

PROBLEM FORMULATION

The first step in development of a regional plan is definition of the region. This may be a state, a subset of a state (such as a regional governmental unit), or perhaps a large county. Waste is considered to be generated at a discrete number of locations. A second set of locations represents candidate sites for treatment or storage facilities (which may include transfer stations). A third set of locations represents disposal facilities.

System Constraints

During a given year, generation location i is assumed to generate G_i tons/year of waste. Treatment facility j, if built, is designed to process up to T_j tons per year of waste, perhaps resulting in a reduction d_j in waste mass (perhaps due to volume reduction or incineration). The variable Y_j will be reserved as a binary integer variable (value of either 0 or 1) depending on whether it is decided to build treatment facility j. Disposal facility k can process D_k tons/year of waste, and the integer variable Z_k (value 0 or 1) describes whether this candidate facility is actually built. The quantities G_i, T_j and D_k are regarded as problem inputs—their values are stipulated for the specific region under question. However Y_j and Z_k are to be determined from the model solution. The other primary variables which must be determined by solving the model are the tons/year flowing from generation location i to treatment location j, designated as R_{ij}, and the shipments from treatment location j to disposal location k, designated as S_{jk}. In the problem as formulated here, it is assumed that all wastes must first pass through to treatment. However, if a given waste can be directly disposed, the model can be easily modified.

Under the above description, a number of constraints must be obeyed by any solution to the problem. First, all the waste of each generating location must be adequately managed. This is stated mathematically by:

$$G_i - \sum_j R_{ij} = 0 \text{ for each } i \tag{9.1}$$

Second, the waste being received at any single treatment location which is included in the solution must not exceed its capacity. This is stated by:

$$Y_j T_j - \sum_i R_{ij} \geq 0 \text{ for each } j \tag{9.2}$$

Third, the residual from treatment must all be shipped to disposal facilities, as given by:

$$d_j \sum_i R_{ij} - \sum_k S_{jk} = 0 \text{ for each } j \tag{9.3}$$

Fourth, the shipments to each disposal site which is included in the solution must not exceed its capacity:

$$Z_k D_k - \sum_j S_{jk} \geq 0 \text{ for each } k \tag{9.4}$$

Thus, there are $i + 2j + k$ constraints involving the following variables (these are termed the decision variables):

R_{ij}	ij variables
S_{jk}	jk variables
Y_j	j variables
Z_k	k variables

for a total of $j(i + 1) + k(j + 1)$ variables. Therefore, providing that either $j > 1$ or $k > 1$, there are more unknowns than equations. Therefore, there are sufficient degrees of freedom to permit optimization (provided that the problem is feasible and admits of a solution in which the R's and S's are nonnegative).

Formulation of the Objective Function

The objective function describes the property of the solution that we seek to optimize. It may be formulated in several ways. In the first approach, the costs of a solution are minimized and the resulting impact (e.g., health and nonhuman effects) are analyzed; based on this consideration, next-best solutions might be sought and a qualitative tradeoff made between the costs and the perceived acceptability of the risk. In the second approach, risk might be incorporated directly into the objective function (either instead of, or in addition to economic factors). A variety of limitations exist which render the latter approach more difficult:

- It is difficult, if not impossible, to place risk factors on commensurable terms with economic factors (the value of a life issue, for example).
- To minimize risk, one would need to know the incremental risks associated with an incremental change in waste management practices under a variety of scenarios. There would be extreme difficulty in computing such marginal risks, and there would be great variability (far greater, for example, than the marginal cost of waste treatment). This would make this approach much more of an intensive data gathering and estimation effort.
- There are actually many components of risk which are intrinsically noncommensurable, and thus it is easier to perform a tradeoff on discrete alternatives that are identified from economic optimization.

Hence, in the formulation of the problem, a strict cost viewpoint is taken recognizing that the ultimate solution that is accepted may not cost the least. However, the objective of the optimization should be to generate ranked alternatives and thereby compute the cost differentials that might be incurred in choosing an option of lower perceived risk.

Costs associated with a particular alternative (combination of values for the R's, S's, Y's, and Z's that satisfy the solution constraints) may be divided into three particular forms:

- fixed costs for a given treatment or disposal location which are independent of the amount of waste processed. These will be denoted by A_j and B_k, respectively.
- variable costs for a given treatment or disposal facility. It will be assumed that the variable costs are linear functions of the waste flows processed, and that a_j and b_k represent the cost/ton of waste received at treatment facility j and disposal facility k, respectively.
- transportation costs from generator to treatment facility and from treatment facility to disposal site. These will be treated as solely variable, and linear functions of the waste shipped. The constant q_{ij} is the cost per ton to ship waste from generator i to treatment facility j and r_{jk} is the cost per ton to ship waste from treatment facility j to disposal facility k. Very frequently these

coefficients can be represented as simple functions of the distance between the two points.

The numerical values for the cost coefficients (A's and B's) and multipliers (a's, b's, q's, and r's) may be in terms of first-year costs (including amortization of capital costs) or in terms of net present value over the useful life of the facility. Particularly in the latter case, the time horizon, interest rate, and inflation rate become assumptions of the model and may affect the final optimal cost as well as the solution itself.

With the previous definitions, the objective function (Q) may be written as follows:

$$Q = \sum_j Y_j A_j + \sum_k Z_k B_k + \sum_j \sum_i (a_j + q_{ij}) R_{ij} + \sum_k \sum_j (b_k + r_{jk}) S_{jk} \qquad (9.5)$$

The optimal solution to the problem is obtained by finding the values of the decision variables that minimize the cost. To find the next to optimal solution, and successive suboptimal solutions, additional constraints are written involving the Y's and Z's. For example, if the optimal solution is found to contain treatment sites one and three and disposal sites two and four, then the following constraint should be appended to find the next most optimal solution:

$$Y_1 + Y_3 + Z_2 + Z_4 \leq 3 \qquad (9.6)$$

In a similar manner, successive constraints should be appended to find next best solutions.

SOLUTION METHODOLOGY

The planning problem, as posed, is formally a mixed integer-linear programming problem. A variety of software exists that can solve this problem, including personal-computer based software (e.g., GAMS). In addition, special-purpose optimization programs are also available to solve the planning problem (e.g., WRAP). Given input data on a problem, the optimal solution and successive less-than-optimal solutions can be found. An example of the necessary input data and the solution results will be illustrated using a hypothetical analysis of waste flow patterns in the State of Illinois.

Five locations for the generation of hazardous waste are considered. These are assumed to generate the following amounts in thousand tons/year (G_i):

Metropolitan Chicago	200
East St. Louis	75
Rockford	50
Champaign	10
Springfield	10

Five potential locations for treatment facilities will be considered. The following table gives the maximum capacity, fixed cost (dollars per year), waste reduction ratios and variable costs for each site:

TABLE 9-1 TREATMENT SITE ASSUMPTIONS

Site	Treatment capacity T_j (1,000 tons/year)	Fixed cost A_j ($/yr)	Residue r_j	Unit treatment cost ($/1,000 tons)
Cook County	250	$1,000,000	0.5	$4,000
Will County	250	$800,000	0.6	$1,500
Peoria	300	$600,000	0.9	$900
Belleville	100	$500,000	0.85	$300
Decatur	100	$500,000	0.7	$400

Table 9-2 provides distances between the generation and treatment sites. It is assumed that the transportation costs are equal to $0.50/ton-mile between generation and treatment site.

Four disposal locations were considered. Table 9-3 provides their maximum disposal capacity, the fixed cost of the site, and the unit cost of disposal. Table 9-4 provides the mileage between the treatment facilities and each of the disposal facilities. For determining transportation costs between treatment facility and disposer, the identical value of $0.50/ton-mile was assumed.

The optimal, and successive suboptimal solutions, are summarized in Table 9-5. Shown in this table are the receipts of each of the treatment and disposal facilities. The absence of a numerical value indicates that that site is not built in the given solution set. The first ten solutions are all within a 10% annual cost. These may be compared risk or impact criteria to assess the minimum cost solution that is acceptable when risks and environmental impacts are considered.

TABLE 9-2 MILEAGE BETWEEN GENERATORS AND TREATMENT SITES

	Miles to treatment location				
Generator	Cook	Will	Peoria	Belleville	Decatur
Metro Chicago	10	20	130	280	160
East St. Louis	280	260	160	10	120
Rockford	80	90	120	290	180
Champaign	130	100	90	160	40
Springfield	180	170	60	100	40

TABLE 9-3 DISPOSAL SITE ASSUMPTIONS

Site	Disposal capacity D_k (1,000 tons/year)	Fixed cost A_k ($/yr)	Unit treatment cost ($/1,000 tons)
Will County	200	$1,000,000	$1,000
Moline	200	$750,000	$500
Cairo	300	$500,000	$400
Monticello	200	$500,000	$600

TABLE 9-4 MILEAGE BETWEEN TREATMENT AND DISPOSAL SITES

	Miles to disposal location			
Treatment site	Will	Moline	Cairo	Monticello
Cook County	20	160	380	140
Will County	5	165	370	130
Peoria	125	80	280	80
Belleville	260	210	130	140
Decatur	150	150	210	30

TABLE 9-5 OPTIMAL AND NEAREST SUB-OPTIMAL SOLUTIONS (1,000 tons per year handled at each location)

Cost of optimal solution	Treatment facility					Disposal facility			
($/yr)	Cook	Will	Peoria	Belleville	Decatur	Will	Moline	Cairo	Monticello
13,925,400		250			95	150			66.5
14,350,400	250				95	125			66.5
14,488,750		250		75	20	164		63.75	
14,913,750	250			75	20	139		63.75	
14,934,550		250		95		150		80.75	
14,947,050		250	20	75		150			81.75
15,288,250	20	250		75		160			63.75
15,354,450		250		95		150			80.75
15,359,550	250			95		125		80.75	
15,372,050	250		20	75		125			81.75

PROBLEM GENERALIZATIONS

The mixed integer linear programming model for waste planning is a powerful one since it is fairly rapidly solved and can be used to explore a variety of "what if" scenarios. However, it has a number of limitations as presented. Fortunately, many limitations can be circumvented by modifications to the modeling approach presented above. These will be described below.

Nonlinearity of Cost Functions

In the previous formulation, the costs are treated as linear functions. In reality, most cost functions are nonlinear, generally showing economies of scale. There are two approaches to treating this problem. In the first, which is computationally more difficult, the costs are used in their nonlinear forms. When this is done, however, the mixed integer programming framework cannot be employed. It is generally more computationally intensive to solve the resulting nonlinear mixed integer programming problem.

Alternatively, it is possible to subdivide a nonlinear cost curve into a series of line segments, which can then be used in the linear mixed integer framework. This process is termed piecewise linearization. A graphical example is shown in Figure 9-1. In this example, three lines are used to describe the cost curve over the range up to 80 tons/day. The intercept of each line is the apparent fixed cost (in dollars per day) and the slope of each line is the variable cost (in dollars/ton)—these can be incorporated directly into the objective function. The overall solution process then

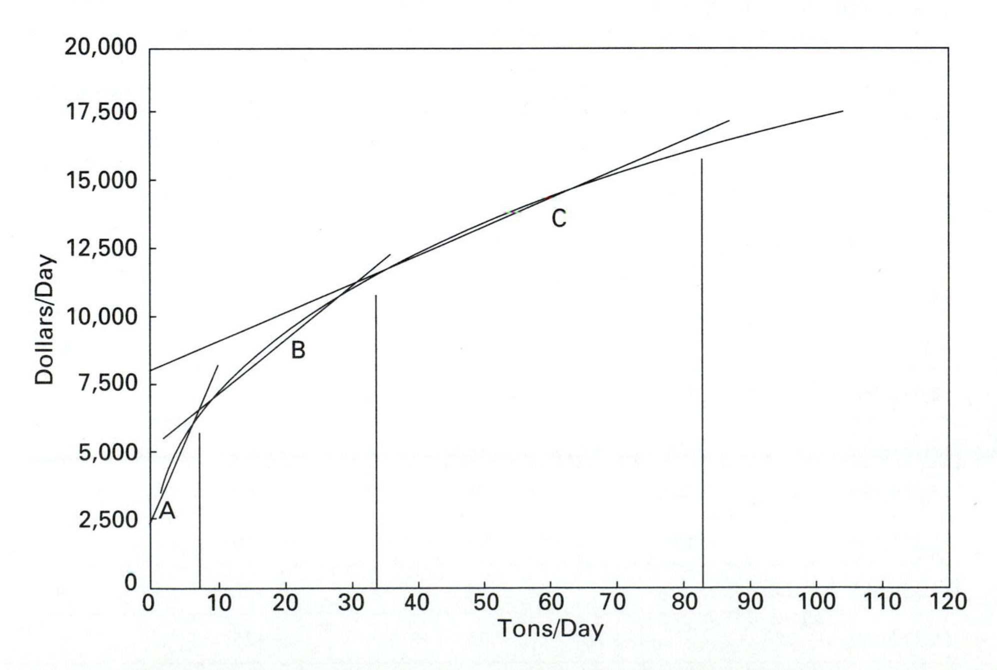

Figure 9-1 Example of piecewise linearization of nonlinear cost functions.

requires that an assumption be made as to the scale at which each process or location will operate so that the appropriate line can be used. The resulting solution is checked for consistency with this assumption. If necessary, the model is rerun with a different assumption as to the scale of operation until consistency of the results with the cost assumptions are achieved.

Multiple Waste Types and Treatment/Disposal Facilities

The above analysis treated all hazardous waste flows as identical. As noted by Jennings and Sholar (1984), if it is desired to compare multiple processing and disposal alternatives, it is necessary to consider that wastes of different types and processing options have different costs and capacities. This problem is solved by using another subscript (m) to denote waste type. Hence, the following variables are redefined:

R_{ijm} = tons per year of waste type m shipped from generator i to treatment facility j

S_{ijm} = tons per year of waste type m shipped from treatment facility j to disposal facility k

G_{im} = tons per year of waste type m generated at site i

T_{jm} = maximum treatment capacity at site j for waste type m

D_{km} = maximum disposal capacity at site k for waste type m

In addition, it is necessary to write each of the capacity constraints (Equations 9.1 through 9.4) for each waste type. The data requirements for this model may include waste type specific cost coefficients and reduction proportions (as well as possibly waste specific transportation costs). Clearly this is a larger model, but it is structurally similar to the simple model presented earlier and may be solved in a similar manner.

The separation of waste types appears to be highly desirable for the consideration of management alternatives, since it can be readily coupled with a risk evalua-

TABLE 9-6 MANAGEMENT ALTERNATIVES IN THE STATE OF KENTUCKY. (Modified from Jennings and Sholar, 1983.)

Option	Cost	Remarks
Optimal solution	$65,476,200	Solvent recovery underutilized Incineration unused
Mandate incineration of organic liquids	$71,626,200	Relative risk is actually increased over optimal case
Either incineration or recycle of recoverable organics allowed; $0.40/gallon credit for recoverable organic liquids	$66,776,200	

tion procedure. As an example of this, Jennings and Sholar (1983) performed a systems analysis for the hazardous waste flows in the state of Kentucky. The following is a synopsis of their results. Their computations show the interesting result that an apparently reasonable regulatory strategy (mandating incineration of organic liquids) can result in both an increase in cost and an increase in risk.

Multiperiod Problems

Another expansion of the planning framework would allow for the projection of waste generation and capacities over time. This would permit a decision to be made about the optimal time for changes in management practice (i.e., facility expansion or closure) to occur. While such multi-period or "capacity expansion" models have found use in other applications (Liebman, 1975), they have not yet appeared in the hazardous waste area. Note that these models will be even more data intensive than the multi-waste models described immediately above.

Impact Assessment of Alternatives

Given a set of alternatives ranked by the above procedure in terms of cost, it is desirable to evaluate the environmental and risk implications of the possibilities to determine whether the incremental excess cost that would be associated with a given alternative may be justifiable. The general framework for accomplishing such analysis has been described by Klee (1971, 1976) and implemented for the hazardous waste regional planning problem by Jennings and Sholar (1983). The objective is to develop an overall ranking Λ_i for each alternative i that we wish to evaluate.

In the first step of the procedure, a set of attributes that will characterize the risk is defined. This should be consistent so that increasing values represent more desirable alternatives, and the relationship between the value and the preference is roughly linear. It is also necessary that the attributes be defined so as to be mutually independent as possible. Possible attributes for a siting problem might include (Wood et al., 1984):

- shortest travel time to nearest aquifer
- shortest travel time to nearest surface water
- distance to nearest occupied residence
- distance to nearest earthquake fault or abandoned mine
- archaeological/historical significance
- impact to terrestrial organisms

Note that some of these factors (e.g., travel times) are objective in nature, but some are subjective. The subjective criteria might usefully be scored with significant input from the political body responsible for the ultimate decision, and/or a representative community body. The attributes that are scored are assumed not to be exclusionary —that is, no attribute is assumed to be infinitely more or less desirable than another

TABLE 9-7 HYPOTHETICAL FACTOR SCORES FOR SEVERAL ALTERNATIVES

Alternative	Score on attribute					
	1 (yrs)	2 (yrs)	3 (mi)	4 (mi)	5 (scale of 1–5)	6 (scale of 1–5)
1	25	0.5	3	2	2	3
2	75	0.2	2	2	3	1
3	100	0.3	1	4	4	3
4	50	0.6	2	5	1	5

attribute. If such attributes do exist, they should be used in a preliminary fashion to screen out alternatives subject to the described analysis.

Each alternative is given a "factor score" for each of these attributes, designated as FS'_{ij}—the score for alternative i on attribute j. For a hypothetical set of four alternatives, these scores might look as indicated in Table 9-7. The ratings for attributes 5 and 6 are performed on a subjective scale from 1 (highly significant adverse impact) to 5 (minimal adverse impact).

The second step of the process is to transform the scores in their natural units to dimensionalized scores on the scale 0–1. This is done for each attribute j by subtracting the minimum value of the attribute from all scores of that attribute and dividing by the range for that attribute. The resulting dimensionalized scores are termed FS_{ij}. For the hypothetical problem, the result is given in Table 9-8.

The third step of the procedure is to develop a set of proportionality factors between attributes. Klee (1971, 1976) developed the DARE (Decision Alternative Ratio Evaluation) for this process. For each one of the attributes considered, the importance of a change from the most desirable to least desirable value must be compared against all other attributes. It is sufficient to compare the attributes in a pairwise manner. Table 9-9 illustrates this procedure for the hypothetical. The last element is arbitrarily given a relative importance ranking of 1. In other words, it is determined that it is twice as important to increase the groundwater travel time from

TABLE 9-8 HYPOTHETICAL DIMENSIONLESS FACTOR SCORES FOR SEVERAL ALTERNATIVES

Alternative	Dimensionless score on attribute					
	1	2	3	4	5	6
1	0.00	0.75	1.00	0.00	0.33	0.50
2	0.66	0.00	1.00	0.00	0.66	0.00
3	1.00	0.25	0.00	0.66	1.00	0.50
4	0.33	1.00	0.50	1.00	0.00	1.00

TABLE 9-9 DEVELOPMENT OF DARE FACTOR WEIGHTS

Attribute	Lowest value (least desirable)	Highest value (most desirable)	Relative importance with respect to next attribute	Derived factor weight	Normalized factor weight
Groundwater proximity	25 years	100 years	2	1.5	0.252
Surface water proximity	0.2 year	0.6 year	0.5	0.75	0.126
Distance to residence	1 mile	3 miles	1.5	1.5	0.252
Distance to faults/mines	2 miles	5 miles	5	1.0	0.168
Archaeological significance	1	4	0.2	0.2	0.034
Terrestrial impact	1	5	1	1	0.168

25 years to 100 years as it is to increase the surface water travel time from 0.2 year to 0.6 year. These relative importance rankings can be used in a systemic manner to obtain public or decision maker input. The derived factor weight is obtained backwards, from the last factor evaluated (which had been given a weight of 1.0). By successive multiplication by the relative importance ratios, a factor weight can be obtained for each attribute. For convenience, these factor weights are normalized so that they sum to unity. The last column should be inspected to ensure that it represents the true evaluations of the importance of each attribute. The normalized factor weights will be denoted as FW_j for attribute j.

The final step is to apply the factor weights to the attribute scores for each alternative. The resulting overall score is obtained as follows:

$$\Lambda_i = \sum_j FW_j FS_{ij} \tag{9.7}$$

Applying this equation to the hypothetical results in the overall scores given in Table 9-10. Based on this evaluation, alternative 4 is ranked as the most desirable, followed by alternatives 3, 1, and 2, respectively. These rankings should now be compared against the costs of each alternative in order to produce a final decision.

Sensitivity Analyses

It should be obvious that the use of a quantitative approach for comparison of alternatives is data intensive. A large number of assumptions, such as costs, interest rates, relative impacts, etc., need to be employed in order to perform the necessary computations. Prior to finalization of a decision, it is necessary, therefore, to determine that any uncertainties in the input information do not drastically alter the final solution. The formal process of examining these effects is termed a sensitivity analysis. For each variable subject to uncertainty, its range of uncertainty should be examined. The analysis (including both the optimization and the DARE analysis)

TABLE 9-10 OVERALL SCORES FOR HYPOTHETICAL ALTERNATIVES

Alternative	Λ_i
1	0.44172
2	0.44076
3	0.51238
4	0.67116

should then be rerun using the range of parameter values. For a large number of parameters, it may suffice to run the models many times to explore solution sensitivity.

The invariance of an optimal solution (at least in terms of a choice of treatment and disposal locations) is termed its robustness. It is desirable to have a solution that is robust to a wide range of values of the input parameters. It may be necessary to use less-than-optimal solution to achieve this robustness property. However, it is necessary to evaluate this property prior to decision formalization.

REFERENCES

BROOKE, A., KENDRICK, D., AND MEERAUS, A. *GAMS: A Users Guide.* Scientific Press, Redwood City, Cal. (1988).

JENNINGS, A. A., AND SHOLAR, R. L. "Hazardous Waste Disposal Network Analysis." *Journal of Environmental Engineering* 110(2), 325–342 (1984).

KLEE, A. J. "The Role of Decision Models in the Evaluation of Competing Environmental Health Alternatives." *Management Science* 18, B52–67 (1971).

———. "Models for Evaluation of Hazardous Wastes." *Journal of the Environmental Engineering Division, ASCE* 102, 111–125 (1976).

MITRE Corporation. "WRAP: A Model for Regional Solid Waste Management Planning: Users Guide." EPA/530/SW–574 (1977).

PIERCE, J. J., AND DAVIDSON, G. M. "Linear Programming in Hazardous Waste Management." *Journal of the Environmental Engineering Division, ASCE* 108(5), 1014–1026 (1982).

WOOD, E. F., FERRARA, R. A., GRAY, W. G., AND PINDER, W. F. *Groundwater Contamination from Hazardous Wastes.* Prentice Hall, Englewood Cliffs, N.J. (1984).

Index

A

acclimation, in biological processes, 285
acid
  definition, 120
  dissociation constant, 121, 123
acids, cost and capacity compared, 129
activated carbon
  addition to biological processes, 286
  preparation, 106
  regeneration using solvents, 89–90
activated sludge
  mass balance, 260–261, 266
  recycle ratio, 266
  typical design variables, 252
activity, solid phase, of precipitates, 136
activity coefficients
  dissolved ions, 138
  effect on relative volatility, 98
  in ion exchange, 198
  Van Laar equation, 102
adaptation, in biological processes, 285
adsorbents, polymeric, 106
adsorption, 105–112
  before air stripping, 93
adsorption isotherm
  Freundlich, 107
  Langmuir, 107
advanced oxidation processes, 167–181
aeration power, 280
air stripping, 90–94
  adsorption of off-gas from, 109
  gasoline in groundwater, 113
alkaline wastes, neutralization, 129
alkalinity
  definition, 124
  effect on anaerobic processes, 282
ammonia removal, steam stripping, 105
anaerobic contact process, 253
anaerobic processes
  fixed film, 253–254
  microbiology, 252
AOP (advanced oxidation processes), 167–181
arc furnace incineration, 317
area source, 41–42
Arochlor 1242, 42
ash, incinerator, 318

assessment of alternatives, 350–353
atmospheric emissions, plume rise, 41
audits, waste, 331–339
azeotropes, 99

B

backwashing
filters, 72
ion exchangers, 225
barometric pressure, effect on volatilization through soil caps, 40
basicity constant, 122, 123
batch distillation, 98
batch microbial growth, kinetics, 247
bicarbonate, radical scavenger in ozonation, 171, 173
biodegradation
qualitative generalizations, 35
second order kinetics, 35–36
biofilm process, schematic, 258
biological decomposition, effect on volatilization, 40
biological treatment
minimum attainable substrate level, 262
washout, 263
bisulfate, for chrome reduction, 162
borohydride
for metal recovery, 330
precipitation, 148
boundary conditions
open, 26
semi-infinite, 24
breakpoint, adsorption, 106
breakthrough, 106, 212
Bronsted–Lowery model, 120
burnout time, rotary kiln incinerators, 312

C

cadmium
hydroxide precipitation, 136–137
solubility, effect of pH on, 142
cake, filter, 77
capacity loss, regeneration of adsorbents, 109
Caputo Dump (NY), 42
carbon, activated. *See* activated carbon
carbon dioxide
as a neutralization agent, 129
supercritical, in solvent extraction, 89
carbon monoxide
destruction in incineration, 308–309
use of, in incinerator monitoring, 322
carbon tetrachloride, 110
carbonate
radical scavenger in ozonation, 171, 173
use in precipitation, 151–152
Carmen–Kozeny equation, 71. *See also* Kozeny-Carmen equation
categorization, by property, 12
caustic agents, comparison by cost and capacity, 126
cement kilns, 314
Census of Manufacture, 2
centrifugation, 67–69
centrifuge, equivalent area, 67
CFR (Code of Federal Regulations), 3
chemical oxidation and reduction, 152–181
chlorine, cyanide destruction using, 160–162
chromate reduction, 162
chromate removal by sulfide precipitation, 150
chromic acid, replacement by peroxide-sulfuric acid, 328
chromium (VI)
reduction, 162
sulfide precipitation, 150
cleaning, reverse osmosis and ultrafiltration, 80
coagulation-flocculation, 144–145
coal liquefaction wastewater, membrane treatment, 85

Code of Federal Regulations (CFR), 3
combustion
  equilibrium constant, 304
  incomplete, products of, 310
combustion efficiency, defined, 309
complexation of metals, effect on precipitation, 141–144
complexing agents, interference with precipitation of metals, 143
conditioning, thickening, importance of, 77
control, neutralization systems, 130–131
coprecipitation, 145
corrosivity, definition, 118
costs, piecewise linearization, 348
cyanate, intermediate in cyanide oxidation, 161
cyanide
  chemical oxidation, 160–162
  danger of neutralization by equalization, 128
  effect complexation on chemical oxidation, 162
  ion exchange treatment of, 227
  treatment using ozone, 180
  wet air oxidation, 162
cyanogen chloride, intermediate in cyanide oxidation, 161

D

Darcy's law, 46
Debye–Hückel equation, 139
decay rate, biological, 247–248
decision alternative ratio evaluation, 351
deep biofilm model, 260
deep shaft processes, supercritical oxidation, 165
density gradients, effect on dispersion, 30
"derived from" rule, 4, 58, 110
design, conceptual, 58
dewatering, filterability of sludges, 77
dinititrotoluene wastewater, supercritical oxidation, 166
dioxin
  incineration performance, 314
  requirements for incineration, 292
disincentives for waste minimization, 339
dispersion, 41
  axial
    effect on hydraulic conductivity variations, 48
    groundwater, 48
    rivers, 25–26
  longitudinal. *See* dispersion, axial
  transverse rivers, 30
  vertical water, 30
dissociation equilibrium, acid, 120–121
dissociation water, 119
distillation, 96–104
  in solvent recovery, 330
  steam, 104–105
Dulong correlation, 298

E

Eckenfelder equation, 271
economics
  advanced oxidation processes, 179–180
  ion exchange, 226–229
  waste minimization, 335–339
eddy diffusion. *See* dispersion
electrode potential, 155
electron acceptor, 244
electron donor, 244
electroplating, waste reduction opportunities, 327–328
electroplating wastes, conventional treatment flowchart, 134
enthalpy, effect of temperature on, 297–298
enthalpy balance, incinerators, 296–298
EP toxicity, 135

equalization, use for neutralization, 128
equilibrium, precipitation, 133–136
equilibrium constant
acidity and basicity, 123
effect of temperature on, in combustion, 303
ion exchange, 183, 197–198
redox reactions, 155–156
equilibrium stages, solvent extraction, 88
erfc, 26–27
error function, complementary, 26–27
ethanol regeneration of adsorbents, 109
excess air, defined, 294
exposure routes, 21
extraction, solvent. *See* solvent extraction

F

Fenton's reagent, 177
ferrous sulfate, chrome reduction, 162
ferrous sulfide precipitation, 151
fetch, 32
filter press, 77–78
filtration
molecular, 78–86
particulate, 70–78
pressure, 77–78
pressure drop vs. particle size, 70
pretreatment for adsorption, 110
vacuum, 75–77
fire brick, characteristics, 321
flocculation basins, design of, for precipitation processes, 145
flooding, air stripping, 93
flue gas, as a neutralization agent, 129
fluidization, of filters, 72
fluidized beds
incineration in, 316
regeneration of adsorbents, 109
flux
solute, 81
water, membrane filtration, 81
fouling, membrane filtration, 80
fractioning column, 98
free radicals, 33
importance in ozonation processes, 171–177
See also hydroxyl radicals
Freundlich equation, 107
Froude number, 95

G

gas-liquid mass transfer, 32, 92
gasoline, air stripping, 113
Gaussian curve, 27
generation, per capita, 1
generation factors, 5–7
generators, size distribution, 9
groundwater
adsorption treatment, 109
air stripping, 94
cost of treatment using advanced oxidation, 179
membrane treatment, 85
groundwater flow, governing equations, 46–47
group contribution methods, 31
growth rate
biological, 247
first order constants, 249
Monod constants, 249
gypsum, formation during lime precipitation, 150

H

Haldane equation, 250
half reactions, biological, free energies, 245
half-cell potential, 155
Hausbrand diagram, 105
hazardous waste
household, 13
sources, 1–4
by SIC, 8
head, defined, 46
head loss, granular filtration, 71

Henry's law, 31, 90–91, 98, 105
adsorption, for, 29
Henry's law constant
effect on volatilization in biological treatment, 286
ozone, 168
hexavalent chromium reduction, 162
hollow fiber membranes, 84
humic materials, effect on adsorption, 108
Hyde Park (NY) landfill, 251
hydraulic conductivity, 46
effect on axial dispersion, 48
estimation of, 47
hydraulics, air stripping, 93
hydrogen peroxide, 33
absorbency of UV light, 176
cyanide oxidation with, 162
importance in ozonation, 174
metals, effect on reaction with organics, 157
use in supercritical oxidation, 165
use in waste minimization, 328
hydrogen sulfide
removal by membranes, 85
see also sulfides
hydrolysis, 33–34
half-lives at neutral pH, 34
hydroxide precipitation
compared to sulfide precipitation, 150
metals, 148–150
hydroxyl radicals
oxidation potential, 172
production by ozone-UV process, 175
rate of reaction with organics, 173
see also free radicals

I

ideal adsorbed solution theory, 107–108
impact assessment, 350–353
incineration
air pollution control of, 317–318
carbon monoxide destruction, 308–309
equilibrium in, 304
heat loss, 300
infrared, 317
required destruction and removal, 292
stoichiometry, 293–294
velocity to assure turbulence, 308
incinerator exhaust, atmospheric dispersion, 41
incinerators
liquid injection, 311–312
rotary kilns, 311–314
incomplete combustion, products of, 310
infrared incineration, 317
inhibition
anaerobic processes, 282
biological growth, 249
integer programming, 345
ion activity product, as a solubility criterion, 135
ion exchange, 181–230
economics, 226–229
effect of cross linking on selectivity, 206
effect of flowrate on capacity, 220–221
equilibrium constant, 197–198
functional groups, effect on selectivity, 196–197
materials, physical properties, 187–189, 192–193
regeneration, 220–223, 225
selectivity sequences, typical, 199
ion exchangers, classification, 190
ionic strength, defined, 138
isotherm. *See* adsorption isotherm

K

kinetic limitations
adsorption, 108
precipitation, 136

kinetics
biodegradation of xenobiotics, 283
biological growth, 246–250
organic destruction in incineration, 306–308
redox reactions, 157
Kozeny–Carmen equation, 47
Kremser equation, 88

L

Langmuir equation, 107
Laplace equation, 46
Leva equation, 93
Lewis acid, 122
lime kilns, 314
lime slurry neutralization systems, 127–128
limestone
bed neutralizing system, 127
relative performance in neutralization, 127
linearization of cost functions, 348
liquid extraction. *See* solvent extraction
liquid injection incinerators, 311–312
local equilibrium model for ion exchange, 214–218
longitudinal dispersion, 25–26
Love Canal (NY), 48

M

manufacture, value added by, 6
mass balances, 21–23
activated sludge, 260–261, 266
batch suspended growth, biological, 255–256
biological film process, 258–259
plug flow biological process, 268–271
mass transfer, ozone, 170–171
mass transfer coefficients
air stripping, 92
in biological packed beds, 279
gas film, 32, 92
liquid film, 32, 92
in ozonation reactors, 171
in rotating biological contactors, 276
soil gas, 39–40
McCabe–Thiele diagram, 102–103
membrane filtration, 78–86
deterioration of membranes, 80
metal hydroxides, solubility, 149
metal removal by precipitation, 147–152
metal sulfides, solubility, 151
metal wastes, amenability to treatment by ion exchange, 227
methanol, regeneration of adsorbents, 109
minimum fluidization velocity, 72
Model City (NY), incineration, 317
molecular filtration, 78–86
monitoring systems, incinerators, 321–322
Monod
constant, 249
equation, 248
parameters, experimental determination, 256–257, 262–265
multicomponent adsorption, 107–108

N

Nernst equation, 155
net heating value (NHV)
defined, 297
supplemental fuels, 295
neutralization, 118–131
process design and control, 130–131
NHV. *See* net heating value
nitrate respiration, effect on biological kinetics, 281
nonconservative pollutants, 28
nonideality. *See* activity coefficients
NO_x, formation in incineration, 309–310
nucleation during precipitation, 137–138
nutrients, biological growth, required for, 241–243

O

octanol-water partition coefficient, 30
Onda correlation, 92
organic carbon, 30
organic compounds
  biodegradation, related to structure, 35
  chemical oxidation, 163–181
  destruction by UV-ozone process, 176
  hydrolysis, 34
  kinetics of destruction in incineration, 306–308
  photolysis, 34
  reactivity with hydroxyl radical, 173
  reactivity with ozone, 170
  relationship between structure and biodegradation, 284
  specific enthalpies, effect of temperature, 298
osmosis, reverse. *See* reverse osmosis
osmotic pressure, 81
overflow velocities in sedimentation, 65
oxidation potential
  hydroxyl radicals, 172
  ozone, 169
  standard, for oxidizing chemicals, compared, 163
oxidation state, definition, 153
oxygen requirement in biological treatment, 280
ozonation, cyanide oxidation, 162
ozone, reactivity with organics, 170
ozone decomposition, effect of water quality, 169
ozone generator, schematic, 167
ozone processes, 167–181

P

packed tower stripper, 91
packing materials, flooding properties, 93
PACT process, 286
paper plant wastewaters, steam stripping, 105
particle removal, granular filters, in, 73–75
partition coefficient
  octanol-water, 30
  sediment-water, 30
  solvent extraction, 86
Pasquill Stability Categories, 43
PBBs, 21
PCBs, 42–45
  incineration using plasma arcs, 317
  for production of fuel, 330
  requirements for incineration, 292
  solvent extraction, 89
  treatment using advanced oxidation, 179
pE-pH diagram, example, chlorine, 158
Peclet number, 26
permselection, 85
pH
  effect on ion exchange capacity, 191
  redox reactions, influence on, 157
phenol removal
  solvent extraction, 89
  using ion exchange, 181
phenolphthalein end-point, 124
photocatalytic oxidation, cyanide, 162
photolysis, 33–35
physical treatment processes
  classification, 61
  definition, 60
PIC (products of incomplete combustion), 310
pickle liquor, recovery of iron salts, 330
pickling, changes in, for waste minimization, 328
plasma arc furnace incineration, 317
plasmids, 285
plating baths, ion exchange treatment, 227
plume dispersion
  one-dimensional slug, 25
  three dimensional slug, 27

plume rise, 41
POHC, 292
pollution prevention, 325–339
polybrominated biphenyls, 21
polychlorinated biphenyls. *See* PCBs
porosity, 29
  granular filters, 71
precipitation, chemical, 132–152
precipitation of metals, economics, 229
present value, 335
pressure
  osmotic, 81
  spreading, 108
pressure filters, 77–78
principal organic hazardous constituent (POHC), 292
priority pollutants
  biodegradation, 283
  solvent extraction, 89
products of incomplete combustion, 310

Q

quantitative structure-activity relationships, biodegradation, 284
quantum yield, 34–35

R

Raoult's law, 40–41
rapid mix, criteria, in precipitation, 145
rate constants, ozone-organic reactions, 172
RAYOX process, 179
RBC. *See* rotating biological contactors
recycle ratio, activated sludge, 266
recycle/recovery/reuse, classification, 329
redox potential, 155
  measurement, 156–157
redox processes, 152–181
reduction, chromium, 162
reflux
  constant, 99
  variable, 99
refractory materials, characteristics, 321
regeneration
  of adsorbents, 109
  ion exchange, 220–223, 225
regulations
  "derived from" rule, 4, 58, 110
  exclusions, 3
rejection, during membrane filtration, 82
relative volatility, 98
residence time, rotary kiln incinerators, 312
respiration rate in biological treatment, 280
retardation, 29
reverse osmosis, 78–86
  fouling, 80
Reynolds number, 32, 47, 279
rinse processes, changes in, for waste minimzation, 328
risk, incorporation in systems analysis, 344
rivers and streams, transport properties, 25–26, 30
robustness of optimal alternatives, 353
rotary hearth, regeneration, of adsorbent, 109
rotary kiln incinerators, 311–314
rotating biological contactors, RBC, 254–255
  removal vs. loading, empirical relationships, 272

S

safety, redox processes, 159
SBR. *See* sequencing batch reactor
Schmidt number, 32, 82
sedimentation, 62–67
  flocculent (Type II), 64

selectivity, ion exchange
  coefficient of, 198
  effect of cross linking, 206
  typical values, 204
selenium
  ion exchange treatment, 228
  precipitation, 147
sequencing batch reactors (SBR), 250
  addition of activated carbon to, 286
SIC (standard industrial classification), 1–4
sludges
  initial settling velocity, 64–65
  resistance to filtration, 77
soil cap
  barometric pressure, effect of, 40
  effect on volatilization, 39, 45
solidification, 230
solubility
  metal hydroxides, 149
  metal sulfides, 151
  partitioning, relationship to, 31
solubilty product, defined, 135
solvent extraction, 86–90
  carbon regeneration, 89–90
  regeneration of adsorbents, 109
  solvent classes, 87
solvent recovery, 99
solvent vapors, adsorption, 109
solvents
  recovery by membranes, 85
  recycling, 329
  replacement, for waste reduction, 328–329
sorption, 28
specific energy gradient, 26
specific resistance of sludges, 77
sphericity, granular filters, 71
spiral wound membrane processes, 85
spreading pressure, 108
stability, anaerobic processes, 281–283
stabilization, chemical, 230
standard industrial classification (SIC), 1–4
steam distillation, 104–105
steam stripping, 104–105
  regeneration of adsorbents, 109
still bottoms, 99
stoichiometry
  biological processes, 241–243
  combustion processes, 293–294
stripping, air. *See* air stripping
stripping, steam. *See* steam stripping
structure, effect on biodegradation, 284
substrate inhibition kinetics, 250
substrate utilization rate, biological, 248
Sulfex process, 151
sulfide precipitation, 150–151
sulfide removal, steam stripping, 105
sulfides, effect on anaerobic processes, 283
sulfur dioxide, chrome reduction, 162
sulfuric acid neutralization, 129
supercritical fluids, solvent extraction, 89
supercritical oxidation, cyanide oxidation with, 162
supercritical water oxidation, 164–166
superoxide anion in ozonation, importance of, 173
supplemental fuels, properties, 295

T

TCE (trichloroethylene), 110
TCLP (toxicity criteria leaching procedure), inorganic criteria, 135
temperature, effect on air stripping, 94
thermal NO_x, defined, 309
thermal regeneration, adsorbents, 109
thermodynamics, biological growth, 244–246
thickening
  centrifugal, 68
  filtration, 78
  gravitational, 64–67
Times Beach (MO), 21

titanium dioxide, use in photocatalytic cyanide oxidation, 162
titration curve
  ion exchangers, 189
  strong acids, 125
toluene, 94
tortuosity, 39, 71
toxic materials, critical levels for anaerobic processes, 282
transfer unit, in air stripping, 92
transport equation, 21–23
  one-dimensional, steady with reaction, 28
  one-dimensional, unsteady state, 24
  three-dimensional, unsteady state, 27
trial burns, incinerator, 318–320
trichloroethylene, 110
  ozone reactions, 170
  treatment using advanced oxidation processes, 181
trickling filters, 253
  design using Eckenfelder equation, 271–272
tubular membrane processes, 84
turbulence, importance in incineration, 308

U

ultrafiltration, 78–86
ultraviolet light. *See* UV
Ultrox Process, 175–178, 181
UV, 33
  absorbence by hydrogen peroxide, 176
  in advanced oxidation processes, 174–181
UV/Ozone/Hydrogen Peroxide process, 175

V

vacuum filters, 75–77
value added by manufacture, 6
Van Laar equation, 102
vapor pressure
  Antoine's equation, 40, 102
  temperature, effect of, 102
velocity, filtration, 71
Velz equation, 272
volatile materials, adsorption, 109
volatile solvents, recovery by membranes, 85
volatility, relative, 98
volatilization, 31–33
  filtration, 78
  wind, effect on, 32

W

waste audits, 331–339
waste management hierarchy, 58
waste minimization, definition, 326
wastewater, per capita generation, 4
water, reaction with ozone, 174
Weber number, 95
wet air oxidation, 163–164
  cyanide oxidation, 162

X

xenobiotics, biodegradation, 283–284

Z

Zimmerman process, 163–164
Zimpro process, 163–164